Awdhesh Kumar Choudhary · Somenath Mondal ·
Subhadeep Metya · G. L. Sivakumar Babu
Editors

Advances in Geo-Science and Geo-Structures

Select Proceedings of GSGS 2020

Springer

Editors
Awdhesh Kumar Choudhary
Department of Civil Engineering
NIT Jamshedpur
Jamshedpur, Jharkhand, India

Somenath Mondal
Department of Civil Engineering
NIT Jamshedpur
Jamshedpur, Jharkhand, India

Subhadeep Metya
Department of Civil Engineering
NIT Jamshedpur
Jamshedpur, Jharkhand, India

G. L. Sivakumar Babu
Department of Civil Engineering
Indian Institute of Science
Bengaluru, Karnataka, India

ISSN 2366-2557 ISSN 2366-2565 (electronic)
Lecture Notes in Civil Engineering
ISBN 978-981-16-1992-2 ISBN 978-981-16-1993-9 (eBook)
https://doi.org/10.1007/978-981-16-1993-9

This Springer imprint is published by the registered company Springer Nature Singapore Pte Ltd.
The registered company address is: 152 Beach Road, #21-01/04 Gateway East, Singapore 189721, Singapore

Preface

In the era of the twenty-first century, India has experienced with expeditious industrialization, aggressive urbanization and population explosion resulting in rapid change in the socio-economic structure of the country. Thus, the need of the hour involves the evolution of cutting-edge researches and application of the innovative technologies in combination with novel strategies on both the construction of new and extension of existing infrastructures. Nevertheless, the complex subsoil conditions along with ever-increasing socio-economic costs, scarcity of good construction material and construction site, coupled with environmental constraints, posed major challenges to the geotechnical engineering professionals. Consequently, engineers and practitioners start modifying and reinforcing the "weak soil" into suitable one in order to meet the demand of civilization. Furthermore, the amount and rate of generation of solid waste has become menace to the progress of civilization. In this context, geotechnical researchers and engineers have been putting enormous effort to convert wastes into value-added products or to generate energy out of waste, which seems to be the only remedy to such issues. All these paved the way for the successful organization of the National Conference on "Geo-Science and Geo-Structures" (GSGS 2020) under the aegis of the Technical Education Quality Improvement Programme (TEQIP-III), MHRD, Government of India, by the Department of Civil Engineering, National Institute of Technology, Jamshedpur.

Needless to mention, the success of a conference depends to a great extent on the availability of contributions from persons of eminence. We are deeply moved by the spontaneous response we have received in this regard from not one but several such personalities from almost all IITs, NITs and other reputed institutes in India. The organizers also render their heartfelt thanks to the contributory authors, delegates and reviewers from whom the response is overwhelming. GSGS 2020 has been a confluence of classical and contemporary geotechnical engineering issues which encompasses applications in the realm of environmental geotechnology, soil reinforcement and ground improvement, solid waste management, slope stability analysis and so on. Altogether, nearly 150 abstracts have been received from authors from different parts of India, from which only 34 full papers have been selected after final review. Besides those from educational institutions, a substantial number of delegates have also attended from the Industry.

It is our fervent hope that the deliberations made during the conference and the records thereof in the proceedings book will help to enrich the knowledge of the geotechnical fraternity with the current state of research and practices.

Jamshedpur, India

Awdhesh Kumar Choudhary
Somenath Mondal
Subhadeep Metya
Organizing Secretary(s)/Chair(s)

Contents

About the Editors

Dr. Awdhesh Kumar Choudhary is presently working as an Assistant professor at the Department of Civil Engineering, National Institute of Technology Jamshedpur, Jharkhand. He obtained his B.E. (Civil) from North Maharashtra University, Jalgaon; M.Tech. from Indian Institute of Technology Guwahati and Ph.D. from Indian Institute of Technology Kharagpur. After completing his Ph.D., Dr. Choudhary also worked as a Post-doctoral fellow at Indian Institute of Science Bangalore. His major areas of research interests include the anchor foundation system, ground improvement, reinforced soil structures, and the stability of the buried structures. He has published 9 papers in reputed international journals as well as 9 conference papers in the proceedings. Dr. Choudhary is also reviewer of several reputed journals such as Geotextile and Geomembrane, Geosynthetics International, Soils and Foundation, and European Journal of Environmental and Civil Engineering.

Dr. Somenath Mondal is presently working as an Assistant Professor in the Department of Civil Engineering, National Institute of Technology, Jamshedpur. He received his B.Tech. degree in Civil Engineering from West Bengal University of Technology in 2011. Then he joined Bengal Engineering and Science University Shibpur (presently known as IIEST Shibpur) for pursuing Master of Engineering in Geotechnical Engineering and he graduated in 2013 from there. In 2014, he joined IIT Bombay to pursue doctoral study in the department of Civil Engineering with the specialization in Environmental Geotechnology and in 2017, he graduated with Ph.D. degree from IIT Bombay. During his Ph.D. tenure he got a couple of research grants from European Commission to conduct his research in Paristech France (October–November 2015; October–January 2016–2017). Dr. Mondal's prime research interests are thermal characterization of geomaterials, utilization of industrial by-products, energy geotechnics, heat induced moisture migration, thermo-hydro-mechanical behaviour of clay. He has published 6 articles in well reputed international journals.

Dr. Subhadeep Metya is presently an Assistant Professor in the Geotechnical Engineering Division, Department of Civil Engineering, National Institute of Technology (NIT) Jamshedpur, India. He obtained his B.Tech. from the West Bengal University of Technology, followed by M.Tech. and Ph.D. from Indian Institute of Engineering

Science and Technology (IIEST), Shibpur. During the tenure of his doctoral research at IIEST, Shibpur, Dr. Metya has been a recipient of the prestigious INSPIRE Fellowship of the Department of Science and Technology (DST), Government of India, from whom he has also received the prestigious Newton-Bhabha Ph.D. Placement Grant 2015–2016 by virtue of which he has completed six months' research internship at the Zienkiewicz Centre for Computational Engineering (ZCCE) at the Swansea University, UK. As for his research contribution so far, Dr. Metya has co-authored one book, four book chapters, ten papers in peer-reviewed journals, as well as fourteen conference papers. His awards and honours include the University Gold Medal in both B.Tech. and M.Tech., the IGS-Shri R. N. Prasad Biennial Prize (2015) and the IGS-HEICO Young Geotechnical Engineer Award (2018) by the Indian Geotechnical Society. Besides, he has gained a rare and invaluable experience and exposure on landslide related issues while attending the internationally well-known school on landslides "LARAM" in University of Salerno, Italy in 2014. His current research interests include slope stability and landslides, geotechnical earthquake engineering, reinforced soil structures, rock mechanics, foundation engineering, computational geotechnics and risk and reliability applications in civil engineering.

Prof. G. L. Sivakumar Babu is a Professor, Geotechnical Engineering Division, Department of Civil Engineering, Indian Institute of Science, Bengaluru, India. He served as the President, Indian Geotechnical Society for four years during 2017–2020 and as Governor, ASCE Board of Governors, Region 10 during 2014 to 2020. He is an awardee of John Booker Medal, International Association for Computer Methods and Advances in Geomechanics (IACMAG), USA, for the year 2017. His areas of research include risk and reliability applications in geotechnical engineering, geosynthetics and reinforced soil structures, environmental geotechnology, fibers in geotechnical engineering, earthquake geotechnical engineering and geotechnics for disaster mitigation. A total of over 150 papers in journals and 100 in conferences have been published. He is the editorial board member and reviewer of several reputed journals such as *American Society for Civil Engineers* (ASCE), *Canadian Geotechnical Journal*, *Geotechnique*, *Computers and Geotechnics*, *Georisk*, *Engineering Structures* and national journals such as *Indian Geotechnical Journal*, *Sadhana*, etc. He has also served as Chairman/Co-chairman/expert for technical sessions in various international/national conferences in India and abroad.

Estimation of Landfill Settlement Using Different Models

Choudhary Aniket Aman, Sunita Kumari, and Sufyan Ghani

Abstract Estimation of settlement is a matter of prime importance for the maintenance of municipal solid waste (MSW) landfills. Settlement of MSW landfill used to take place for a long duration, and the total displacement can approach up to 30% of original landfill height. The main phases of settlement in MSW landfills are initial compression, primary compression, and long-term secondary compression. The mechanisms involved in the waste settlement are densification, placement of finer particles in the voids of larger size particles, physical–chemical change, and bio-chemical decomposition. The factors which affect the settlement values are initial density or void ratio of the solid waste, placement sequence, decomposable materials, leachate levels, and landfill operation methods. A large value of settlement after non-functioning of the landfill is undesirable from management angle as it can cause surface ponding, development of cracks in clay liners (such as compacted clay liners), and tear in the geomembrane layer. Therefore, understanding of total landfill settlement is necessary for landfill design and the future construction over it. Different prediction models are used to determine the displacement of a typical MSW landfill. These models are based on soil mechanics-based model, rheological model, empirical model, and biodegradation-induced settlement model. The model parameters are taken from data available in the published literature. Measured settlement data from each model is displayed as settlement-versus-time graph. In the study, it can be seen that the predicted settlement magnitude differs notably depending on the prediction model used and the selection of parametric values. Marques (2001) model was found out to predict the settlement more closely as it considers all the mechanisms involved in the settlement in good proportion.

Keywords Landfills · Municipal solid wastes · Settlement · Models · Predictions

C. A. Aman (✉) · S. Kumari · S. Ghani
Department of Civil Engineering, NIT Patna, Patna, Bihar, India

S. Kumari
e-mail: sunitafce@nitp.ac.in

A. K. Choudhary et al. (eds.), *Advances in Geo-Science and Geo-Structures*,
Lecture Notes in Civil Engineering 154,
https://doi.org/10.1007/978-981-16-1993-9_1

1 Introduction

In the past few decades, the world is facing a variety of rapid changes in environmental and humanitarian issues. The environmental issues include climate change, shortage of resources, etc., and the humanitarian issues include the rapid spike in population, problems related to urbanization, etc. These rapid unexpected changes lead people to take fast reactions to settle, which further resulted in issues due to unequal distribution of resources and lack of planning. The fast urbanization in cities causes a large influx of people to the cities in search of occupation. The residents of the city are nearer to the landfill site than before, and in many cases, the landfill site gets enclosed within the residential zone of the city. Urbanization contributes to increase in solid waste generation and unhygienic disposal of wastes which results in degradation of urban environment and causes health hazards. The engineered landfill is the answer to these problems as it will improve the soil, ground water, and air standards. Both active and after-closure management of landfills are important in an environmental point of view. Post-closure landfill maintenance can be economically favorable if the landfill site can be reused. Post-closure landfill settlement analysis helps in designing of the final cover system and landfill ancillary facilities, such as system for collection of gases that liberate from the landfill, drainage pipes, and leachate injection pipes to protect the human as well as the environment. It also helps in designing of post-closure developments on closed landfills and in formulating post-closure maintenance plan. Several experiences for reuse of capped landfill areas are livestock farming, installation of solar power plants, green capping for recreational use, or industrial reuse of waste utilization plant is a common practice in Europe. The main mechanisms involved in waste settlement are: (a) mechanical compression: It includes processes like densification, distortion, bending, crushing, and reorientation. Compression caused by the weight of the waste itself and applied loads occur in the form of initial compression and delayed compression, (b) raveling: It includes the movement of finer wastes into the voids or cavities of large size wastes, (c) physical-chemical change: Reactions like redox reaction and corrosion leads to volumetric decrease of wastes, (d) bio-chemical decomposition: It includes a reduction in the waste mass due to aerobic and anaerobic decomposition. In this paper, the time-dependent settlement of the municipal solid waste landfill is examined using different methods. Settlement-versus-time relationships are investigated and compared.

1.1 Settlement Estimation Methods for MSW Landfills

The settlement estimation models which are used for the prediction of MSW landfill settlement are described below:

(a) **Soil mechanics-based models**

Sowers method

Sowers [1] equations for the prediction of MSW landfill settlement are similar to those used for the estimation of initial and long-term settlement of the soil. Total displacement (ΔH) is taken equal to the sum of primary settlement (ΔH_c) and secondary settlement (ΔH_α). The equations used are Eqs. (1), (2), and (3)

$$\Delta H = \Delta H_c + \Delta H_\alpha \tag{1}$$

$$\Delta H_\alpha = C'_\alpha \times H_o \times \log\left(\frac{t_2}{t_1}\right) \tag{2}$$

$$\Delta H_c = C'_c \times H_o \times \log\left(\frac{\sigma_i}{\sigma_o}\right) \tag{3}$$

where H_0 denotes the thickness of each layer of waste; Cc' denotes the modified in initial compression index; C_α' denotes the modified long-term compression index; σ_0 denotes the previously applied pressure in the waste layer; σ_i denotes the total overburden pressure applied at the mid-level of the waste layer under consideration which is equal to (γ_{waste} * height of the waste fill above the mid-level of the layer under consideration); t_1 denotes the starting time of the time period for which long-term settlement of the layer is desired; and t_2 denotes the time period for which settlement of the layer is desired.

Bjarngard and Edgers' model

According to Bjarngard and Edgers [2], the total settlement is equal to the summation of immediate compression displacement, displacement due to reorientation of the solid wastes, and the displacement due to degradation. Considering all these three phases the final equation is Eq. (4)

$$S/H = \left[\mathrm{CR} \times \log \frac{(\overline{P}_o + \Delta P)}{(\overline{P}_o)}\right] + \left[C_{\alpha(1)} \times \log\left(\frac{t_{(2)}}{t_{(1)}}\right)\right] + \left[C_{\alpha(2)} \times \log\left(\frac{t_{(3)}}{t_{(2)}}\right)\right] \tag{4}$$

where S denotes the displacement; H denotes the thickness of the landfill waste layer; $\overline{P}_o$ denotes the initial effective stress; ΔP denotes the stress increment; t_1 denotes the time of completion of initial compression; t_2 denotes the time duration for intermediate secondary compression; t_3 denotes the time for total period of time considered in modeling; CR denotes the compression index; $C_{\alpha(1)}$ denotes the intermediate secondary compression coefficient; and $C_{\alpha(2)}$ denotes the long-term secondary compression coefficient.

Hossain and Gabr model

Hossain and Gabr [3] focus on the secondary displacement with three terms as given in Eq. (5)

$$\frac{\Delta H}{H} = \left[C_{\alpha i} \times \log\left(\frac{t_2}{t_1}\right)\right] + \left[C_{\beta} \times \log\left(\frac{t_3}{t_2}\right)\right] + \left[C_{\alpha f} \times \log\left(\frac{t_4}{t_3}\right)\right] \tag{5}$$

where $C_{\alpha i}$ denotes the compression index; C_{β} denotes the biodegradation index; $C_{\alpha f}$ denotes the creep index; t_1 denotes the time of completion of initial compression; t_2 denotes the time duration for which compression is to be evaluated; t_3 denotes the time for completion of biological compression; and t_4 denotes the time for creep at the end of biological degradation.

(b) **Empirical models**

Power creep law model

Equation proposed by Edil et al. [4] is expressed as Eq. (6)

$$S(t) = [H \times \varepsilon(t)] = \left[H \times \Delta\sigma \times m \times \left(\frac{t}{t_r}\right)^n\right] \tag{6}$$

where m denotes the reference compressibility; n denotes the coefficient of compression; t_r denotes the reference time; $\Delta\sigma$ (kPa) denotes the loading pressure; t denotes the time duration of interest; $\varepsilon(t)$ denotes the strain; H denotes the initial thickness of MSW landfill; and $S(t)$ denotes the settlement.

Hyperbolic function model

Equation proposed by Ling et al. [5] is expressed as Eq. (7)

$$S = [\varepsilon(t) \times H_o] = \left[\frac{S_{\text{ult}}}{1 + S_{\text{ult}}/(\rho_o \times t)}\right] \tag{7}$$

where S denotes the settlement; H_o denotes the initial height of MSW landfill; $\varepsilon(t)$ denotes the strain; t denotes the time duration of interest; and S_{ult} denotes the ultimate strain.

(c) **Rheological model**

Gibson and Lo's model

Equation proposed by Gibson and Lo's [6] is used mainly for the prediction of secondary displacement and is expressed as Eqs. (8) and (9)

$$\varepsilon(t) = \Delta\sigma \times [a + b\{1 - \exp(-(\lambda/b) \times t)\}] \tag{8}$$

$$\varepsilon(t) = (\Delta\sigma \times a) + \left[\Delta\sigma \times b\{1 - \exp((\lambda/b) \times t)\}\right] \tag{9}$$

where ε(t) denotes strain; $\Delta\sigma$ denotes compressive stress; *a* denotes coefficient of primary compression rate; *b* denotes coefficient of secondary compression rate; λ/b denotes Newtonian dashpot; $\Delta\sigma \cdot$ a denotes initial amount of compression; and $\Delta\sigma \cdot b$ denotes long-term secondary compression.

(d) **Settlement models incorporating biodegradation**

Marques model

Marques [7] considered all the mechanisms involved in the settlement of MSW landfill. Total displacements equal to the sum of instantaneous displacement, secondary mechanical creep, and displacement due to biodegradation. It is expressed as Eq. (10)

$$\frac{\Delta H}{H} = \left[C_c' \times \log\left(\frac{\sigma_o + \Delta\sigma}{\sigma_o}\right)\right] + \left[\Delta\sigma \times b \times \left(1 - e^{-ct'}\right)\right] + \left[E_{dg} \times \left(1 - e^{-dt''}\right)\right] \tag{10}$$

where ΔH denotes settlement; *H* denotes initial height of waste; C_c' denotes primary compression ratio; *b* denotes coefficient of secondary mechanical compression; *c* denotes secondary mechanical compression rate; E_{dg} denotes total compression due to waste degradation; *d* denotes secondary biological compression rate; t' denotes time elapsed since loading application; and t'' denotes time elapsed since waste disposal.

2 Methodology

Prediction of time-settlement relationship

To determine the displacement of municipal solid waste landfill by different methods, a typical landfill is considered. Select the moist unit weight and saturated unit weight of refuse as accurately as possible from the field results or published data. Decide the range of time over which settlement is required. Calculate the overburden pressure at the mid-layer of each waste layer using the principles of geotechnical engineering. Select the appropriate value of parameters from either tests or published literatures. For each time estimate, calculate the settlement using different proposed models.

MSW landfill conditions

An MSW landfill is selected, and the total height of the landfill is 40.2 m. It is filled in 6 layers, each having different thickness in the time period of 8 months. Different

Table 1 Filling sequence of MSW landfill (Geotechnical aspects of landfill design and construction by Qian et al. 9

Time period	Height of the solid waste filled (m)
1st month	7.5
2nd month	9.3
3rd month	5.4
4th month	0
5th month	0
6th month	2.4
7th month	7.5
8th month	8.1

models are used to predict the displacement of the MSW landfill for the period of 30 years. The original applied pressure on the solid waste is 48 kN/m^2, and the unit weight of solid waste is 10.2 kN/m^3.

Values of various model parameters are taken from the by Babu et al. [8] and published literature. For comparison of settlement values from different methods, the values of relevant parameters are considered constant. For example, the value of the compression index (C_c) for the soil mechanics-based model is 0.28, and the same value is used for other models too.

Table 1 shows the filling sequence for the MSW landfill site considered. The site is taken from Geotechnical aspects of landfill design and construction by Qian et al. [9].

3 Results

Using different proposed models and parameters, calculations have been performed. Figure 1 clearly shows the settlement of different layers (H1–H6) obtained from the soil mechanics-based Sowers method as a function of time. The total displacement of the landfill is obtained by the addition of the settlement values of each layer.

Comparative study

A comparative study of the different models helps in gaining the knowledge reliability and limitations of the model for specific site conditions. Table 2 presents a summary of the values of the parameters used for comparative study. Table 3 summarizes the values of the total settlement obtained from different models.

The maximum settlement at the end of 10,950 days by Sowers method [1] which considers primary and secondary settlement is given as 13.54 m.

Similarly, maximum settlement at the end of 10,950 days by Bajarngard and Edgers model [2] which does not take into consideration the bio-decomposition displacement in great depth is predicted as 23.67 m.

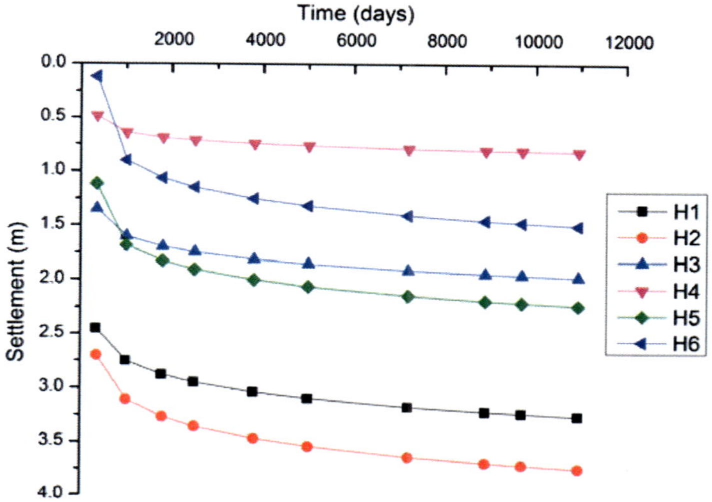

Fig. 1 Time-settlement relationship of different layers of MSW landfill based on Sowers method [1]

Fig. 2 Comparison of settlement response using different models

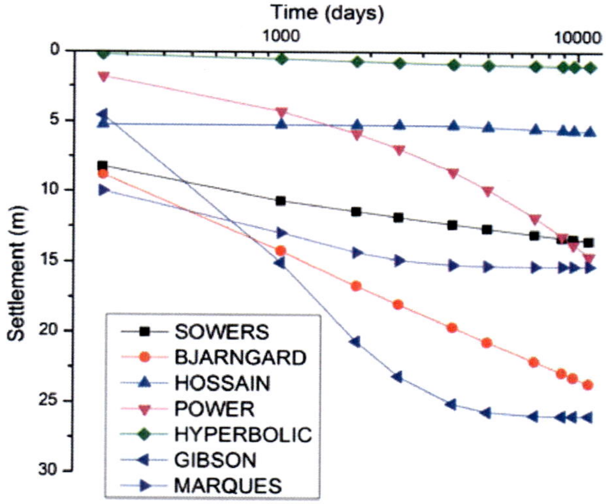

The maximum settlement at the end of 10,950 days by Hossain and Gabr model [3] which does not consider the compression due to imposed load but take consideration of bio-decomposition-induced displacement in great depth is given as 5.67 m.

Empirical models like the hyperbolic model and power creep model predicted settlement equal to 1.03 m and 14.62 m, respectively.

The maximum settlement of different points at the end of 10,950 days by Gibson and Lo model [6] is predicted equal to 25.96 m.

Table 2 Model parameters and their values

Prediction methods	Parametric values	Source of parametric values
Sowers method [1]	$C_c = 0.28$ $C'_\alpha = 0.065$	Geotechnical aspects of landfill design and construction by Qian et al. 9
Bjarngard and Edgers method [2]	$C_{\alpha 1} = 0.035$, $C_{\alpha 2} = 0.215$, $t_1 = 1$–25 days, $t_2 = 200$ days, $C_C^* = 0.106$	El-Fadel et al. [10] Bjarngard and Edgers [2]
Hossain and Gabr model [3]	$C_{\alpha i} = 0.03$, $C_\beta = 0.19$, $C = \alpha f = 0.022$, $t_1 = 10$–15 days, $t_2 = 100$–2000 days, $t_3 = 3500$ days, $t_4 = 0$	Hossain and Gabr [3]
Power creep function model	$N' = 0.50$, $t_r = 1$ days, $M' = 1.7 * 10^{-5}$	El-Fadel et al. [10]
Hyperbolic function model	$\rho_o = 0.001$ m/day, $S_{ult} = 1.140$, $t_{c*} = 1$ month	Ling et al. [5]
Gibson and Lo model [6]	$a = 0.0001$, $b' = 0.00305$, $\lambda/b' = 0.0009$	El-Fadel et al. [10]
Marques model [7]	$E_{dg} = 0.1585$, $d = 1.14 * 10^{-3}$, $C_{c'} = 0.106$, $c = 1.79 * 10^{-3}$, $b = 5.72 * 10^{-4}$	Marques [7]

Table 3 MSW landfill predictions after 30 years using different models

Prediction methods	Total settlement at the end of 30 years (m)
Sowers method [1]	13.54
Bjarngard and Edgers method [2]	23.67
Hossain and Gabr model [3]	5.67
Power creep function model	14.62
Hyperbolic function model	1.03
Gibson and Lo model [6]	25.96
Marques model [7]	15.27

The maximum settlement of different points at the end of 10,950 days by Marques [7] model which considered all the components of the displacement of landfill waste is predicted as 15.27 m.

The predicted total settlement by different models differs on the basis of the model used. Some of the models ignore the mechanical compression as it occurs during or suddenly after placing the waste. But the time-dependent settlement due to creep and biodegradation cannot be ignored as they affect the clay liners (such as compacted clay liners), cover system, and ancillary landfill facilities. It even affects any structure constructed over the landfill.

Figure 2 clearly shows the comparison of displacement response using different models.

4 Conclusions

Estimation of MSW landfill displacement is important to plan cover system and for the protection of appurtenant structures that are set up in the waste, such as gas collection and leachate collection pipes, and structures constructed over the waste. Settlement estimation is important for post-closure maintenance of landfill. Various methods have been published for the computation of MSW landfill settlement. Each model varies on the basis of the assumptions taken and the type of settlement they are computing. A step-by-step procedure is suggested to estimate the settlement of proposed site, and the total settlement is estimated by different methods. The settlement predicted by different models is found to vary significantly. This happens because some models ignore the mechanical compression or the biodegradation-based settlement. However, the settlement occurs due to mechanical compression and biodegradation is the matter of concern as they are going to affect the cover, the appurtenant structures, and the structure to be constructed over the landfill. After analyzing various models based on the mechanisms of settlement they consider, it is concluded that the Marques [7] model considers all the three (primary compression, secondary mechanical compression, and secondary biological compression) components of settlement in good proportion and, hence, gives a fair value of the total settlement. Models like the hyperbolic model and Hossain and Gabr model severely underestimates the settlement. Soil mechanics-based Sowers model predicts the settlement prediction closer to that of the Marques model [7]. There were differences in the initial settlement values of power creep law model compared to Marques [7] settlement values, but the total displacement values at 10,950 days are closer. Bjarngard and Edgers model and Gibson and Lo's model severely overestimate the settlement values.

References

1. Sowers GF (1973) Settlement of waste disposal fills. Proc., 8th Int. Conf. on Soil Mechanics and Foundation Engineering, Moscow 2(2):207–210
2. Bjarngard A, Edgers L (1990) Settlement of municipal solid waste landfills. Proc., 13th Annual Madison Waste Conf., Univ. of Wisconsin, Madison, Wis., pp 192–205
3. Hossain SM, Gabr M (2005) Prediction of municipal solid waste landfill settlement with leachate recirculation. Proc., Geo-Frontiers, Austin, Tex., vol 168, ASCE, 50
4. Edil TB, Ranguette VJ, Wuellner WW (1990) Settlement of municipal refuse. Geotechnics of waste fills: theory and practice, STP1070. In: Landva A, Knowles GD (eds) ASTM, West Conshohocken, Pa., pp 225–239
5. Ling HI, Leschchnsky D, Mohri I, Kawabata T (1998) Estimation of municipal solid waste landfill settlement. J Geotech Geoenviron Eng 124(1):21–28

6. Gibson RE, Lo KY (1961) A theory of soils exhibiting secondary compression. Acta Polytech Scand C (10):1–15
7. Marques ACM (2001) Compaction and compressibility of municipal solid waste. Ph.D. thesis, Sao Paulo Univ., Sao Carlos, Brazil
8. Babu GS, Reddy KR, Chouskey SK, Kulkarni HS (2010) Prediction of long-term municipal solid waste landfill settlement using constitutive model. Pract Periodical Hazard Toxic Radioactive Waste Manage 14(2)
9. Qian X, Koerner RM, Gray DH (2002) Geotechnical aspects of landfill design and construction. Prentice Hall, p 474
10. El-Fadel M, Shazbak S, Saliby E, Leckie J (1999) Comparative assessment of settlement models for municipal solid waste landfill applications. Waste Manage Res 17(5):347–368

Prediction of Liquefaction Using Reliability-Based Regression Analysis

Sufyan Ghani and Sunita Kumari

Abstract The liquefaction susceptibility of soil is generally associated with some soil parameters which are usually measured by laboratory tests on distributed and undistributed samples under different test conditions. This study uses Idriss and Boulanger's (Soil Dyn Earthq Eng 26(2–4):115–130, [6]) simplified procedure based on standard penetration test to evaluate liquefaction criteria to estimate the liquefaction susceptibility for soil deposits of zone IV of Bihar. To overcome the shortcomings of these empirical methods, a regression model has been developed using multi-linear regression analysis to predict liquefaction. The proposed model is a function of the plasticity index, liquid limit, water content, and some other geotechnical parameters. Reliability index (β) and probability of liquefaction (P_L) have also been determined for both the proposed methods for better understanding of their accuracies and robustness. First-order second-moment (*FOSM*) reliability analysis has been adopted in the present paper. The observation drawn from the study demonstrated a reliable and decent prediction rate of the regression model as compared to Idriss and Boulanger method. A robust regression model for evaluating the liquefaction susceptibility which is based on field test data for preliminary prediction would be of great help in the field of geotechnical engineering.

Keywords Liquefaction · Multi-linear regression · Plasticity index · Reliability analysis

1 Introduction

Soil liquefaction is one of the most concerning natural hazards that arises of earthquake and needs to be properly addressed for better safety and serviceability of engineering structures. Liquefaction occurs when a saturated or partially saturated soil transforms to a liquefied state and loses its shear strength and stiffness due to the applied cyclic loads. The excessive pore water pressure builds up to the extent that it

S. Ghani (✉) · S. Kumari
Department of Civil Engineering, National Institute of Technology Patna, Patna, Bihar 800005, India

A. K. Choudhary et al. (eds.), *Advances in Geo-Science and Geo-Structures*, Lecture Notes in Civil Engineering 154,
https://doi.org/10.1007/978-981-16-1993-9_2

exceeds the force between the soil grains that keep them in contact. This contact force between grains holds and withstands the weight of the substructure and superstructure standing on the soil. This loss of contact force in soil structure leads to the loss of shear strength, and the soil is observed to flow like a liquid, causing great damages and creating several problems such as terminating the bearing capacity of structure, settlement of soil, boil-sand condition, and deformation or lateral displacement. The study of liquefaction behavior of soil has been explained by many researchers based on different parameters. In the early stages of the research, it was believed that liquefaction occurs only in saturated loose sand during application of dynamic and seismic loads. After the occurrence of Alaska and Niigata earthquakes in 1964, Seed and Lee [1] and Seed and Idriss [2] developed a methodology known as 'Simplified Approach' for assessing liquefaction potential of cohesionless soil during an earthquake, which uses laboratory data of cyclic triaxial loading tests. This method takes into consideration soil type, relative density, initial confining pressure, intensity, and duration of ground shaking. This simplified procedure has been modified and updated by various researchers [3–7]. This study focuses on a revised and updated semi-empirical approach developed by Idriss and Boulanger [7] for assessment of seismic behavior of soil deposits based on standard penetration test. This empirical equation became a popular practice in many projects, for assessing liquefaction of soil deposits. The liquefaction results evaluated using these empirical approaches are based on large number of equations which require large number of laboratory-based datasets. This extended demand for dataset needs more laboratory testing which makes these empirical approaches a bit resource and time-consuming. Therefore, a realistic equation for quick evaluation of liquefaction susceptibility of soil deposits which uses fewer datasets and based on its high-end training predicts a broader area, would be extremely helpful for preliminary prediction. Many researchers have promoted the use of multi-linear regression analysis in place of conventional empirical equations in geotechnical engineering [8–13]. Regression analysis is one of the most reliable statical analysis tools for identifying the impact of input variables on the liquefaction behavior of soil. Performing a regression analysis allows you to assertively determine which soil parameters influence the liquefaction susceptibility of the soil deposit the most along with which parameters can be ignored during this study and how these parameters influence each other. Therefore, an attempt has been made to develop a simplified equation based on regression analysis which uses field test data obtained from four investigation sites of Patna districts, which falls in seismic zone IV in Bihar, India, such as liquid limit, plasticity index, fine content, standard penetration test blow count number, bulk density, and water content for evaluating cyclic resistance ratio (*CRR*) which in return determine the liquefaction susceptibility of soil deposits.

Both the methods adopted in this study to evaluate liquefaction susceptibility of soil deposits are robust and appropriate methods, but there are many possible factors that may cause uncertainty in determining liquefaction potential. To justify the uncertainties in different methods and models and to obtain the most reliable method out of the two, a reliability analysis has been performed. Reliability-based design methods have been active for most of the civil engineering works such as building structure, dam, analysis of slope, and settlement of foundation. [14–23]. This study adopts Idriss and Boulanger [7] method to evaluate liquefaction susceptability of soils and develops a regression model to overcome the major shortcomings of the above-mentioned approch. Further, a reliablity analysis has been performed using first-order second-moment (FOSM) method to obtain a most appropriate and reliable method among the two. Probability of liquefaction has aslo been determined for better understanding and clarity of the results.

2 Methodology

2.1 Idriss and Boulanger [6] Method

For liquefaction assessment of soil deposits, the cyclic stress ratio (*CSR*) that illustrates the loading that is imposed by earthquakes on the soil has been proposed as:

$$(\mathrm{CSR})_{7.5} = 0.65\left(\frac{\sigma_{\mathrm{vo}} a_{\max}}{\sigma'_{\mathrm{vo}}}\right)\frac{r_d}{\mathrm{MSF}}\frac{1}{K_\sigma} \tag{1}$$

where σ_{vo} and σ'_{vo} are the total and effective vertical overburden stress, respectively, at some specified depth z; $a_{\max}$ is the peak horizontal ground acceleration in $g's$; r_d is a stress reduction factor; *MSF* represents the magnitude scaling factor which is applied to alter the induced *CSR* to the reference earthquake magnitude of 7.5; and K_σ presents the correction factor for effective overburden.

Stress reduction coefficient (r_d) is calculated using analytical procedures and is adequately expressed as a function of depth (z) and earthquake magnitude (M). The following expression for r_d is applicable to a depth where $z \leq 34$ m, in which z is depth in meters and M is earthquake magnitude.

$$Ln(r_d) = \alpha(z) + \beta(z) M Ln(r_d) = \alpha(z) + \beta(z) M \tag{2}$$

$$\alpha(z) = -1.012 - 1.126\sin\left(\frac{z}{11.73} + 5.133\right) \tag{3}$$

$$\beta(z) = 0.106 + 0.118\sin\left(\frac{z}{11.28} + 5.142\right) \tag{4}$$

whereas the following expression is applicable for $z > 34$ m;

$$r_d = 0.12\exp(0.22M) \tag{5}$$

Equations used to determine cyclic resistance ratio (*CRR*) and corresponding factor of safety (*FOS*) generated from the corrected blow count $(N_1)_{60}$ have been listed below:

$$(\text{CRR} = \exp\left[\frac{(N_1)_{60\text{cs}}}{14.1} + \left\{\frac{(N_1)_{60\text{cs}}}{126}\right\}^2 - \left\{\frac{(N_1)_{60\text{cs}}}{23.6}\right\}^3 + \left\{\frac{(N_1)_{60\text{cs}}}{25.4}\right\}^4 - 2.8\right] \tag{6}$$

where term $(N_1)_{60\text{cs}}$ depends on $(N_1)_{60}$ and given as:

$$(N_1)_{60\text{CS}} = (N_1)_{60} + \Delta(N_1)_{60} \tag{7}$$

The following equation is used to evaluate factor of safety (FOS) against liquefaction, which is described as:

$$\text{FOS} = \frac{\text{CRR}}{\text{CSR}} \tag{8}$$

2.2 *Multiple Regression Analysis:*

Multi-linear regression analysis is a statistical technique which can be used to examine the relationship between a dependent variable (input parameters) and a set of independent variables, i.e., output parameter. For a multiple linear regression analysis, it is assumed that the relationship between the input variables and output variables is linear and has the following form:

$$y = \beta_0 + \beta_1 X_1 + \beta_2 X_2 + \cdots \beta_p X_p \tag{9}$$

where y is the predicted or expected value of the dependent variable, X_I through X_P are p distinct independent or predictor variables, β_0 is the value of y when all of the independent variables (X_1 through X_P) are equal to zero, and β_1 through β_p are the estimated regression coefficients.

A multiple linear regression analysis of the results obtained from the conventional methods was carried out, considering all the significant parameters such as liquid limit (LL), plasticity index (PI), moisture content (w_C), fine content (FC), SPT blow count (N_{60}), and bulk density (Y). An equation has been proposed to evaluate CRR_model [Eq. (13)] which further evaluates the modeled factor of safety FOS_model [Eq. (14)] to predict liquefaction susceptibility. Various statical performance parameters have been established to evaluate the performance of regression model. In this study, coefficient of determination (R^2) also known as model fit value and correlation coefficient (R) have been used. R^2 ranges from 0 to 1. The closer it is to 1, the better the fit. If R^2 equal to 1, it means that perfect linear relationship exists between the dependent variable and independent variables, while R^2 equal to 0 indicates independent variables have no impact on the dependent variable.

2.3 Reliability Analysis

The factor of safety evaluated from the above-mentioned analysis cannot be directly considered as a reliable output for risk evaluation as considerable amount of uncertainties may be involved, such as error in describing the soil properties and errors associated with the adopted analytical techniques. Thus, soil deposits with a factor of safety greater than unity can still have some probability of failure. Hence, the need of reliability analysis comes into action. First-order second-moment (FOSM) approach has been used to determine the reliability index in this study. The FOSM method applies a Taylor series elaboration for the function to be estimated and is generally used to imprecise the ambiguities available in the input variables. According to FOSM method, if μ_Z and σ_Z are the mean value and the standard deviations of the performance function Z, respectively, reliability index (β) can be calculated by the following equation:

$$\beta = \frac{\mu_Z}{\sigma_Z} = \frac{\mu_R - \mu_S}{\sqrt{\sigma_R^2 + \sigma_S^2}} \tag{10}$$

The probability of liquefaction needs to be dependent on mean and variance of obtained factor of safeties; therefore, reliability index (β) in terms of factor of safety can be obtained by the following equation:

$$\beta = \frac{\mu_F - 1}{\sigma_F} \tag{11}$$

where μ_F are the mean values of factor of safety and σ_F are the standard deviations of factor of safety. Assuming that the all the random variables are normally distributed, the probability of failure (P_F) can be evaluated by:

$$P_F = 1 - \Phi(\beta) \tag{12}$$

where $\Phi(\beta)$ is the standard normal cumulative probability.

3 Result and Discussion

The proposed sites in the present paper are situated near the banks of Ganga River of Patna district of Bihar, which in an active zone of earthquakes. These selected sites are highly populated, and the subsoil strata is of clayey nature with the presence of plastic fines. Hence, a study has been carried out for developing a regression model to evaluate modeled cyclic resistance ratio (CRR_model), considering liquid limit (LL), plasticity index (PI), SPT blow count value (N_{60}), fine content ($F \cdot C$) and water content (wc), and bulk density (γ).

The liquefaction susceptibility of the proposed sites is presented in Tables 1, 2, 3, 4, and 5. **Depths with FOS < 1 are liquefiable layers, whereas soil depths with FOS ≥ 1 are non-liquefiable layers.**

Due to uncertainty and ambiguities observed in the conventional methods, a multi-linear regression analysis was performed and an equation has been proposed that can facilitate the predication of liquefaction susceptibility. The developed regression model determines the modeled cyclic resistance ratio (CRR_{model}) **for soil profile. An equation has been formed based on the regression coefficient present in Table** 5. The results obtained from the equation will help in determining the FOS_{model} which will indicate the liquefaction potential of the concerned soil deposits. The following section provides an insightful data obtained from the regression model.

$$\mathrm{CRR}_{\mathrm{model}} = -1.2024 + 0.0067\mathrm{LL} - 0.0073\mathrm{PI} + 0.0163wc$$

Table 1 Evaluation of FOS as per Idriss and Boulanger method for site-1

Depth	Soil type	TS	ES	N_{60}	$F{\cdot}C$	CSR	CRR	FOS = CRR/CSR
1.5	CI	29.28	14.56	6	28.9	0.20	0.16	0.77
3.0	CI	58.56	29.13	9	28.5	0.20	0.21	1.03
4.5	CI	88.73	44.58	15	27.8	0.20	0.37	1.81
6.0	CI	118.30	59.44	14	27.4	0.20	0.25	1.25
7.5	CI	148.62	75.04	9	26.7	0.20	0.16	0.79
9.0	CI	178.34	90.05	15	26.4	0.20	0.21	1.07
10.5	CI	208.07	105.06	15	26.3	0.23	0.20	0.99
12.0	CI	237.79	120.07	16	26.2	0.22	0.19	0.98
13.5	CI	267.51	135.08	16	25.4	0.22	0.18	0.92
15.0	CI	297.24	150.09	26	25.0	0.25	0.30	1.51

Table 2 Evaluation of FOS as per Idriss and Boulanger method for site-2

Depth	Soil type	TS	ES	N_{60}	$F{\cdot}C$	CSR	CRR	FOS = CRR/CSR
1.5	CL	28.69	13.98	4	85.4	0.21	0.13	0.64
3.0	CL	56.80	27.37	7	79	0.21	0.18	0.83
4.5	CL	88.73	44.59	13	99	0.20	0.29	1.41
6.0	CL	118.31	59.45	16	87.7	0.20	0.32	1.58
7.5	CL	147.89	74.31	13	97.9	0.20	0.21	1.04
9.0	CL	177.46	89.17	14	94	0.20	0.21	1.03
10.5	CL	208.07	105.07	19	97.8	0.20	0.27	1.31
12.0	CL	237.79	120.07	21	95.4	0.20	0.28	1.36
13.5	CL	268.84	136.41	22	92.2	0.20	0.26	1.31
15.0	CL	298.71	151.56	19	96.7	0.20	0.21	1.03

Table 3 Evaluation of FOS as per Idriss and Boulanger method for site-3

Depth	Soil type	TS	ES	N60	*F·C*	CSR	CRR	FOS = CRR/CSR
1.5	CL	27.92	13.20	5	40.2	0.32	0.21	0.65
3.0	CL	55.84	26.41	3	40.2	0.32	0.14	0.46
4.5	CL	83.76	39.62	3	40.2	0.31	0.14	0.45
6.0	CL	103.02	44.16	2	40.2	0.32	0.11	0.35
7.5	CL	128.78	55.20	6	40.2	0.32	0.19	0.60
9.0	CL	171.80	83.51	17	83.8	0.31	0.48	1.53
10.5	CL	200.43	97.43	11	83.3	0.29	0.27	0.91
12.0	CL	229.07	111.35	11	83.3	0.28	0.25	0.90
13.5	CL	266.61	134.17	12	89.7	0.26	0.26	0.97
15.0	CL	296.22	149.07	12	89.7	0.25	0.24	0.98

Table 4 Evaluation of FOS as per Idriss and Boulanger method for site-4

Depth	Soil type	TS	ES	N60	*F·C*	CSR	CRR	FOS = CRR/CSR
1.5	CI	20.93	11.12	2	31.9	0.33	0.18	0.55
3.0	CI	41.86	32.05	6	31.9	0.32	0.32	0.97
4.5	CI	62.75	52.94	3	40.2	0.32	0.15	0.45
6.0	CI	83.67	73.86	4	40.2	0.32	0.16	0.49
7.5	CI	104.59	94.78	2	40.2	0.32	0.11	0.34
9.0	CI	141.00	131.19	5	89.2	0.31	0.16	0.51
10.5	CI	164.49	154.68	13	89.2	0.30	0.32	1.07
12.0	CI	188.97	179.16	9	90.3	0.28	0.22	0.77
13.5	CI	212.59	202.78	14	90.3	0.27	0.30	1.12
15.0	CI	240.61	230.80	15	84.6	0.25	0.30	1.20

Table 5 Regression coefficient for Mithapur site

Regression coefficient	Values
Intercept	−1.2024
LL	0.0067
PI	−0.0073
wc	0.0163
γ	0.0610
N_{60}	0.0161
$F \cdot C$	−0.0023

$$+ 0.0610\gamma + 0.0161 N_{60} - 0.0023\text{FC} \quad (13)$$

$$\text{FOS}_{\text{model}} = \frac{\text{CRR}_{\text{model}}}{\text{CSR}} \quad (14)$$

where N_{60} is SPT blow count value, LL is liquid limit (%), PI is plasticity index, *wc* is natural moisture content, F · C is fine content, and γ is bulk density The regression statistics of the above $\text{CRR}_{\text{model}}$ are correlation coefficient (R) = 0.87 and coefficient of determination (R^2) = 0.77. The value of R^2 signifies that the accuracy of model developed is 77%. Figure 1 presents the factor of safety evaluated using regression model and empirical method. Idriss and Boulanger method uses empirical relations and Eqs. (1) and (6) to determine *CSR* and *CRR* and later uses Eq. (8) to calculate *FOS* which predicts the liquefaction susceptibility of the soil deposits. These empirical equations demand higher datasets for liquefaction evaluation of soil layers which needs high-end laboratory and in situ field tests which tend to make this approach a bit time taking and costly. Therefore, to overcome the drawbacks of empirical methods and to make liquefaction prediction fast and cost efficient, a regression model was proposed to evaluate $\text{CRR}_{\text{model}}$ which further evaluates $\text{FOS}_{\text{model}}$ to predict liquefaction behavior of soil deposits. Based on the observation drawn after the analysis, it is quite difficult to bifurcate from the obtained result that out of the two methods which is more reliable. Therefore, a first-order second-moment (*FOSM*) reliability analysis has been performed to determine the most reliable method.

The reliability index (β) and probability of liquefaction (P_{L}) have been evaluated for both the methods using Eqs. (11) and (12). Figures 2 and 3 show that the reliability index (β) values of the developed regression model are better as compared to Idriss and Boulanger [7] method. Higher reliability index value indicates a reliable result as compared to lower reliability index values which can be used for designing and construction purposes. In few cases, it has been observed that β is less than 1 which indicates triggering of liquefaction. So, ground improvement techniques must be used in such sites before designing of the structure. For Idriss and Boulanger approach,

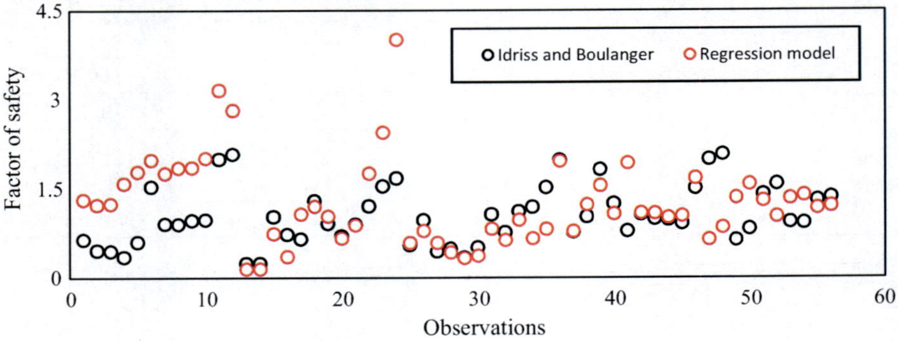

Fig. 1 Factor of safety of regression model and Idriss and Boulanger method

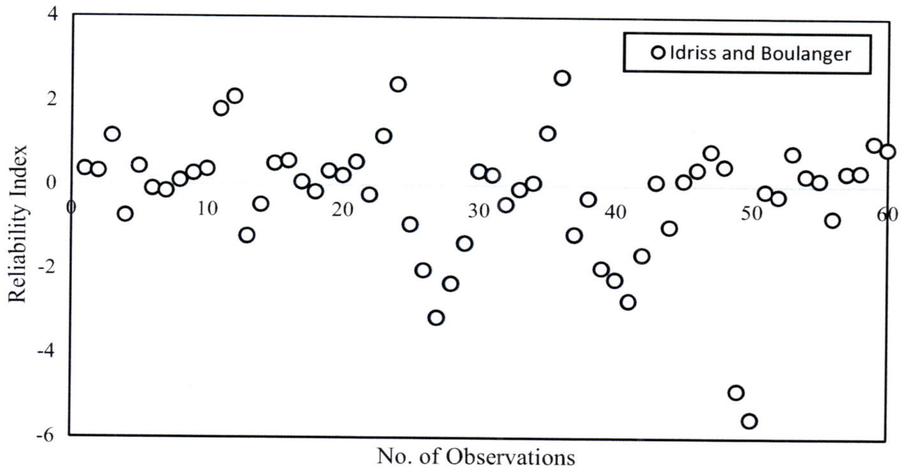

Fig. 2 Reliability index (β) of Idriss and Boulanger method

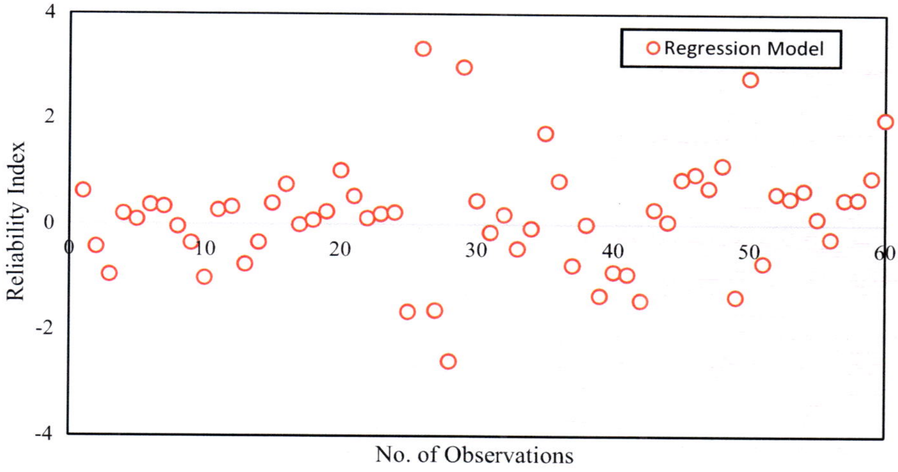

Fig. 3 Reliability index (β) of developed regression model

the majority of reliability indices values are negative indicating high probability of liquefaction as compared to MLR model.

Figure 4a and b presents the trend of probability of liquefaction along with the evaluated factor of safety for Idriss and Boulanger method and regression model adopted in the study. The accuracy of calculated probability depends on the accuracy with which β is calculated, PL = 0.5 at $\beta = 0$. Based on the observation drawn from both the figures, it is concluded that the overall probability of failure for the developed regression model is less as compared to Idriss and Boulanger method. The average liquefaction probability for Idriss and Boulanger method is around 61%, whereas for

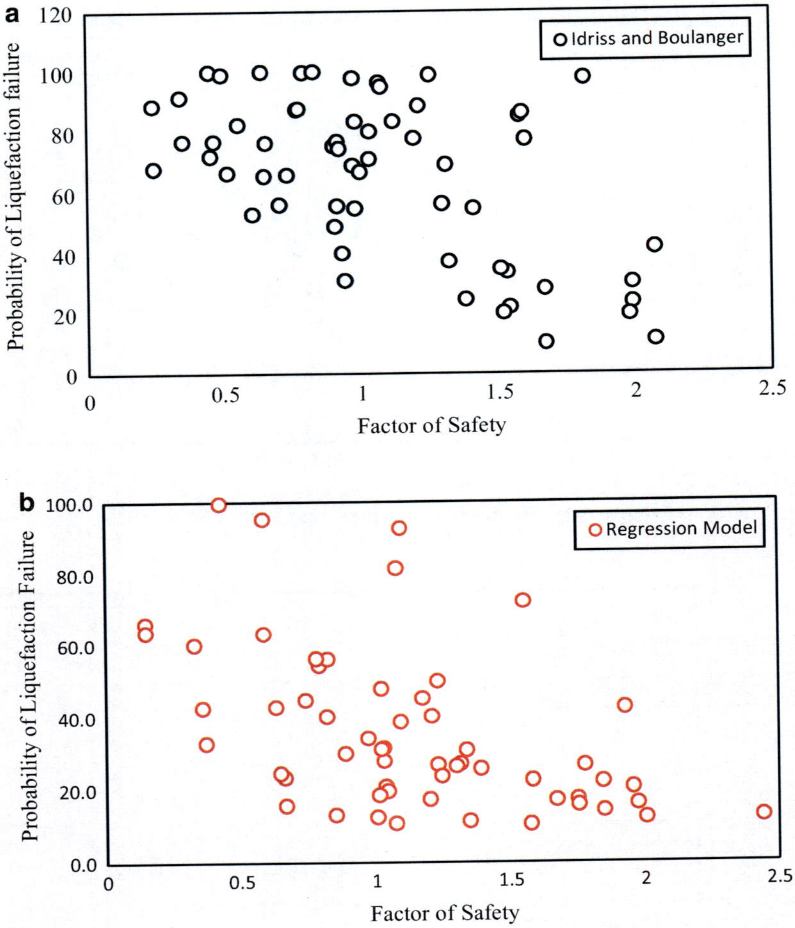

Fig. 4 **a** Probability of liquefaction failure corresponding to factor of safety evaluated using Idriss and Boulanger method. **b** Probability of liquefaction failure corresponding to factor of safety evaluated using developed regression model

the developed regression model, this failure probability reduces to 36%, nearly half in comparsion with the conventional method.

Table 6 presents the comparison of liquefiable and non-liquefiable soil layer between Idriss and Boulanger [7] method and developed regression model. It can be observed that for the developed regression model, the probability of liquefaction failure in soil decreases as compared to Idriss and Boulanger method. The reason for these differences and variations in the output of regression model is due to the consideration of different input parameters. Consideration of plasticity index (PI) and fine content (FC) in evaluating CRR_{model} tends to increase the liquefaction resistance of a soil due to the dilative nature of plastic fines—a finding which is in agreement with the

Table 6 Success rate of Idriss and Boulanger method and developed regression model

Methodology	Success rate	
	Liquefiable (%)	**Non-liquefiable (%)**
Idriss and Boulanger empirical method	61.33	38.67
Regression model	36.32	63.68

results of other researchers. The relation of PI with liquefaction potential of soil has been tested and well understood by many researchers [24]. Gratchev [25] suggested that the use of PI for liquefaction analysis is a reliable and appropriate approach, and soils with higher plasticity indices are less likely to undergo liquefaction. Sridharan [26, 27] explained the relation and influence of PI on pore water pressure which itself is one of the root causes for liquefaction. Fines can also influence a soil's ability to develop excess pore pressures, which is significantly accountable for increasing the liquefaction potential of the soil deposits. The present study concludes that the developed multi-linear regression model is an effective and reliable approach when compared with Idriss and Boulanger method, which is a conventional and empirical approach involving detailed calculation and skilled knowledge for evaluating liquefaction. It was also perceived that the consideration of plasticity index, liquid limit, and fine content for evaluation of liquefaction susceptibility is an appropriate approach.

4 Conclusion

The aim of the study was to develop a CRR_{model} based on multi-linear regression analysis of the data obtained from laboratory tests to estimate the liquefaction susceptibility of soil deposits. The CRR_{model} developed is a function of different geotechnical properties of soil. Reliability indices and probability of liquefaction were obtained to define the liquefaction potential in a more effective manner. The difference in the probabilities of soil liquefaction (PL) obtained for both the methods can be observed from the developed β and PL-graphs. It clearly indicates that the developed regression model is a better reliable method and possesses lesser probability to liquefaction. The developed regression model in the present study could be used for the preliminary design purposes. Construction of a structure on liquefiable zone requires a huge amount of investment, and if the probability of liquefaction is higher, liquefaction mitigation will add on the investment. Therefore, the robustness of the developed regression model and the enhanced success rate as compared to the conventional method will certainly help in minimizing the use of available natural resources and will also contribute toward economical design. This study could be also used as a practical means for the engineers and designers to measure liquefaction potential of the sites and selecting proper mitigation techniques.

References

1. Seed HB, Lee KL (1966) Liquefaction of saturated sands during cyclic loading. J Geotech Eng ASCE 92(6):105–134
2. **Seed HB, Idriss I (1971) MSimplified procedure for evaluating soil liquefaction potential. J** Soil Mech Found Divis 92(6):1249–1273
3. Seed HB (1979) Soil liquefaction and cyclic mobility evaluation for level ground during earthquakes. J Geotech Eng Divis ASCE 105(2):201–255
4. **Seed HB, Idriss IM, Arango I (1983)Evaluation of liquefaction potential using field perfor-** mance data. J Geotech Eng Div ASCE 109(3):458–482. https://doi.org/10.1061/(ASCE)0733-9410(1983)109:3(458)
5. **Iwasaki T, Arakawa T, Tokida K (1984) Simplified procedure for assessing soil liquefac-** tion during earthquakes. Soil Dyn Earthq Eng 3(1):49–59. https://doi.org/10.1016/0261-7277(84)90027-5
6. Idriss IM, Boulanger RW (2004) Semi-empirical procedures for evaluating liquefaction potential during earthquakes. In: Proceedings of 11th international conference on soil dynamics and earthquake engineering, vol 1, pp 32–56
7. Idriss IM, Boulanger RW (2006) Semi-empirical procedure for evaluating liquefaction potential during earthquakes. Soil Dyn Earthq Eng 26(2–4):115–130. https://doi.org/10.1016/j.soildyn.2004.11.023
8. Bartlett SF, Youd TL (1992) Empirical analysis of horizontal ground displacement generated by liquefaction-induced lateral spread. Tech. Rep. No. NCEER-92-0021, National Centre for Earthquake Engineering Research, Buffalo, N.Y., p 114
9. Bartlett SF, Youd TL (1995) Empirical prediction of liquefaction-induced lateral spread. J Geotech Eng. https://doi.org/10.1061/(ASCE)0733-9410(1995)121:4(316),316-329
10. Youd TL, Hansen CM, Bartlett SF (2002) Revised multilinear regression equations for prediction of lateral spread displacement. J Geotech Geoenviron Eng. https://doi.org/10.1061/(ASCE)1090-0241(2002)128:12(1007),1007-1017
11. Latha GM, Somwanshi A, Reddy KH (2013) A multiple regression equation for prediction of bearing capacity of geosynthetic reinforced sand beds. Indian Geotech J 43:331–343. https://doi.org/10.1007/s40098-013-0053-7
12. Youd TL (2018) Application of MLR procedure for prediction of liquefaction-induced lateral spread displacement. J Geotech Geoenviron Eng 144(6)
13. Anwar A, Jamal Y, Ahmad S, Khan MZ (2016)'Assessment of liquefaction potential of soil using multi-linear regression modeling. Int J Civ Eng Technol (IJCIET) 7(1):373–415
14. **Uzielli M, Duzgun S, Vangelsten BV (2006) A first-order second-moment framework for** probabilistic estimation of vulnerability to landslides. Engineering Conferences International
15. Babu GLS, Srivastava A (2007) Reliability analysis of allowable pressure on shallow foundation using response surface method. Comput Geotech 34(3):187–194. https://doi.org/10.1016/j.compgeo.2006.11.002
16. Juang CH, Fang SY, Li DK (2008) Reliability analysis of liquefaction potential of soils using standard penetration test. In: Phoon KK (ed) Reliability-based design in geotechnical engineering, Computers and applications. Taylor & Francis, London
17. Juang CH, Ching J, Luo Z, Ku CS (2012) New models for probability of liquefaction using standard penetration tests based on an updated database of case histories. Eng Geol 133:85–93. https://doi.org/10.1016/j.enggeo.2012.02.015
18. Jha SK, Suzuki K (2008) Reliability analysis of soil liquefaction based on standard penetration test. Comput Geotech 36(4):589–596. https://doi.org/10.1016/j.compgeo.2008.10.004
19. Jha SK, Suzuki K (2009) Liquefaction potential index considering parameter uncertainties. Eng Geol 102(1–2):55–60. https://doi.org/10.1016/j.enggeo.2009.03.012
20. Bungenstab FC, Bicalho KV (2016) Settlement predictions of footings on sands using probabilistic analysis. J Rock Mech Geotech Eng 8(2):198–203. https://doi.org/10.1016/j.jrmge.2015.08.009

21. Malkawi AIH, Hassan WF, Abdulla AF (2000) Uncertainty and reliability analysis applied to slope stability. Struct Saf 22(2):161–187. https://doi.org/10.1016/S0167-4730(00)00006-0
22. Johari A, Javadi AA (2012) Reliability assessment of infinite slope stability using the jointly distributed random variables method. Scientia Iranica 19(3):423–429. https://doi.org/10.1016/j.scient.2012.04.006
23. Umar SK, Samui P, Kumari S (2018) Deterministic and probabilistic analysis of liquefaction for different regions of Bihar. Geotech Geol Eng. Springer, Published online 36(5):3311–3321
24. Ghani S, Kumari S (2020) Insight into the effect of fine content on liquefaction behavior of soil. Geotech Geol Eng. https://doi.org/10.1007/s10706-020-01491-30000
25. Gratchev I, Sassa K, Fukuoka H (2006) How reliable is the plasticity index for estimating the liquefaction potential of clayey sands? J Geotech Geoenviron Eng 132(1):124–127
26. Sridharan A, Rao S, Murthy NS (1986) Liquid limit of montmorillonite soils. Geotech Test J 9(3):156–159
27. Sridharan A, Rao S, Murthy NS (1988) Liquid limit of kaolinitic soils. Geotechnique 38(2):191–198

Modeling of Consolidation Considering Coupled Soil-Fluid Interaction

Sunita Kumari and Avinash Arya

Abstract Expulsion of excess pore water pressure caused by load applied for longer time which results in reduction of volume of the soil is termed as consolidation of soil mass. The response in terms of displacement is dependent of soil mass properties, applied loading, pore water pressure, and time. The present paper includes formulation of consolidation problem using finite element method (FEM). The basic matrix equations for coupled finite element are formulated based on Biot's consolidation equations, coupled with two-dimensional equilibrium equation and continuity equation. Shape function is used to discretize the spatial variables, and finite difference time marching scheme is used to discretize the time domain. The FORTRAN 90 program has been used for writing the FEM code. A parametric study has been also carried out for different elastic modulus to obtain the response of soil deposit.

Keywords Consolidation · Saturated soil · Finite element method (FEM) · Excess pore pressure · Settlement · Depth

1 Introduction

Soil is a three-phase system with water and air filling the voids of solid particles. When the air voids are occupied by water completely, the soil becomes a two-phase system called as saturated soil [1, 2]. This water filled in voids causes occurrence of pore water pressure due to the stress applied on the soil. When stress is applied to the soil, initially the stress is carried out by the soil skeleton, and with passage of time, the stress is transferred to the pore water present in between the soils resulting in development of pore water pressure and consolidation occurs. The concept of three-dimensional deformation of saturated soil skeleton was first developed by Biot [3]. Eventually, this coupled formulation was extended to implement different developed constitutive models in a two-dimensional finite element program [4–10].

S. Kumari · A. Arya (✉)
National Institute of Technology Patna, Patna, India

A. K. Choudhary et al. (eds.), *Advances in Geo-Science and Geo-Structures*, Lecture Notes in Civil Engineering 154,
https://doi.org/10.1007/978-981-16-1993-9_3

2 Finite Element Formulation

A fully coupled analysis accounts for the dynamic interaction between the solid and fluid phase. Resulting dynamic equilibrium and flow continuity equation are solved simultaneously. The governing coupled equation for consolidation process is obtained by discretizing the equilibrium and continuity equation. These equations are as follows:

$$[S]u - [G]P_e = [f_u] \tag{1}$$

$$[G]^T\{u\} - [H][P_e] = [\{n_G\} + G] \tag{2}$$

In which, $[S]$, $[G]$, and $[H]$ are known as stiffness, coupling, and permeability matrix, respectively. Also, f_{u} and n_{G} are known as force matrix for soil and fluid phase, respectively. $u - P_{\mathrm{e}}$ are defined as nodal displacement and excess pore water pressure.

In the present study, a mixed element having eight-noded element for displacement and four-noded element for pore pressure has been considered. Vertical settlement, u_z, and Horizontal settlement, u_x, at any node in the soil element can be expressed in terms of nodal settlements as given below:

$$u_z = \sum_{i=1}^{n} N_i(u_z)_i;\ u_x = \sum_{i=1}^{n} N_i(u_x)_i \tag{3}$$

The pore pressure in the four-noded soil element is given as:

$$P = \sum_{i=1}^{4} N_i^P P_i \tag{4}$$

Finite difference time marching scheme has been used to discretize Eqs. 1 and 2 which can be given as:

$$\frac{\partial u}{\partial t} = \frac{u_i - u_{i-1}}{\Delta t} \tag{5}$$

3 Validation

The present formulation is validated with Terzaghi one-dimensional consolidation theory. For this analysis, a clay layer of thickness 5 m subjected to a vertical pressure of 100 kPa and maintained constant with time is considered. Drainage is allowed from the top and bottom surfaces. The clay layer is divided into 20 uniform elements. And

Table 1 Soil properties for validation

Modulus of elasticity, E	4131 kPa
Coefficient of permeability, K_v	1.38×10^7 m/s
Coefficient of consolidation, C_v	5×10^{-5} m^2/s
Depth of soil, z	5 m

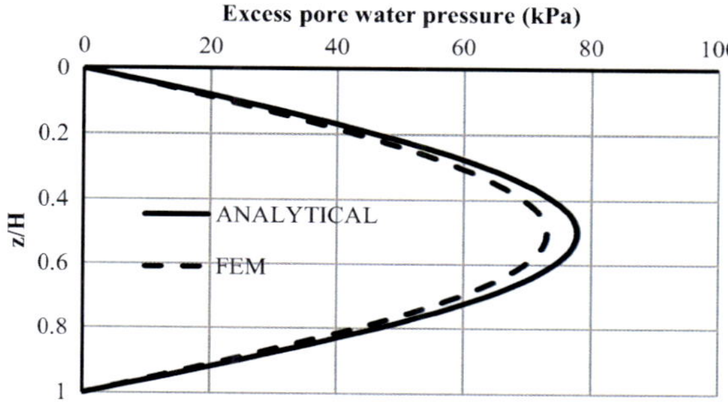

Fig. 1 Comparison of excess pore pressure with respect to depth for 50% consolidation

the time step of 0.01 s was used. The material properties used for validation purpose are given in Table 1.

The value of pore pressure is zero at the top and bottom of the assumed problem domain which shows a double drainage case. Figure 1 also simulates a consolidation behavior, and this figure shows the good agreement for 50% consolidation. For excess pore water pressure response, it is observed that the variations are not much in this case also. The maximum excess pore pressure at mid depth is observed 73.2 kPa by FEM approach, and the same is 77.95 kPa using analytical approach. The results show very good agreement between the numerical analysis and the exact solution. These figures show that it is approximately simulating the consolidation phenomena.

4 Result and Discussion

A homogenous soil domain of finite 10 m depth in vertical direction is considered as shown in Fig. 2. Also, to find the influence of modulus of elasticity on clayey soil under consolidation, parametric study has been carried out at different modulus of elasticity varying as 5000, 6000, 8000, and 10,000 kPa while maintaining other parameters constant. The present study is carried out for a soil domain having double drainage condition and permeability of 8.64×10^{-7} m/d. The variations of excess pore water pressure (EPWP) and displacement in horizontal and vertical direction with depth factor at different modulus of elasticity and 30% consolidation stages are shown in Figs. 3, 4, and 5.

From the EPP response as in Fig. 3, it is observed that pore pressure increases with decrease in modulus of elasticity. Since analysis has been done for double drainage,

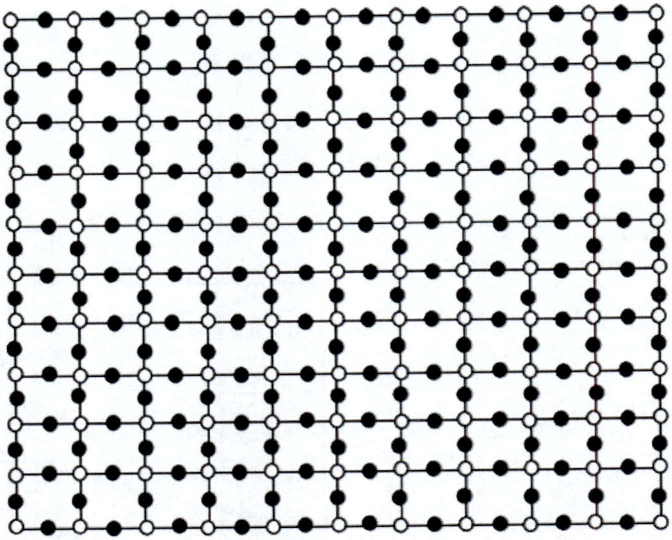

Fig. 2 Soil domain (10 m depth)

Fig. 3 Comparison of excess pore pressure for different modulus of elasticity (kPa) with respect to depth

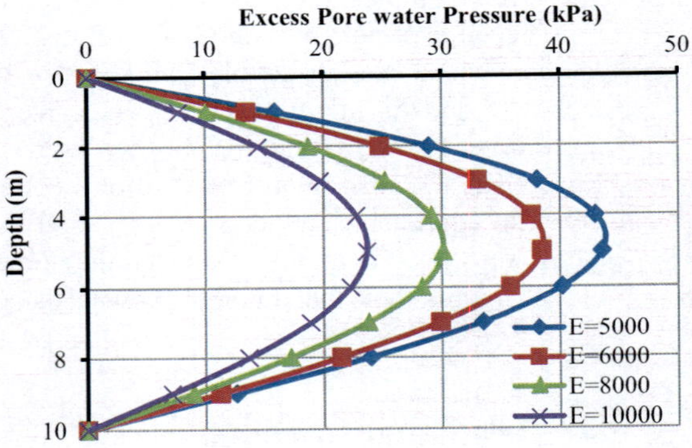

this value is seen maximum at middle of soil layer and consequently decreases at bottom layer. At modulus of elasticity (E) is 5000 kPa, the pore pressure at mid-depth is achieved as 43.60 kPa, whereas this value decreases to 23.70 kPa for 10,000 kPa. It is also noticed that with increasing value of modulus of elasticity, loads will be distributed among soil skeleton; thus, a moderate behavior of excess pore pressure at each step is noticed.

Figure 4 shows the settlement response in vertical direction. It is observed that with decrease in modulus of elasticity, the settlement increases and vice versa. Keeping the permeability constant, it is noticed that at $E = 5000$ kPa, settlement is 14.7 cm at the top, and as this value increases to 6000 kPa, settlement is observed as 12.6 cm

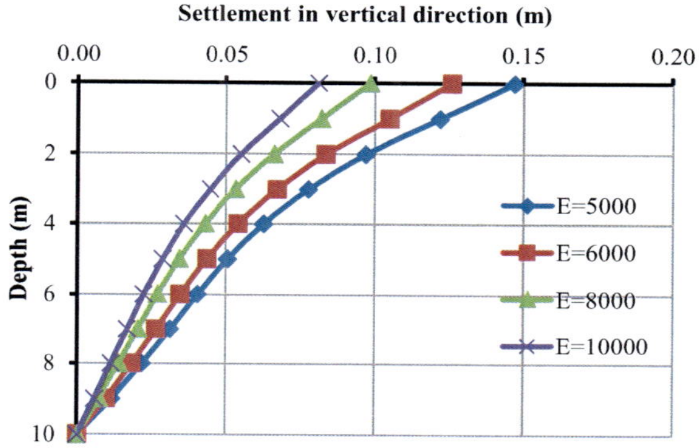

Fig. 4 Comparison of settlement in vertical direction for different modulus of elasticity with respect to depth

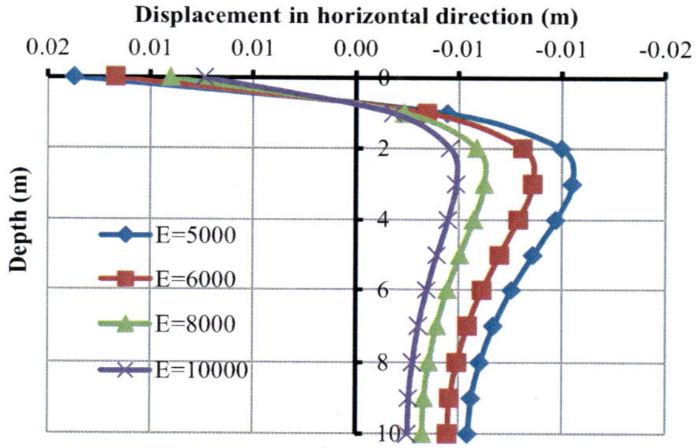

Fig. 5 Comparison of displacement in horizontal direction for different modulus of elasticity with respect to depth

at the top surface. These values are decreased up to 9.80 and 8.10 cm for $E = 8000$ kPa and $E = 10{,}000$ kPa, respectively. From Fig. 4, it is observed that with increase in depth, the settlement reduces, and it approaches minimum value at the bottom of the concerned soil domain.

In Fig. 5, displacement in horizontal direction is plotted with depth for different modulus of elasticity. From response, it is observed that the displacement is very small as compared to vertical settlement at the top which is 1.37 cm. At the depth of 2–3 m, it gives maximum displacement of 1.19 cm in opposite direction to that obtained at the top. With increase in modulus of elasticity, the displacement decreases as the soil becomes stiffer due to which load is carried by the soil skeleton.

5 Conclusion

The present formulation is capable of modeling the consolidation behavior of the soil mass. It gives the simultaneous response of pore pressure and displacement with time which considers the continuous degradation of stiffness of soil mass under loading. Therefore, this formulation is able to measure the real behavior of consolidation behavior. It is also concluded that excess pore pressure, vertical settlement, and horizontal displacement show uniform decrease with increase in modulus of elasticity.

References

1. Booker KJR, Small JC (1977) Finite element analysis of primary and secondary consolidation. Int J Solids Struct 13:137–149
2. Terzaghi K (1943) Theoretical soil mechanics. Wiley, New York
3. Biot MA (1941) General theory of three-dimensional consolidation. J Appl Phys 12:155–164
4. Desai CS, Johnson LD (1972) Evaluation of two finite element formulations for one-dimensional consolidation. Comput Struct 2:469–486
5. Sandhu RS, Wilson EL (1969) Finite element analysis of flow in saturated porous elastic media. J Eng Mech Divis ASCE 95(3):641–652
6. Carter JP, Small JC, Booker JR (1977) A theory of finite elastic consolidation. Int J Solids Struct 13:467–478
7. Menendez C, Nieto PJG, Ortega FA, Bello A (2009) Mathematical modelling and study of the consolidation of an elastic saturated soil with an incompressible fluid by FEM. Math Comput Model 49:2002–2018
8. Vermeer PA, Verruijit A (1981) An accuracy condition for consolidation by finite elements. Int J Numer Anal Meth Geomech 5:1–14
9. Abid MM, Pyrah IC (1988) Guidelines for using the finite element method to predict one-dimensional
10. Huang J, Griffiths DV (2010) One-dimensional consolidation theories for layered soil and coupled and uncoupled solutions by the finite element method. Geotechnique 60(9):709–713

Comparative Assessment of Kurukshetra City Waste Dumping Sites Using RIAM Analysis: A Case Study

Sanjeev Kumar and **Surinder Deswal**

Abstract Kurukshetra, a historical city in India, is dealing with major environmental issues regarding the final disposal of municipal solid waste (MSW). The waste of urban localities is directly dumped in the landfills situated in nearby villages. The selection of a new disposal site and technique also requires much attention. Environmental impact assessment or EIA is needed for establishing such waste management facilities for the city waste. The whole procedure of EIA is time-consuming, but it is urgent to consider all the different components affecting the environment. Rapid impact assessment matrix (RIAM) is a proven method to investigate and evaluate the physical, ecological, economic, and social-cultural impacts due to the developmental projects. The same tool is applied for the assessment of two different MSW landfill sites. Both of the disposal sites are on different stages of operations. Thus, analysis based on RIAM is used to find out the concerned areas which are affected due to dumping operations on these locations.

Keywords EIA · Sustainability · MSW · RIAM · Landfill

1 Introduction

Every developing nation is now dealing with issues of poor solid waste management. India is no exception, here, as both urban and rural cities are not doing well with the overall municipal solid waste (MSW) produced daily. Various factors are responsible for the considerable gap between the generation and collection of total MSW generated in India [1–3]. Urbanization, high consumer needs, lack of eco-friendly options, and lack of funds are some of the important reasons which need urgent attention. While suggesting and deciding the process of disposal of MSW, there is a need to address the issue of environmental impact assessment or EIA. The process of EIA is time-consuming, but it surely prevents the ill effects of such developmental projects [4–6]. EIA is being used in taking serious decisions regarding the type of project and

S. Kumar · S. Deswal (✉)
NIT Kurukshetra, Kurukshetra, India

A. K. Choudhary et al. (eds.), *Advances in Geo-Science and Geo-Structures*, Lecture Notes in Civil Engineering 154,
https://doi.org/10.1007/978-981-16-1993-9_4

process required for MSW management. The EIA-2016 rules ask for serious actions if the guidelines are not met.

The process of EIA should be scientific and transparent as it involves prediction, identification, and mitigation of social, biophysical, and additional relevant effects of proposals [7]. Rapid impact assessment matrix which is also known as RIAM analysis has been successfully used to analyze a project in the least possible time [8]. This tool has been successfully used in various case studies [9, 10]. Among various rapid assessment techniques, RIAM methodology is helpful for fast and accurate analysis of specified components of EIA [11, 12].

The study targets the sanitary landfill sites of Kurukshetra which is an important city in Haryana, India. On average, the city generates 70 tons of MSW per day, and a single municipal corporation in Thanesar handles the collection, segregation, and disposal of this MSW. The collected waste is then directly dumped to the landfill sites which are situated at two different locations, in the outskirts of the city. One of the landfill sites is now non-operational, and the other landfill site is newly allotted in the village Muqimpura. Impact analysis of the selected disposal sites was done and compared by their impacts on specific areas. This study will be useful in creating awareness that is necessary while dealing with the issues of MSW management.

2 Methodology

2.1 Collection of Baseline Data

Site 1 is the old dumping site located at the village Mathana, and Site 2 is situated in the agricultural fields of Muqimpura village. A GIS-based analysis is used to identify the accurate locations of both the sites. Figure 1 depicts the location of these two sites.

- Site 1- Mathana Dumping Site (29.9802° N, 76.9542° E)
- Site 2- Muqimpura Dumping site (29.9815° N, 76.6730° E)

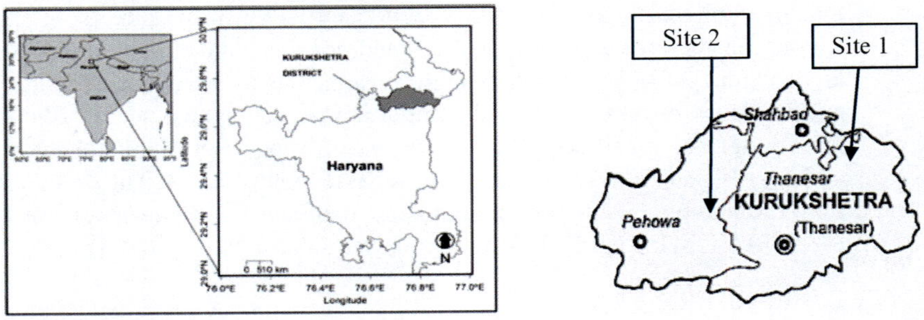

Fig. 1 Location of landfill sites

Site 1. Mathana dumping site (29.9802° N, 76.9542° E)
Site 2. Muqimpura dumping site (29.9815° N, 76.6730° E).

2.2 *RIAM*

To collect the responses, a team of 15 people from NIT Kurukshetra was formed to interact with the villagers. The team members are full-time students of the institute and are aware of the basic concepts and components of EIA and the role of MSW in environmental degradation. The overall evaluation of possible environmental impacts of disposal sites was done based on given points [13]:

1. Criteria A: have significance to the state and exclusively adequate to alter the obtained score.
2. Criteria B: significant to condition but independently incapable of altering the obtained score.

Selected individual components are then evaluated, and the environmental score or ES is calculated using the following equations (Tables 1 and 2).

Table 1 Evaluation criteria for RIAM

	Criteria	Scale	Description
A_1	Importance of condition	4	Important to national/international Interest
		3	Regional or national interests are related
		2	Significant to outside areas
		1	Only related to local conditions
A_2	Magnitude of change/effect	3	Major positive benefit
		2	Significant improvement
		1	Improvement in status quo
		0	No impact
		−1	Negative transformation
		−2	Substantial negative outcome
		−3	Major consequence
B_1	Permanence	1	Not applicable or neutral
		2	Provisional
B_2	Reversibility	1	Not applicable or neutral
		2	Changeable
		3	Unalterable
B_3	Cumulative	1	No developments or not applicable
		2	Single
		3	Synergistic

Table 2 Range bands for respective ES values

ES range values	RB codes	Explanation of RB
+72 to +108	E	Foremost positive impact
+36 to +71	D	Substantial optimistic impact
+19 to +35	C	Temperate positive impact
+10 to +18	B	Constructive impact
+1 to +9	A	Slightly optimistic impact
0	N	Neutral/not valid
−1 to −9	−A	Slightly negative impact
−10 to −18	−B	Negative impact
−19 to −35	−C	Temperate adverse impact
−36 to −71	−D	Substantial adverse impact
−72 to −108	−E	Major consequence

$$A_1 \times A_2 = A_{\mathrm{T}} \tag{1}$$

$$B_1 + B_2 + B_3 = B_{\mathrm{T}} \tag{2}$$

$$A_{\mathrm{T}} \times B_{\mathrm{T}} = \mathrm{ES} \tag{3}$$

2.3 Analysis Using Rapid Impact Assessment Matrix

Data were collected in the form of answers to the questionnaire prepared from the people living in the vicinity. Questions based on different environmental aspects were considered by keeping the following vital areas in mind.

1. PC: physical and chemical components
2. BE: biological and ecological components
3. SC: sociological and cultural components
4. EO: economical and operational components.

To conduct this study systematically and scientifically, the mentioned components were further divided into subcomponents. The included subcomponents are as follows:

Physical/chemical components

- PC1 Proximity to human settlement
- PC2 Land recommendation
- PC3 Proximity to the water body
- PC4 Impact on surface water
- PC5 Odor emissions

- PC6 Soil fertility
- PC7 Groundwater pollution
- PC8 Landscape and topography
- PC9 Quality of affected land
- PC10 Closeness to the forest area.

Biological/ecological components

- BE1 Greenhouse gases emissions
- BE2 Impact on health of the soil
- BE3 Ecological risks due to chemicals in use
- BE4 Impact on nearby forest cover
- BE5 Forest cover for CO_2 sink
- BE6 Land pollution
- BE7 Impact on biodiversity and wildlife
- BE8 Implication on human well-being.

Sociological/cultural components

- SC1 Risk of communicable ailments
- SC2 Trouble due to dust emissions
- SC3 Trouble due to odor emission
- SC4 Noise pollution due to operation
- SC5 Disturbance to the community
- SC6 Public acceptance
- SC7 Implications due to site operation.

Economical/operational components

- EO1 Economic impact on agriculture land
- EO2 Water treatment cost
- EO3 Cost of land development
- EO4 Financial gains.

3 Result and Discussions

Open dumping is a common practice used by authorities to get rid of MSW, and hence, the lack of proper facilities is a cause of concern for both the sites. The following are the important results of this assessment:

1. Although Site 1 is now officially closed, its negative impacts on PC and BE components are very much similar to Site 2.
2. The newly allotted land in Site 2 is now a significant source of income to its owners, and it also creates jobs for the people living in the locality; but in the case of Site 1, the dumping site is not generating revenue anymore.

3. The emission of GHG, leachate collection, and odor control are the major unaddressed issues for both of these sites as these unplanned landfill sites have no provision for collecting and controlling these liquids and gases.
4. Comparing the ES scores and RB's confirms that the soil and water pollution in the area is a big concern for the environment at both the locations.
5. The SC and BE components for both the sites have no positive RB values. It interprets that both the sites have no positive contribution to these components and the negative RB values imply potential damage caused to these components.
6. The results mentioned in Figs. 2 and 3 depict that both the landfill sites are

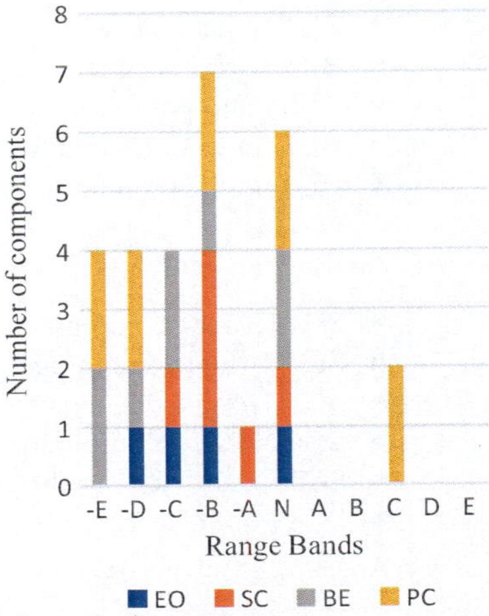

Fig. 2 RIAM analysis of Site 1

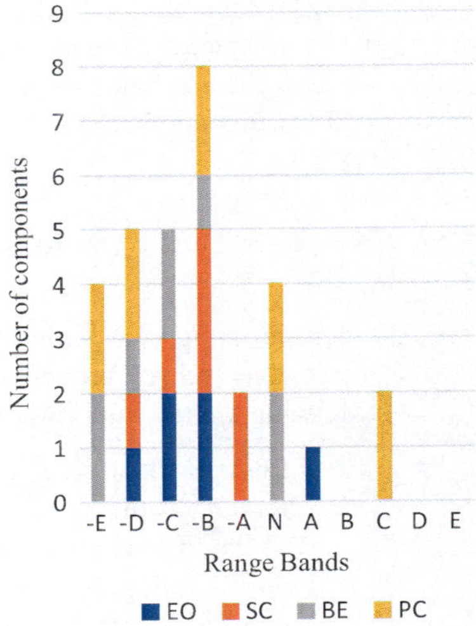

Fig. 3 RIAM analysis of Site 2

not environmentally sound. There is an urgent need to opt for better disposal techniques.

7. The study suggests that both the landfill sites are not sustainable. The authorities are deciding in a haste, and it is advised that the dumping sites be relocated to another location. A proper impact assessment of new disposal sites can help to avoid environmental damage.
8. As most of the collected MSW are wet and biodegradable, therefore, the direct disposal should be prohibited. Other modes of disposal are required depending upon the type of organic content and the availability of land and manpower.

4 Conclusion

The study of both these disposal sites suggests that open dumping of MSW in nearby village areas is not a sustainable solution. The final obtained range bands are almost identical which confirms that Site 2 will meet the same fate as Site 1. The final assessment using RIAM suggests that both these dumping locations need urgent attention from the authorities and the people living nearby. The study suggests the implication of vermicomposting and other bioconversion techniques will help to reach the sustainable aim of MSW management. RIAM can further be used for analyzing the future impacts of such facilities to make the decisions in record time. RIAM cannot be a shortcut to the whole EIA process, but it surely can contribute as a major tool for such impact assessments by the authorities and the decision-makers.

References

1. Kumar S, Bhattacharyya JK, Vaidya AN, Chakrabarti T, Devotta S, Akolkar AB (2009) Assessment of the status of municipal solid waste management in metro cities, state capitals, class I cities, and class II towns in India: an insight. Waste Manag 29:883–895. https://doi.org/10.1016/j.wasman.2008.04.011
2. Bhalla B, Saini M, Jha M (2012) Characterization of Leachate from Municipal Solid Waste (MSW) Landfilling Sites of Ludhiana, India: A comparative study. Int J Eng 2:732–745
3. Rawal N, Rai S, Duggal SK (2017) An approach for the analysis of the effects of solid waste management in slum areas by rapid impact assessment matrix analysis. Int J Environ Technol Manag 20:225–239. https://doi.org/10.1504/IJETM.2017.089652
4. Paliwal R (2006) EIA practice in India and its evaluation using SWOT analysis. Environ Impact Assess Rev 26:492–510. https://doi.org/10.1016/j.eiar.2006.01.004
5. Panigrahi JK, Amirapu S (2012) An assessment of EIA system in India
6. Sainath NV, Rajan KS (2015) Meta-analysis of EIA public hearings in the state of Gujarat, India: its role versus the goal of environmental management. Impact Assess Proj Apprais 33:148–153. https://doi.org/10.1080/14615517.2014.964085
7. Zamorano M, Garrido E, Moreno B, Paolini A, Ramos A (2005) Environmental diagnosis methodology for municipal waste landfills as a tool for planning and decision-making process. Sustain. Dev Plan II, 12.84:545–554

8. Mondal MK, Dasgupta BV (2010) Resources, conservation and recycling EIA of municipal solid waste disposal site in Varanasi using RIAM analysis. Resour Conserv Recycl 54:541–546. https://doi.org/10.1016/j.resconrec.2009.10.011
9. Suthar S, Sajwan A (2014) Rapid impact assessment matrix (RIAM) analysis as decision tool to select new site for municipal solid waste disposal: a case study of Dehradun city. India Sustain Cities Soc 13:12–19. https://doi.org/10.1016/j.scs.2014.03.007
10. Afroosheh F, Shahrashoub M, Toosi MG, Saffari M (2018) A field study of the environmental effects of marginalization in the 19th district of Tehran using rapid impact assessment matrix (RIAM). Environ Energy Econ Res, 123–135. https://doi.org/10.22097/eeer.2018.149024.1043
11. Aliakbari-beidokhti Z, Ghazizade MJ, Gholamalifard M (2017) Environmental impact assessment of municipal solid waste disposal site using rapid impact assessment matrix (RIAM) analysis in Masshad city. Iran Environ Eng Manag J 16:2369
12. Valizadeh S, Hakimian H (2019) Evaluation of waste management options using rapid impact assessment matrix and Iranian Leopold matrix in Birjand. Iran Int J Environ Sci Technol 16:3337–3354. https://doi.org/10.1007/s13762-018-1713-z
13. Pastakia CMR, Jensen A (1998) The rapid impact assessment matrix (RIAM) for EIA. Environ Impact Assess Rev 18:461–482. https://doi.org/10.1016/S0195-9255(98)00018-3

Effect of Soil Structure Interaction on Seismic Response of Buildings

Sudhanshu Shekhar, Shivam Mani Tripathi, and Shri Ram

Abstract Large human causalities and huge economic loss occur due to structural damages which are the result of the earthquake. The greater part of the structural designing structures includes basic component of the structure which has direct contact with ground. When the earthquake forces strike on the structures, we saw that both ground and structural displacement is related to each other. The procedure in which reaction of the soil impacts the movement of structure and the movement of structure impacts the reaction of the soil is known as soil structure interaction. Our main aim is to study about the seismic demand of structure which lies in zone IV. For superstructure (G+3) basic normal building is considered for seismic analysis. To study the effect of soil structure interaction medium type soil is used. The aggregate work is partitioned into two sections. In the initial segment, the analysis is done manually in which some techniques are utilized in assessment of seismic demand of moment resisting frame buildings and in the second part the analysis is performed by the E.TAB software. The extra need has given on manual earthquake investigation. Results show the difference in displacement values for both with soil model and without soil model. The displacements are lesser in without soil condition. The model requires more seismic demand in with soil condition due to more seismic response.

Keywords Soil-structure interaction · Seismic response · Seismic demand · Displacement · E.TAB

1 Introduction

The technique in which the reaction of the soil influences the movement of the structure influences the reaction of the soil is named as soil structure interaction (SSI). There are two types soil structure interaction inertial and kinematic soil structure

S. Shekhar (✉) · S. M. Tripathi · S. Ram
Madan Mohan Malaviya University of Technology, Gorakhpur, Uttar Pradesh, India

A. K. Choudhary et al. (eds.), *Advances in Geo-Science and Geo-Structures*,
Lecture Notes in Civil Engineering 154,
https://doi.org/10.1007/978-981-16-1993-9_5

interaction. Inertial interaction discusses the vibration of the superstructure, rotation, and displacement at the foundation level of the structure that comes due to inertia drove forces like moment and base shear. These rotations and displacements are the sources of energy dissipation in the soil structure system. Kinematic soil structure interaction causes motions at the foundation to diverge from free field motion. Gorakhpur lies in zone IV. Zone IV lies in the active seismic zone. It is high-causality hazard zone which covers zone responsible to MSK VIII. IS Code: 1893-2002 recommended zone factor of 0.24 for zone IV. In Gorakhpur region, there is random development and rapid growth of population occurs day-by-day. Due to the earthquake, there is some damage occurs such as structure failure, pipeline failure, highways and railways damage, etc [1]. During an earthquake, the structures bear the lateral forces of significant magnitude as well as the vertical load acting on the structure [2]. It is essential to determine these forces in a practical manner and use in design for safety of structure due to failure from the earthquake. The framed structures are generally analyzed with their bases considered to be either completely hinged or rigid. Though, the foundation resting on deformable soil also undergoes deformation relying on the relative rigidities of the foundation, soil, and superstructure. Therefore, the soil structure interaction analysis is necessary for the precise estimation of the response of the superstructure. In this research paper, a G+3 building which is lies in Gorakhpur is considered for seismic analysis. For the SSI analysis purpose, we considered the cohesive soil. The objective of this analysis is to calculate the effect of SSI on several static and dynamic properties of R.C. frame for base shear, fundamental natural period, natural frequencies, mode shapes, time period etc. To understand the effectiveness and serviceability of these models, the effort can be done. The other aim of this research work is to study the SSI effect in the analysis of a residential building G+3 and constructed as an RC frame using E.TAB 2016. In this paper firstly, we performed the analysis manually with the help of equivalent static load method and after that we use E.TAB software for the analysis purpose. In the software analysis firstly, we developed the model of building structure with soil and without soil and find displacements occurs in the building.

2 Idealization of the System

2.1 Structural Idealization

Generally, in real life, the precise analysis of a structure cannot perform easily. The idealization of a structure is a phenomenon in which the simplification of structural components can be done, while its behavior under loading cannot be changed. For simplifying the calculations, we did idealization of the structure. In designing the structure, the time taken is large if idealization is not done. For a structural engineer, it is necessary to develop techniques to idealize a structure for performing analysis easily. For the study, I take a four-storey (G+3) residential building which is in

Gorakhpur region. This building lies in zone 4 according to IS-1893 (Part-1):2002. This building is a 3-bay structure, and each bay is of 5 × 5 in a plan supported on the isolated footing. The storey height of this building is 3 m and the foundation depth is 1.0535 m. The dimension of the column is 300 mm × 300 mm is taken. The dimension of the beam is 250 mm × 400 mm is taken. The thickness of the slab is 140 mm and the thickness of brick masonry wall is 120 mm considered only at the periphery of the building.

2.2 Idealization of Soil Continuing Medium

When the earthquake comes then the soil- structure response generally depends on the size of the building, soil profile, dynamic characteristics, and nature of excitation [3]. The analysis of SSI is also performed by FEM. In FEM, we consider soil as an elastic continuing medium beneath the foundation [4, 5]. The finite soil mass is taken, with boundary which is very far from the location where there is no loading effect comes on the structure. For modeling with finite boundary, it is assumed that the plan dimension of soil is taken as 5 times of the length of the structure. Therefore, the soil mass block is (50 × 50) and 8 m depth is considered.

3 Methodology

3.1 Estimation of Seismic Design Force by Static Analysis Using Equivalent Static Force Method

Step 1: (Table 1)

Step 2: Seismic weight calculation:

Take unit weight of concrete is 25 kN/m^2 and unit weight of masonry is 22.5 kN/m^2.

(a) Slab:

D.L. due to self-weight of slab = (15 × 15 × 0.14) × 25 = 787.5 kN.

(b) Beams:

Self-weight of beam per unit length = (0.25 × 0.4 × 25) = 2.5 kN/m.
Total length = 4 × 15 × 2 = 120 m.
D.L. due to self-weight. Of beam = (2.5 × 15) × 4 × 2 = 300 kN.

(c) Columns:

Self-weight of column per unit length = (0.3 × 0.3 × 25) = 2.25 kN/m.
D.L. due to self-weight of column = (16 × 2.25 × 3) = 108 kN.

Table 1 Design parameters

Number of storey	4
Z	0.24 [Table 2, Clause 6.4.2 IS 1893(Part-1):2002]
Live load	3 kN/m^2
Columns	300 mm × 300 mm
Beams	250 mm × 400 mm
Thickness of slab	140 mm
Thickness of wall	120 mm
Importance factor	1.0 [Table 6, clause 6.4.2 IS 1893(Part-1):2002]
Response reduction factor	3 [Table 7, clause 6.4.2 IS 1893(Part-1):2002]

(d) Walls:

Self-weight of wall per unit length = (0.12 × 3 × 22.5) = 8.1 kN/m.
Total length of wall = (4 × 15 × 2) = 120 m.
D.L. due to self-weight of wall = (8.1 × 15 × 4) = 486 kN.

- Load on all floors = W_1 = W_2= W_3= (787.5 + 168.75 + 300 + 108 + 486) = 2450.25 kN
- Load on roof slab (live load on slab is zero).

W_4 = 787.5 + 0 + 300 + (108/2) + (486/2) = 1384.5 kN.

Total Seismic Weight, $W = W_1+W_2 + W_3+W_4$= {(2450.25 × 3) + 1384.5} = 8735.25 kN.

Step 3: Fundamental natural period of vibration (T_a):

According to clause 7.6.2 of IS 1893 (Part-1):2002

$$T_a = \frac{0.09h}{\sqrt{d}}$$

whereas

h = Height of building in meter, d = Base dimension of the building at the plinth level, in m, along the considered direction of the lateral force.

$$T_a = \frac{0.09h}{\sqrt{d}} = \frac{0.09 \times 12}{\sqrt{15}} = 0.278\,\text{s}$$

The building is situated on medium soil type II (Fig. 1).

Spectral Acceleration:
$\frac{S_a}{g} = 2.50, Z = 0.24, I = 1.0, R = 3.$

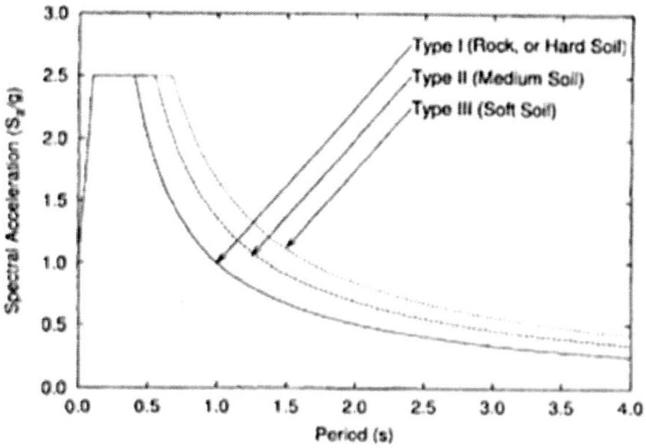

Fig. 1 Graph between spectral acceleration and period

According to clause 6.4.2 IS 1893 (Part 1): 2002,

Design horizontal seismic coefficient

$$(A_h) = \frac{Z}{2} \times \frac{S_a}{g} \times \frac{I}{R}$$

where,

Z Zone factor
$\frac{S_a}{g}$ Average response acceleration coefficient
I Importance factor

R Response Reduction Factor.

Now we calculate the value of A_h

$$A_h = \frac{0.24}{2} \times 2.50 \times \frac{1}{3} = 0.1$$

According to clause 7.5.3 IS 1893 (Part 1): 2002,

Design Seismic Base Shear

$$V_B = A_h \times W$$

where,

A_h = Design horizontal seismic coefficient.
W = Seismic weight of the building as per clause 7.4.2 IS 1893 (Part 1): 2002.
Now we calculate the value of V_B,

$$V_B = 0.1 \times 8735.25 = 873.525\,\text{kN}.$$

According to clause 7.7.1 IS 1893 (Part 1): 2002, vertical distribution of Base Shear to different floor level (Q_i) (Figs. 2 and 3; Table 2).

$$Q_i = V_B\left[\frac{W_{(i)}h_i^2}{\sum\left(W_{(i)}h_i\right)^2}\right]$$

$$V_4 = 342.42\,\text{kN},\ V_3 = 683.98\,\text{kN},\ V_2 = 835.62\,\text{kN},\ V_1 = 873.53\,\text{kN}$$

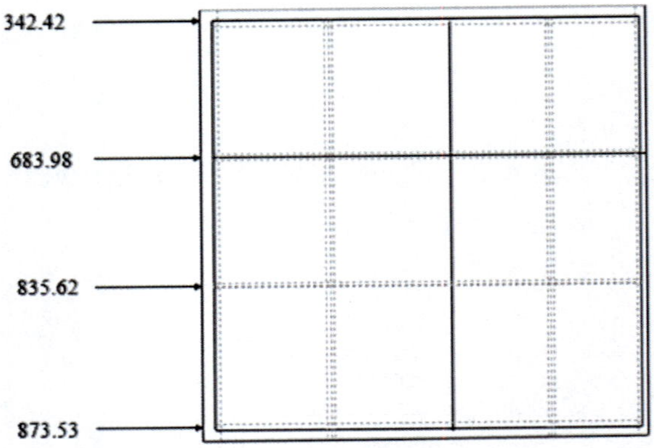

Fig. 2 Distribution of lateral force direction along the height of structure (kN)

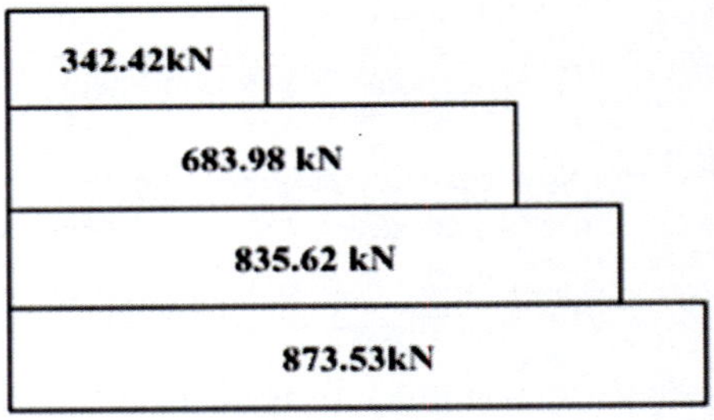

Fig. 3 Distribution of shear force direction along the height of structure (kN)

Table 2 Lateral load distribution with height by the equivalent static method

Floor level	W_i	h_i	$W_i h_i^2$	$\frac{W_i h_i^2}{\sum(W_i h_i)^2}$	Q_i	Storey shear force
4	1384.5	12	199,368	0.3924	342.77	342.64
3	2450.25	9	198,470.25	0.3906	341.21	683.98
2	2450.25	6	88,209	0.1736	151.64	835.62
1	2450.25	3	22,052.25	0.0434	37.91	873.53
Σ			508,099.5			

3.2 Modeling in E.TAB 2016 Software

Step—1: Firstly, we set the standard code such as IS 456: 2000, IS 1893: 2002 and other codes.
Step—2: We open ETABS and after this, we select a new model, therefore, a window opens and in this, we enter the dimension of the grid and storey dimension of the building.
Step—3: In this step, we define the property. We select the define menu material properties and then define the property of the material. We include new material for our basic components of the structure such as a beam, column, and slab according to the specified details. After this, we select the frame section and define these sections and we add the necessary section for a beam, column, and slab.
Step—4: In this step, we assign the property and draw the structural components by command menu. We draw lines for beams and generate the column in the specified region in which assigning of the property is done for beams and columns.
Step—5: In this step, we assign the supports. Firstly, we select all columns and base of the structure and then we fixed the support by assigning menu for joint or frame.
Step—6: In this step, we defined the all load consideration in ETABS and then we assign the load. In ETABS loads are defined in the defined menu by using static load cases command.
Step—7: In this step, we assign the dead load. In ETABS, dead load is assigned automatically such as dead load for the external and internal wall.
Step—8: In this step, we assign the live load. Live loads are assigned to an overall structure as well as floor finishing.
Step—9: In this step, we assign the wind load. Wind load is provided according to IS 875 1987 PART 3 by assigning wind angle and wind speed. For analysis purpose, we design a G+3 building which height is less than 12 m, so we do not need to provide wind load.
Step—10: In this step, we assign the seismic loads. According to IS 1893:2002, we define and assign the seismic load by providing response reduction factor in both X and Y direction.
Step—11: In this step, we assign the combination of the load. We use load combinations command in define menu for assigning the load combination.
Step—12: In this step, we perform analysis and check for errors after completion of all the above step.
Step—13: In this step, we design the structure according to IS 456:2000 after completion of analysis. From ETABS, we design each structural element.

4 Result and Analysis

Assign earthquake load according to IS 1893 (Part-1): 2002, Soil type: Medium soil or Type II soil according to soil classification and IS 1893 (Part-1): 2002, Height of

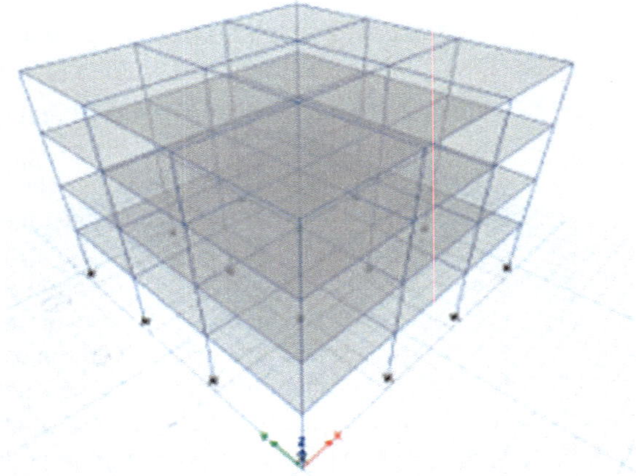

Fig. 4 Actual model of building without soil in 3D

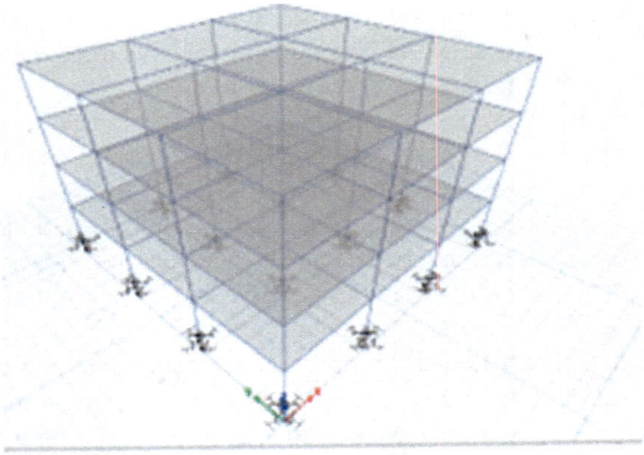

Fig. 5 Actual model of building with soil in 3D

storey = 3 m, Foundation depth below soil = 1.0535 m, Building type: Residential (Figs. 4, 5 and 6; Tables 3, 4 and 5).

5 Importance of Soil Structure Interaction

Analysis of SSI is very important for the structures. In the seismic code, it is mentioned that ignore the SSI phenomenon for the structure on rock or rock-like material, but there is no provision for soft soil. 2015 Nepal earthquake is very hazardous [6]. There was a huge loss occurs due to this earthquake. When we consider the SSI effect, the structure becomes more flexible and the natural period as well as damping ration increases. The SSI effect is different on high-rise and low-rise

Fig. 6 Model of building after deformation in 3D with soil

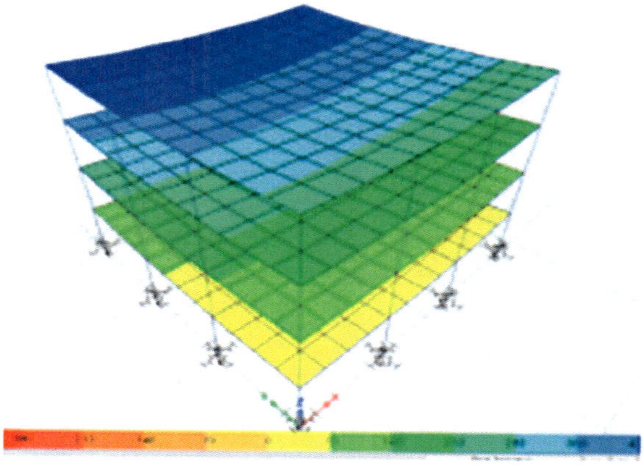

Table 3 Geometric properties of building

Thickness of medium soft soil layer	8 m
Thickness of slab	140 mm
Size of columns	300 mm × 300 mm
Size of beam	250 mm × 400 mm

Table 4 Material properties of building

	Elastic modulus (kN/m^2)	Poisson's ratio	Density (kN/m^3)
For medium soft soil layer	3500	0.4	18
For R.C.C. building	22,360	0.2	24

Table 5 Comparison of result between maximum displacement in building without considering soil condition and maximum displacement building with considering soil condition

Storey	Maximum displacement without soil in mm	Maximum displacement with soil in mm
4	39	52.5
3	34.1	50
2	26	41.2
1	12.5	29

building. The base shear coefficient reduces more in low-rise building in comparison with high-rise building. When we consider the SSI effect the force produced is greater than ESM. When we consider the SSI phenomenon the storey drift and displacement occurs more [7]. The infrastructures as well as life-line facilities are safe during earthquake when we consider SSI effect in design [8].

6 Conclusion

1. Static loading as well as dynamic loading is important for calculating the response of building due to soil structure interaction effect.
2. The accurate idealization of supporting soil is feasible by FEM. It is feasible to merge deviation in the layered soil, boundary conditions and properties of soil etc. Thus, this will generate accurate data.
3. It is noticed that the soil below the building and close to the fixed- boundaries is analogously stable considering the overall behavior of soil.
4. We found that more displacement occurs in the building with soil than building without soil therefore, we see the effect of soil structure interaction from these data of displacement.
5. For analyzing the effect of soil structure interaction, FEM is a very effective method. With the help of FEM, the complexity of the real problem decreases.
6. From above analysis, we found that for the stability of structure it is necessary to consider the effect of soil structure interaction.

References

1. Wolf JP (1985) Dynamic soil-structure interaction. Prentice-Hall, Englewood Cliffs
2. Hosseinzadeh NA (2012) Shake table study of soil-structure interaction effects on seismic response of single and adjacent buildings. Ph.D. thesis, IIEES, Tehran, Iran
3. Abdollahi B (2009) Soil-structure interaction analysis using cone models. J Seismol Earthq Eng 10:167–174
4. Luco JE, Contesse L (1973) Dynamic structure soil structure interaction. Rep Bull Seismic Soc Am 63:89–303
5. Todorovska MI, Trifunac MD (1990) Analytical model for enplane building foundation-soil interaction: incident P-, SV- ABD Rayleigh waves. Report No. CE 90–01, Department of Civil Engineering, University of Southern California, Los Angeles, California, USA, p 218
6. NEHRP Consultants Joint Venture A partnership of the Applied Technology Council and the consortium of Universities for Research in Earthquake Engineering, U.S. Department of Commerce, 2012, Soil-Structure Interaction for Building Structures
7. Stewart JP, Fenves GL, Seed RB (1999) Seismic soil-structure interaction in buildings. Anal Methods J Geotech Geoenviron Eng (ASCE)
8. Worku A (2014) Soil-structure interaction provisions a potential tool to consider for economical seismic design of buildings? J S Afr Inst Civ Eng Kenya
9. Hosseinzadeh N, Davoodi M, Roknabadi ER (2002) Shake table study of soil structure interaction effects in surface and embedded foundations. In: Structural Engineering Research Center, International Institute of Earthquake Engineering and Seismology (IIEES), Tehran, Iran

Bamboo Strips as Eco-friendly Soil Reinforcement Material

Manika Kumari and Somenath Mondal

Abstract Utilization of reinforcement material in enhancing engineering properties of weak soil has gained immense popularity in contemporary geotechnical engineering practices. However, it can be noted that most of the reinforcement products are factory-made synthetic materials. In this regard, bamboo strips are proposed to be used as natural soil reinforcement material. Moreover, the utilization of bamboo strips as reinforcement material endorses the concept of the sustainable solution in geotechnical engineering practices. In this study, an experimental investigation consists of numerous laboratory testing has been conducted to assess the performance of bamboo strips as potential reinforcing material in silty-clayey soils. Particularly, the focus has been kept in analyzing the improvement in CBR values in unsoaked conditions compared to unreinforced soil. Furthermore, the variation in the improvement of CBR values has been studied for different percentage of bamboo strip. From the obtained experimental outcomes, it has been found that the CBR value of soil increases with increase in bamboo strips. Hence, the present study demonstrates the potency of the bamboo strip as an eco-friendly reinforcing material to improve the engineering properties of soil.

Keywords Soil reinforcement · Bamboo strip · Sustainable material · Experimental investigation · Engineering properties

1 Introduction

Approximately 1380 million of people live in nearly 6.60 lakh villages spread across India. The connectivity on rural roads is a major part of rural development as it

M. Kumari (✉) · S. Mondal
Department of Civil Engineering, National Institute of Technology, Jamshedpur, India

S. Mondal
e-mail: smondal.ce@nitjsr.ac.in

A. K. Choudhary et al. (eds.), *Advances in Geo-Science and Geo-Structures*,
Lecture Notes in Civil Engineering 154,
https://doi.org/10.1007/978-981-16-1993-9_6

facilitates the rural sector to reach economic and social services, enhancing agriculture and non- farm productivity, thereby reducing the poverty and enhancing rural development and proper income opportunities [1].

In recent years, synthetic fiber composites have become a specific option as traditional construction material. However, natural fibers can be compared to that of most synthetic fibers, because natural fibers have a cheap price, good specific engineering properties, very easy to handle and easily available [5]. In this context, utilization of natural fibers as reinforcement materials for the subgrade in the rural road construction project can be a most preferred option since the designed maximum load for these roads are not high. Moreover, natural fibers are an economical and non-conventional replacement for man-made and synthetic fibers [2, 3].

Natural fibers have easily contributed to economic prosperity and permanence to our daily lives. Due to environmental concerns, there has been increasing interest in the replacing synthetic fibers composites with natural plant fibers such as jute, coir, bamboo, sisal, and so on. Bamboo can be used because of their cheap price, durability, sustainability, and biodegradability. Natural fibers provide advantages of economical, functional, and environmental as replacements for traditional synthetic fibers [6, 7].

Soil reinforcement by fiber material is considered an effective method of soil improvement due to its cost-effectiveness, ease of adaptation, and reproducibility. Geo-textiles and geo-synthetics are commonly used as soil reinforcements. Nonetheless, in this study, natural materials like bamboo strips are proposed to use as sustainable reinforcement for soil [11].

Bamboo production is increasing in importance as an economic alternative for villagers, and several local governments have developed bamboo development programs. The involvement of villagers in growing bamboo products may serve as a boost to the economic development of the country. It will not only improve the rural livelihood, but will also eliminate unemployment and hence will enhance the economy of the country [4].

In few of the studies [9–15], researchers have investigated the performance of bamboo in different forms. Nonetheless, an effort has been made in the present study to investigate the performance of bamboo grids for the application in construction of rural road.

2 Materials and Methodology

2.1 Soil (Saraikela Kharswana, Jharkhand)

Experiments were carried out on soil taken from Matekumdih in Saraikela Kharswana district, Jharkhand. The soil used in this study was Alfisols soil which is form in semi-arid to humid areas, typically under hardwood forest cover. They have a clay-enriched subsoil and relatively high native fertility.

Fig. 1 Bamboo strips

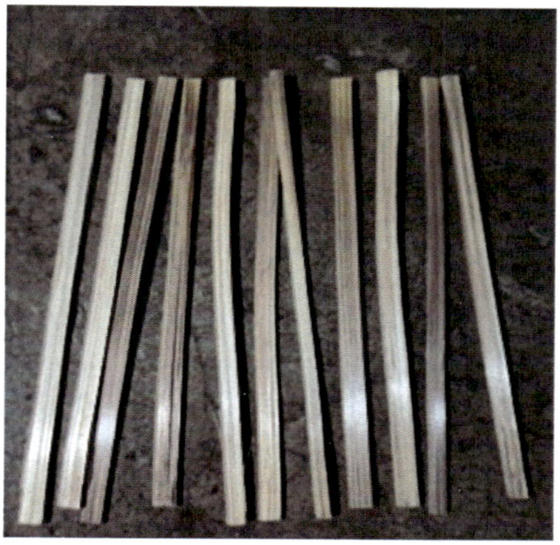

2.2 Bamboo

For this study, the bamboo was collected from Chakulia in East Singhbhum district, Jharkhand, India. Chakulia, Jharkhand, India is located in the Town's place category with the GPS coordinates of 22°28′48.0888″ N and 86°42′59.3424″ E.BambusaBambos species was used (Fig. 1).

2.3 Experimental Work

The present study comprises of laboratory tests for finding out the index properties and engineering properties of soil. As per IS: 2720 for soil and IS: 6874-2008 for bamboo, the geotechnical laboratory tests were conducted.

The soil was collected, air-dried, and pulverized appropriately for compaction and remolding of soil with and without bamboo. All tests were carried out as per the Indian standards specifications published by the Bureau of Indian Standards, New Delhi. To classify of the soil sample, there are five laboratory tests required to be done. Table 1 shows the results of index properties of soil.

Standard proctor compaction test (Light Compaction) was conducted as per IS 2720 (Part 7)-1980 for the determination of maximum dry density (MDD) and optimum moisture content (OMC). Figure 2 shows the lightweight compaction curve of silty clay soil, and it can be observed that the MDD value of the soil is 15.9 kN/m^3, and OMC value is 21% for light compaction.

The bamboo culms were collected and cut into strips (Fig. 1). The strips were cut into distinct shapes and sizes. To conduct the tensile strength test, it was necessary to prepare the bamboo sample. The bamboo strips of 1 m length and 20 mm width were cut and allowed to dry and season for 30 days.

Table 1 Index properties of soil

S. no.	Description	Values
1	Specific gravity,	2.45
2	Liquid limit (%)	35.6
3	Plastic limit (%)	21.2
4	Plasticity index (%)	14.4
5	Soil classification	CL
6	Maximum dry density (MDD, kN/m^3)	15.9
7	Optimum moisture content (OMC, %)	21

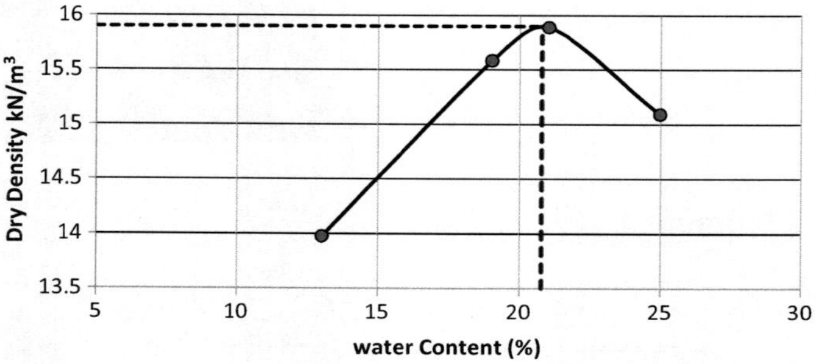

Fig. 2 Compaction test graph

The bamboo strips were cut and trimmed. The smooth surfaces of the bamboo strips were roughened to increase the friction between the strips and the soil. The thickness of the sample varies throughout its length since it is a natural material whose properties cannot be controlled strictly.

The mechanical test for determining the strength of bamboo as soil reinforcement is conducted as per IS: 6874 (2008). Test like index tensile strength, density, compressive strength were conducted. The mechanical test results of the bamboo are given in Table 2.

This study, the soil was compacted with or without bamboo strips. The CBR test was conducted on a soil sample as per IS: 2720(part XVI)-1987. The California bearing ratio (CBR) is a penetration test for evaluation of the mechanical strength of road subgrade and base courses. The obtained CBR values for different percentage of bamboo strips have been showcased in Table 3.

Table 2 Mechanical properties of bamboo specimen

Properties	Value (N/mm^2)
Tensile strength	170
Compressive Strength	89
Static bending strength	143

Table 3 CBR values for soil with percentages of bamboo strips

Percentages of Bamboo strips	CBR values
0.0	1.87
0.25	3.67
0.50	4.08
0.75	4.78

3 Results and Discussion

The CBR test was conducted to determine the optimum amount of bamboo strips in soil. This is done by mixing soil with varying percentages (0.0, 0.25, 0.50, 0.75, and 1.0%) of bamboo strips in soil. The CBR tests were performed unsoaked condition. Separate tests were performed for length of strips (10 cm); the effect of bamboo strips was done by comparing the CBR values under unsoaked condition.

3.1 Effect on CBR Value of Soil With or Without Bamboo Strips

Figure 3 shows that as percentages of bamboo strips in silty clay soil increases, increasing the CBR value of soil. The maximum improvement in CBR was obtained on 0.75% of bamboo strips used.

The graph (Fig. 4) indicates that the maximum CBR value of a reinforced soil was approximately 2.2 times that of unreinforced soil.

Based on CBR test on soil (with strips of length 10 cm) with strips reinforcement of 0.0%, 0.25%, 0.50%, 0.75%, and 1.0% increases in CBR test, values were found to be 1.87, 3.62, 4.087, 4.788, and 4.55, respectively.

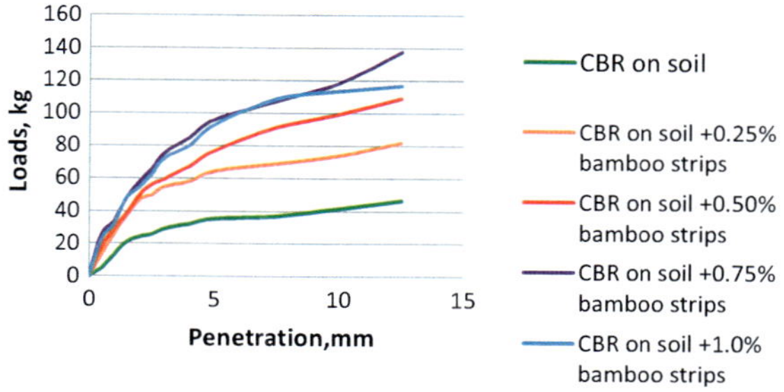

Fig. 3 CBR curve for soil with various percentage of Bamboo strips

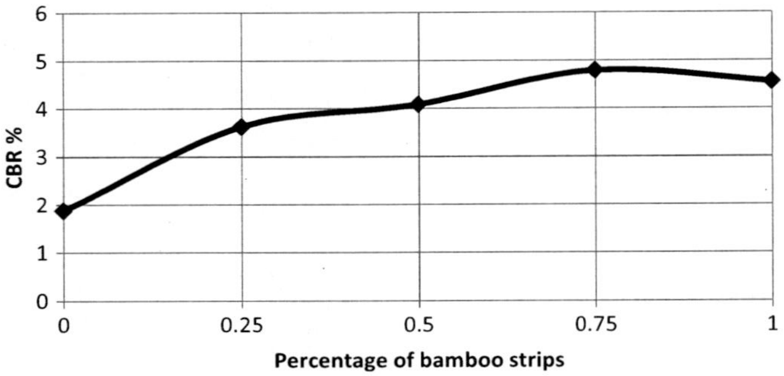

Fig. 4 Relation between CBR value and percentage of bamboo strips

4 Conclusion

In the present study, the experimental outcome successfully demonstrates the enhancement in the CBR value of the soil is because of percentages of bamboo strips, and further, it can be concluded that;

1. The bamboo strips can improve the strength thus increasing the bearing capacity of soil.
2. The maximum CBR value of a reinforced soil was approximately 2.2 times that of unreinforced soil.
3. Based on CBR test on soil, 0.75% of bamboo strips was used for the compaction.
4. The optimum amount of bamboo strips was 0.75% for OMC was 21%.
5. CBR value of soil was increased with the increasing of percentages of bamboo strips.

Having mentioned the above conclusions, it should be notified that the applications of bamboo strips as soil reinforcement should be limited to utilize in the project of low estimated cost. In the present study, rural roads have been considered where durability of the roads for several decades is not necessary; however, within limited budget, a provision of reinforcing soil underneath the roads can be kept. This will enhance the service life of the rural roads without significantly increasing the initial cost of construction.

Studies can be attempted in future to reduce the biodegradability of bamboos with the aid of the state-of-the-art coating technology, thereby increasing the durability.

References

1. Pitroda J, Pankhaniya H, Agravat S, Parmar P (2016) A critical review on innovative utilization of bamboo in rural road construction. Int J Constr Res Civ Eng 2(4):27–32

2. Khalil HA, Bhat IUH, Jawaid M, Zaidon A, Hermawan D, Hadi YS (2012) Bamboo fibre reinforced biocomposites: a review. Mater Des 42:353–368
3. Parashar S, Tomar P (2019) Synergy of sustainable bio-composite bamboo material in green technology–an explicit report. In: Proceedings of international conference on sustainable computing in science, technology and management (SUSCOM). Amity University Rajasthan, Jaipur-India
4. Selvan T, Tripathi KM (2017) Economic potential of bamboo for rural livelihood. Agrobios Newsletter
5. Dittenber DB, GangaRao HV (2012) Critical review of recent publications on use of natural composites in infrastructure. Compos A Appl Sci Manuf 43(8):1419–1429
6. Liu D, Song J, Anderson DP, Chang PR, Hua Y (2012) Bamboo fiber and its reinforced composites: structure and properties. Cellulose 19(5):1449–1480
7. Huda S, Reddy N, Yang Y (2012) Ultra-light-weight composites from bamboo strips and polypropylene web with exceptional flexural properties. Compos Part B Eng 43(3):1658–1664
8. Bergado DT, Bukkanasuta A, Balasubramaniam AS (1987) Laboratory pull-out tests using bamboo and polymer geogrids including a case study. Geotext Geomembr 5(3):153–189
9. Mandal JN, Manjunath VR (1995) Bearing capacity of strip footing resting on reinforced sand subgrades. Constr Build Mater 9(1):35–38
10. Marto A, Othman BA (2011) The potential use of bamboo as green material for soft clay reinforcement system. In: International conference on environment science and engineering, Singapore IPCBEE, vol 8, pp 129–133
11. Hejazi SM, Sheikhzadeh M, Abtahi SM, Zadhoush A (2012) A simple review of soil reinforcement by using natural and synthetic fibers. Constr Build Mater 30:100–116
12. Asaduzzaman M, Islam MI (2014) Soil improvement by using bamboo reinforcement. Am J Eng Res 3(8):362–368
13. Hegde A, Sitharam TG (2015) Use of bamboo in soft-ground engineering and its performance comparison with geosynthetics: experimental studies. J Mater Civ Eng 27(9):04014256
14. Brahmachary TK, Rokonuzzaman M (2018) Investigation of random inclusion of bamboo fiber on ordinary soil and its effect CBR value. Int J Geo-Eng 9(1):10
15. Akhil KS, Sankar N, Chandrakaran S (2020) Surface heave behaviour of sand bed reinforced with woven bamboo mat. Geotech Geol Eng, 1–8
16. Priyadarshee A, Chhotu AK, Kumar V (2014) Use of bamboo in low volume rigid pavement as reinforced material: a review. J Civ Eng Environ Technol 1(3):14–16

Assessment of the Geotechnical Properties of Red Earth Stabilized Using Quarry Dust and Cement

Sitaram Nayak, H. K. Preetham, and S. Deepthi Prakash

Abstract The present investigation aims at improving the engineering properties of red earth by incorporating quarry dust (QD) and analyze the results thus obtained by conducting a comparative study with the basic soil. Red earth is widely spread over a large part of peninsular India. Red earth/red soil could yield better results by the addition of rough-textured granular quarry dust and a hydraulic binder like cement. QD is the output from the rubble crushing units. Red earth was replaced with different proportions of quarry dust by dry mass of the soil: QD (5–30%, with an increment of 5%) with cement (2–6%) for the optimum QD-soil mix. The geotechnical properties like specific gravity, consistency limits, compaction, unconfined compressive strength (UCS), and triaxial compression test were performed on the red earth as well as on the mixes as per Indian Standard Codal provisions. UCS test results conclude that an optimum percentage of replacement of red earth by QD is 10% which yielded maximum strength than other mixes. It was observed that on replacement of red earth by granular quarry dust and cement, the shear strength properties and other geotechnical properties were improved. Thus, the utilization of granular industrial by-products has proven to be socially and economically beneficial.

Keywords Red earth · GBFS · Stabilization · UCS · Shear strength parameters

1 Introduction

The rapid development in the field of infrastructure and urbanization in the present context in a developing country like India has led to the improvement of the geotechnical properties of existing soil to suit a project accordingly [1]. Most of the ideal sites have been over exploited for the cause of massive urban transformation, with the remaining of poor character grounds [2]. These sites may not always be suitable for the intended or planned construction project. Here the concept of soil stabilization comes to the forefront. Soil stabilization is proving to be the optimum solution

S. Nayak · H. K. Preetham (✉) · S. D. Prakash
Department of Civil Engineering, National Institute of Technology Karnataka, Surathkal, Mangalore 575025, India

A. K. Choudhary et al. (eds.), *Advances in Geo-Science and Geo-Structures*, Lecture Notes in Civil Engineering 154,
https://doi.org/10.1007/978-981-16-1993-9_7

whenever the properties of the soil must be altered to make it suitable for further processing [3]. Soil stabilization is the method of enhancing the engineering properties of the poor soil by the addition of superior soil or binder material addition to the or soil [4]. This results in the shear strength improvement and reduction in the compressibility character of the soil [5]. The coarse grained material supplement also improves the geotechnical property of soil and warrants as a best stabilizer. The coarse grained material incorporation improves the frictional resistance of the mix, and thus, the shear strength of the mix enhances [6]. In the present situation, to conserve the coarse grained natural resource (sand), industrial by-products are utilized so as to meet its disposal problem and also to meet the economy of the project [7].

Lime, fly ash, cement, etc., are the conventional and traditional stabilizers normally used in the pavement structures to improve the capacity and to resist the **heavy axel loads. It involves the modification of the soil structure, pozzolanic reac**tions, etc. [8, 9] Cement is used due to its binding property resulting in the enhanced bearing capacity, compressibility, and volume stability to water and frost action. Stabilization with cement is independent of the soil type, and the strength gain is solely due to pozzolanic reactions [10]. Due to hydration of cement, cementitious products like CSH, CAH, and CASH are formed [11].

The process of weathering in the tropical regions moorum, which is also called red earth, is formed [12] which is normally rich in iron. This iron oxide induces the reddish-brown color to the soil, and hence, the soil is termed as red earth [13]. This soil is majorly formed in the peninsular regions [14]. The desirability of red earth as a foundation soil can be improved with stabilization.

Quarry dust (QD) is the by-product of rubble crusher units. The rock must be crushed into different sizes during quarrying activities which results in a large cloud of dust. This cloud of dust settles down into a heap and is known as quarry dust [15]. Quarry dust constitutes nearly 20–25% of the output of each rubble crusher unit. A rough estimate suggests that around 200 million tons of quarry by-products are being generated each year in India [16]. Leaving this in huge heaps in the neighborhood can cause serious health hazards. Besides the space required for disposal is yet another problem. Hence, quarry dust could be effectively used for building purposes resulting in a double advantage of reducing the cost of construction and management of waste disposal. The addition of quarry dust to the soil often serves the purpose **of densification of the soil mass by filling in vacant voids** [12, 17]. Also, it imparts higher frictional properties to the soil due to its inherently high friction angle. With the addition of cement, the absence of cementation associated with soils stabilized only with QD can be eliminated as the hydration products of cement imparts binding properties to the soil [18].

Previous studies carried out to analyze the effects of addition of these by-products have proved advantageous in many aspects. Thus, the study attempts to investigate the use of quarry dust as an effective material in stabilizing red earth.

2 Materials

The red earth was brought from Hassan, Karnataka, India. The soil samples were transported to the geotechnical laboratory to investigate its engineering properties and attempt to improve its geotechnical properties. The soil was dried, pulverized, sieved, and stored in a container for various tests. Quarry dust was collected from Pakshikere, Mangalore, Karnataka, India.

3 Methodology

Initially the study soil was examined (Table 1) for its basic properties like gradation property, specific gravity, consistency limits, compaction characteristics, and strength properties. The strength properties like unconfined compressive strength, shear strength parameters based on triaxial compression test, etc., were also analyzed. The initial characterization of quarry dust was carried out and presented in Table 2. Further red earth was replaced by quarry dust in different percentages (5, 10, 15, 20, and 30%) and analyzed for its performance. The mix yielding maximum UCS is termed an optimum mix. The next phase mainly deals with the various percentage of cement addition to the optimum mix. All the tests performed were in accordance with the Indian standard codal provisions (IS 2727, IS 1727, and SP 36).

The index and shear strength properties were determined. Cylindrical samples compacted to their respective maximum dry density (MDD) and optimum moisture content (OMC) for determination of shear strength properties. As QD is inert, fresh samples were prepared and tested immediately for different percentage addition of

Table 1 Engineering properties of red earth

Sl. No	Properties	Particulars
1	Particle size distribution Clay size % Silt size % Sand size % Gravel size %	14 30 54 2
2	Specific gravity	2.64
3	Liquid limit % Plastic limit % Plasticity index %	31 21 10
	IS Classification	SC
4	Compaction characteristics IS light compaction: MDD (g/cc) OMC (%)	1.83 15.2
5	Unconfined compressive strength (kPa)	230

Table 2 Engineering properties of quarry dust

Sl. No	Properties	Particulars
1	Grain size analysis Fines % Fine sand size % Medium sand size % Coarse sand size % Gravel size %	9 39 43 7 2
2	**Specific gravity**	2.76
3	Strength parameters Cohesion (kPa) Friction angle ϕ (degrees)	4 45

QD. Among the various mixes, the peak UCS value is termed as the optimum mix. For this optimum percentage replacement/optimum mix, different percentages of cement were added and examined for their index properties and strength properties at 7 days and 28 days. All the test results are analyzed and presented systematically in the paper.

4 Results and Discussion

4.1 Stabilization of Soil with QD

The test results on mixes of soil-quarry dust are presented in Table 3. The liquid limit reduces upon addition of quarry, and hence, the plasticity index decreases. The MDD did not show any considerable variation with the addition of QD. This is because of the frictional resistance offered by quarry dust for the compaction effort [17]. OMC decreased due to the addition of a material coarser than red earth.

Table 3 Geotechnical properties of red earth before and after stabilization with QD

Sl. No	Properties	Percentage of QD replacing red earth					
		(Red earth)	5	10	15	20	30
1	Maximum dry density(g/cc)	1.83	1.835	1.84	1.85	1.86	1.88
2	Optimum moisture content (%)	15.2	14.9	14.8	14.6	14.5	14.3
3	Liquid limit %	31	28.3	26.9	26.5	25.9	24.7
4	Plastic limit %	21	20	19.7	19.5	19	18.7
5	Plasticity Index %	10	8.3	7.2	7	6.9	6

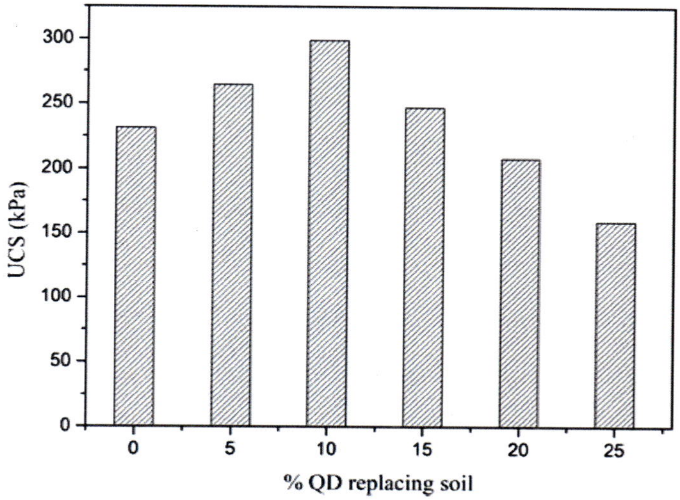

Fig. 1 UCS variation of red earth and quarry dust mixture

The UCS value increases with an increased replacement with QD (Fig. 1). Since QD is a chemically inert material, only the immediate strengths are assessed [12]. Compared to basic soil, the UCS value is increased by 26% for a replacement percentage of 10%. With a further increase in QD%, the percentage increment was found to decrease. It even reduced below the basic soil UCS when the percentage replacement exceeded 20%. This is due to the loss of cohesion and confining capacity with the addition of a coarser material. Therefore, the optimum percentage replacement was concluded as 10%.

Unconsolidated undrained (UU) tests were performed on fully saturated samples of soil and the optimum mix to obtain the shear strength parameters. For QD, the cohesion intercept was found to decrease due to the addition of a coarser material and a consequent reduction in clay content of soil but friction angle showed an increase from 22° to 27°.

4.2 Stabilization of Soil with Optimum QD and Cement

Varying percentages of cement were added (2, 4, and 6%) to the optimum QD replacing the soil (10% QD). The addition of cement initiates the pozzolanic reactions [19]. The increment in strength was found in the case of QD replacement and cement addition. For additions of 2, 4, and 6%, incremental percentages of 253%, 756%, and 1159% of increased unconfined compressive strength were observed. This is essentially due to the cohesion effect provided by hydration products of cement with water and the inherent high friction angle of QD [16]. Likewise, UU tests conducted

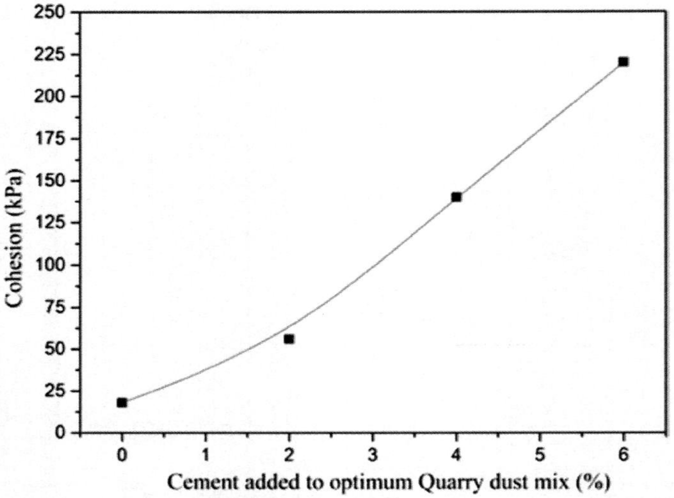

Fig. 2 Variation of cohesion with percentage of cement added to optimum red earth and QD mix

on cement stabilized optimum mix provided beneficial results. Increments of 210, 678, and 1120% were observed in cohesion values (Fig. 2) and of 14, 26, and 44% were found in friction angle values (Fig. 3) when compared with the same properties of untreated soil with the addition of 2, 4, and 6% of cement, respectively.

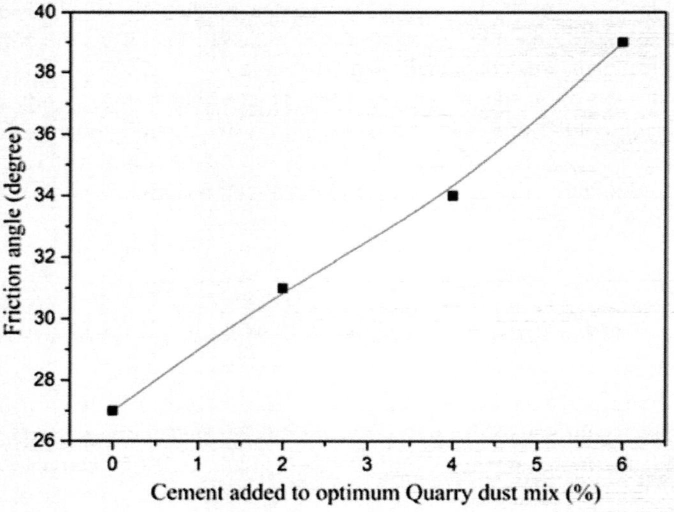

Fig. 3 Variation of angle of internal friction with percentage of cement added to optimum earth and QD mix

5 Conclusion

Having found red earth suitable for stabilization, it was replaced by varying percentages of quarry dust and subjected to the addition of a binder (cement), in varying proportions. Geotechnical investigations were conducted on the resulting mixes. The following conclusions are drawn from the work performed:

- Liquid limit and plasticity index decreases with the addition of QD to soil. Maximum dry density increases, and optimum moisture content reduces upon QD addition to red earth.
- The UCS value of the mix improved with the addition of quarry dust, but after a certain amount of replacement, the value decreased. Thus, an optimum percentage of replacement of red earth by QD is established as 10%.
- An exceptional improvement was found in the UCS and shear strength parameters with the addition of cement to the optimum mix. The high friction angle provided by the coarse particles of QD combined with the cohesive effect contributed by cement hydration can be attributed for these improvements in the case of QD replacement. Overall QD with cement addition to red earth proved advantageous.

References

1. Preetham HK, Nayak S (2019) Geotechnical investigations on marine clay stabilized using granulated blast furnace slag and cement. Int. J. Geosynth. Gr. Eng. 5:1–12. https://doi.org/10.1007/s40891-019-0179-5
2. Nayak S, Balaji M, Preetham HK (2020) A study on the behaviour of stone columns in a layered soil system. Transp. Infrastruct. Geotechnol. 7:85–102. https://doi.org/10.1007/s40515-019-00090-x
3. Sekhar DC, Nayak S, Preetham HK (2017) Influence of granulated blast furnace slag and cement on the strength properties of lithomargic clay. Indian Geotech. J. 47:384–392. https://doi.org/10.1007/s40098-017-0228-8
4. Preetham HK, Nayak S, Surya EV (2019) Experimental investigation on the stabilization of soft clay using granulated blast furnace slag. IOP Conf Ser Mater Sci Eng 561. https://doi.org/10.1088/1757-899X/561/1/012047
5. Mishra S, Sachdeva SN, Manocha R (2019) Subgrade soil stabilization using stone dust and coarse aggregate: a cost effective approach. Int J Geosynth Gr Eng 5. https://doi.org/10.1007/s40891-019-0171-0
6. Nayak S, Sarvade PG (2012) Effect of cement and quarry dust on shear strength and hydraulic characteristics of lithomargic clay. Geotech Geol Eng 30:419–430. https://doi.org/10.1007/s10706-011-9477-y
7. Hamer K, Karius V (2002) Brick production with dredged harbour sediments an industrial-scale experiment. Waste Manag 22:521–530. https://doi.org/10.1016/S0956-053X(01)00048-4
8. Baldovino JA, Moreira EB, Teixeira W, Izzo RLS, Rose JL (2018) Effects of lime addition on geotechnical properties of sedimentary soil in Curitiba Brazil. J Rock Mech Geotech Eng 10:188–194. https://doi.org/10.1016/j.jrmge.2017.10.001
9. González-López JR, Juárez-Alvarado CA, Ayub-Francis B, Mendoza-Rangel JM (2018) Compaction effect on the compressive strength and durability of stabilized earth blocks. Constr Build Mater 163:179–188. https://doi.org/10.1016/j.conbuildmat.2017.12.074

10. Sekhar CD, Nayak S (2017) SEM and XRD investigations on lithomargic clay stabilized using granulated blast furnace slag and cement. Int J Geotech Eng 6362:1–15. https://doi.org/10.1080/19386362.2017.1380355
11. Amulya S, Ravi Shankar AU, Praveen M (2018) Stabilisation of lithomargic clay using alkali **activated fly ash and ground granulated blast furnace slag. Int J Pavement Eng 1–8** https://doi.org/10.1080/10298436.2018.1521520
12. Akinyemi BA, Elijah A, Oluwasegun A, Akpenpuun DT, Glory O (2020) The use of red earth, lateritic soils and quarry dust as an alternative building material in sandcrete block. Sci African 7:e00263. https://doi.org/10.1016/j.sciaf.2020.e00263
13. **Nagaraj HB, Rajesh A, Sravan MV (2016) Influence of soil gradation, proportion and combina-**tion of admixtures on the properties and durability of CSEBs. Constr Build Mater 110:135–144. https://doi.org/10.1016/j.conbuildmat.2016.02.023
14. Okagbue CO, Onyeobi TUS (1999) Potential of marble dust to stabilise red tropical soils for road construction. Eng Geol 53:371–380. https://doi.org/10.1016/S0013-7952(99)00036-8
15. Soosan TG, Sridharan A, Jose BT, Abraham BM (2005) Utilization of quarry dust to improve the geotechnical properties of soils in highway construction. Geotech Test J 28:391–400. https://doi.org/10.1520/GTJ11768
16. Kumar A, Gift S, Siksha BB (2016) Improvement in geotechnical properties of an expansive **soil using fly ash—quarry dust mixes improvement in geotechnical properties of an expansive soil using fly ash—quarry dust mixes. J Electron Eng Geotech 3487–3500**
17. **Amadi AA (2014) Enhancing durability of quarry fines modified black cotton soil subgrade** with cement kiln dust stabilization. Transp. Geotech. 1:55–61. https://doi.org/10.1016/j.trgeo.2014.02.002
18. Eze-Uzomaka OJ, Agbo D (2010) Suitability of quarry dust as improvement to cement stabilized-laterite for road bases. Electron J Geotech Eng 15(K):1053–1066
19. Harikumar M, Nipin S, Shayifa T, Jini K (2016) Experimental investigation on compressed stabilized earth block

Improvement in the Properties of Red Soil Using Granulated Blast Furnace Slag

H. K. Preetham, Sitaram Nayak, and Praveen Jagapur

Abstract In the current study, an attempt is made to improve the geotechnical properties of red soil using industrial by-product: Granulated-Blast Furnace Slag (GBFS). Red soil is distributed over large part of the peninsular region in India. Red soil could be effectively stabilized to yield better strength characteristics. GBFS is the primary by-product of the iron and steel industry. Red soil was replaced with varying percentages of admixture (GBFS) by dry weight of soil (5, 10, 15, 20 and 30%). Basic geotechnical properties like specific gravity, Atterberg limits, compaction, unconfined compressive strength (UCS) and triaxial compression test, were performed on the red soil as well as on the mixes as per Indian Standard Codal provisions. From the UCS test results, the optimum percentage of replacement of 15% was found. It was observed that on replacement with admixtures, the liquid limit was found to decrease which reduces the compressibility. The presence of CaO in GBFS has improved the shear strength and shear strength parameters of soil. Thus utilization of granular industrial by-product (GBFS) has proven beneficial in geotechnical structures.

Keywords Red soil · GBFS · Stabilization · UCS · Shear strength parameters

1 Introduction

It is an accepted fact that urbanization is taking place at an alarming rate in India. With this massive urban transformation, most of the ideal construction sites have already been built upon leaving us with undesirable sites for future use [1]. These sites may not always be ideal for the proposed or planned construction which in turn inhibits sound construction and development of quality infrastructure. This necessitates the use of sites with soils of minimal quality. This is where the idea of soil stabilization comes in [2]. Soil stabilization has made an optimum solution when the soil properties need an alteration to make it suitable for further processing. It is the process of improving the engineering properties of soil by the addition of another soil or a

H. K. Preetham (✉) · S. Nayak · P. Jagapur
Department of Civil Engineering, National Institute of Technology Karnataka, Surathkal, Mangalore 575025, India

A. K. Choudhary et al. (eds.), *Advances in Geo-Science and Geo-Structures*, Lecture Notes in Civil Engineering 154,
https://doi.org/10.1007/978-981-16-1993-9_8

cementitious material [3]. It can lead to the improvement of shear strength as well as the reduction of permeability and compressibility of the soil mass [4]. Among the various methods available for soil stabilization, the most common and widely accepted one is the mixing of a natural coarse-grained soil and a fine-grained soil which results in a mixture having an adequate angle of internal friction and cohesion [5]. But the method gaining popularity nowadays is the mixing of natural soils with a traditional stabilizer and industrial by-products which would otherwise be a serious issue concerning disposal [6].

Red earth also called Moorum is formed in the tropical regions through the process of weathering [7]. It is usually rich in Iron, Aluminum, Manganese and Titanium oxides. Iron and Aluminum oxides which are prominent in red earth results in the reddish-brown color of the soil with the seasonal fluctuation of the water table [8]. Red earth is widely spread throughout the world especially in the regions with high rainfall. In India, it is found majorly in the peninsular regions [9]. The desirability of red soil as a foundation soil can be improved with stabilization.

The traditional stabilizers used are lime, fly ash, cement, etc. Lime is usually used when the paving properties need to be improved and the ability of soil to get compacted must be high. It involves alteration of the soil structure, carbonation, pozzolanic reactions, etc. [10]. For the poor soil for its low baring capacity, volume instability and high compressibility, cement is used. Stabilization with cement is not dependent on the soil type and the strength gain is due to pozzolanic reactions [11]. Cement stabilization occurs with the hydration of cement and the consequent formation of hydration products CSH, CAH and CASH [12].

Granulated-blast furnace slag (GBFS) is generated as the primary by-product from iron and steel industries and they are produced in large quantities. The production of iron involves the supply of a controlled mass of limestone, coke and iron ore into the blast furnace which is operating at a high temperature of 1500 °C. The by-product left behind is slag which floats on the top of molten iron due to its low-specific gravity [13]. This molten slag, comprising of silicates and aluminates from iron ore and limestone in various oxide forms, is quenched with the help of high-pressure jets. This prevents crystallization leading to the formation of glassy granular aggregates [14]. Approximately, around 400 kg of slag is produced for every ton of crude iron. Since they induce no harm to the environment or human life, The National Slag Association recommends its incorporation in various civil engineering activities. Besides, the problem of disposal associated with GBFS can also be solved.

GBFS has latent hydraulic properties and similar chemical composition and hydration products as that of cement. But, the reactivity of GBFS is slow and time-consuming. To speed up the reactivity, activators like cement, lime, are required. Although the strength gain due to hydration reactions of GBFS are slower, the combined effect of the long-term strength of GBFS hydration products and cement hydration products certainly exceeds that due to cement hydration alone [15]. Generally, the amount of improvement which happens is a function of the chemical composition of the soil and the slag [16].

Previous studies on the utilization of these by-products have proven to be advantageous in many respects. Thus, to address the issue of disposal of these industrial remnants and to study how these waste materials can be effectively used for the stabilization of red soil, this study was taken up.

2 Materials

The red soil used in the study was procured locally from Hassan City, Karnataka located at 13.0033° N, 76.1004° E. The disturbed samples obtained were brought to the geotechnical laboratory to establish their physical as well as geotechnical properties. For that purpose, the red soil was dried pulverized and sieved properly as per the requirements for various tests. GBFS was procured from iron industry, Koppal, Karnataka.

3 Methodology

The whole experimental program was subdivided into two phases. It started with the initial characterization of the materials involved, i.e., Red soil and GBFS. The shear strength properties like unconfined compressive strength (UCS), cohesion and friction, etc. were also analyzed. Tables 1 and 2 show the engineering properties of red soil and GBFS, respectively. The second phase mainly dealt with the replacement

Table 1 Physical and engineering properties of red soil

Sl. No	Properties	Particulars
1	Gradation analysis Gravel size particles% Sand size particles % Silt size particles % Clay size particles %	2 54 30 14
2	Atterberg's limits Liquid limit % Plastic limit % Shrinkage limit % Plasticity index %	31 21 16 10
	Indian Standard Classification	SC
3	Specific Gravity	2.64
4	Compaction Characteristics IS Light Compaction: MDU (kN/m^3) OMC (%) IS Heavy Compaction: MDD (kN/m3) OMC (%)	18.3 15.2 19.1 13.8
5	Unconfined compressive strength (kPa)	230

Table 2 Physical and engineering properties of GBFS

Sl. No	Properties	Particulars
1	Specific gravity	1.73
2	Gradation analysis Gravel size particles (%) Coarse sand size particles (%) Medium sand size particles (%) Fine sand size particles (%)	01 14 43 42
3	Strength parameters Cohesion (kPa) Angle of internal friction ϕ (degrees)	4 42
4	pH	9.2
5	Calcium oxide (%)	41
6	SiO_2(%)	35.3

of soil with the industrial by-product (GBFS) in varying percentages of 5, 10, 15, 20 and 30% and finding out the optimum mix for which strength properties showed maximum values.

Geotechnical experiments were conducted on these samples as per IS 2720 and SP 36. The index, as well as strength properties of soil-GBFS mixes, was determined. The various chemical compounds present in GBFS is determined as per the procedure mentioned in IS 1727–1967. For determination of shear strength properties, cylindrical samples of L/D ratio 2 were prepared at maximum dry density (MDD) at their optimum moisture content (OMC) whose values were obtained from compaction tests. Due to the pozzolanic reactivity of GBFS, cylindrical samples were cured for 7 days and 28 days before they were subjected to strength tests. These cylindrical specimens were cured in desiccators for 7 days and 28 days. After the curing periods, these samples were tested for their unconfined compression strengths and triaxial compression strengths under UU (Unconsolidated Undrained) conditions. Fresh samples were prepared by replacing soil with different percentages of GBFS and conducting all the experiments as per IS codal provisions. The percentage replacement wherein the highest value of UCS strength was obtained was recorded as the optimum percentage of replacement. All the test results are analyzed and presented systematically in the paper.

4 Results and Discussions

4.1 Effect of GBFS on Index Properties of Red Soil

The experimental test results obtained after replacing soil with varying percentages of GBFS is presented in Table 3. The plasticity index of the soil reduced after the addition of GBFS due to a reduction in liquid limit and an increase in the plastic

Table 3 Geotechnical properties of red soil before and after stabilization with GBFS

Sl. No	Properties	Percentage of GBFS replacing soil (%)					
		0	5	10	15	20	30
1	Specific gravity	2.64	2.63	2.62	2.61	2.59	2.56
2	Liquid limit %	31	30	30	29	29	28
3	Plastic limit %	21	21	22	22	22	23
4	Plasticity Index %	10	8.8	8.1	7.6	7.1	5.7
5	Maximum dry density (kN/m3)	18.3	17.8	17.6	17.3	17.2	16.4
6	Optimum water content %	15.2	15.8	16.0	16.3	16.6	17.2

limit. The decrement was considered marginal. Further addition of a coarser material (GBFS) will not result in a considerable decrease in liquid limit. The MDD did not show any considerable variation with GBFS addition [17]. It might be due to the frictional resistance offered by GBFS to the compaction effort. The OMC value was found to increase with percentage increase in GBFS [3].This can be attributed to the fact that while compaction, GBFS can break down into smaller particles resulting in an increase in surface area and a consequent increase in OMC.

4.2 *Effect of GBFS on Unconfined Compressive Strength of Red Soil*

The variation of UCS with an increased percentage of addition of GBFS for immediate, 7 days and 28 days curing are shown in Fig. 1. From figure, it is clear that with increase in the percentage addition of GBFS, the immediate UCS values increased marginally up to 15% after which it showed a declining trend [18]. This declining trend may be due to the increase in coarser particles in the mix leading to a low value of cohesion intercept and hence, a low value of shear strength. In mixes subjected to 7 days curing, a remarkable increase in UCS values was observed. For the optimum replacement percentage of 15%, the percentage increase in UCS value was found to increase by 77% (from 193 to 341 kPa). With 28 days curing period, the increment in UCS was highly appreciable. For 15% replacement, an increment of 677% was found in the UCS value. This increase can be explained with the presence of free lime and reactive silica present in GBFS. The hydration reaction which takes place between free lime in GBFS and the soil in the presence of water forms cementitious pozzolanic products similar to those obtained during cement hydration. These products have binding properties resulting in an increase in UCS values. With an increase in the age of curing, there is formation of CSH which results in further increase of strength [19]. But after 15%, the UCS value decreased. This is because, with increase

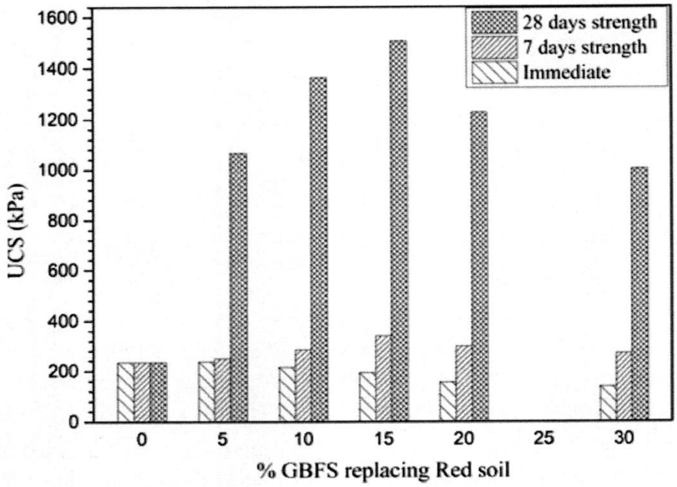

Fig. 1 Variation of UCS of red soil and GBFS blend with curing age

in GBFS content and a decrease in the molding water content, the pozzolanic reactions were hindered. The excess GBFS particles are considered redundant. Thus, the optimum percentage replacement was concluded as 15%.

4.3 *Effect of GBFS on Shear Strength Parameters of Red Soil*

To assess the effect of replacement of GBFS on the shear strength parameters, unconsolidated undrained (UU) tests were conducted on saturated soil samples mixed with optimum GBFS cast at their MDD and OMC. In the case of GBFS, samples cured at 28 days were tested. Both cohesion intercept and friction angle were found to increase with the addition of GBFS. The cohesion intercept was found to rise from 20 kPa (for basic soil) to 78 kPa (Table 4). This occurs due to the pozzolanic reaction

Table 4 Shear strength parameters and permeability of untreated and optimum GBFS treated red soil

Sl. No	Properties	Percentage of GBFS replacing soil (%)	
		0% (untreated soil)	15% (opt mix)
1	Cohesion (kPa)	20	78
2	Friction angle (Degree)	22	29
3	Coefficient of permeability (m/day)	0.012	0.0293

between free lime in slag and the soil [20]. The friction angle also increased from 22° to 29° due to the dominance of coarser particles.

The coefficient of permeability increases by 144% (Table 4) for optimum mix when compared with the untreated soil. The reason behind the increase is that the void space increase with the addition of granular slag.

5 Conclusion

Red soil was replaced by GBFS in different proportions, and the mixes were subjected to geotechnical investigations. The following conclusions can be drawn from the investigations:

- GBFS is a by-products obtained from iron industries and quarries, respectively. Due to the presence of CaO in GBFS, it imparts binding properties to the soil in the presence of water.
- OMC increased, whereas MDD showed only marginal variations with GBFS addition. This is due to the increase in surface area caused due to the breaking down of particles.
- UCS of the mix enhanced with the addition of GBFS, but after a certain proportion of addition, the percentage increase was found to reduce. Thus, an optimum percentage of replacement was found at 15%. With the age of curing, the UCS strength enhanced due to the pozzolanic reactions.
- Shear strength parameters and coefficient of permeability showed a considerable increase for optimum GBFS treated soil.

References

1. Soosan TG, Sridharan A, Jose BT, Abraham BM (2005) Utilization of quarry dust to improve the geotechnical properties of soils in highway construction. Geotech Test J 28:391–400. https://doi.org/10.1520/GTJ11768
2. Basha EA, Hashim R, Mahmud HB, Muntohar AS (2005) Stabilization of residual soil with rice husk ash and cement. Constr Build Mater 19:448–453. https://doi.org/10.1016/j.conbuildmat.2004.08.001
3. Preetham HK, Nayak S, Surya EV (2019) Experimental investigation on the stabilization of soft clay using granulated blast furnace slag. IOP Sonf Ser Mater Sci Eng 561. https://doi.org/10.1088/1757-899X/561/1/012047
4. Nayak S, Balaji M, Preetham HK (2020) A study on the behaviour of stone columns in a layered soil system. Transp Infrastruct Geotechnol 7:85–102. https://doi.org/10.1007/s40515-019-00090-x
5. Mishra S, Sachdeva SN, Manocha R (2019) Subgrade soil stabilization using stone dust and coarse aggregate: a cost effective approach. Int J Geosynth Gr Eng 5. https://doi.org/10.1007/s40891-019-0171-0
6. Okagbue CO, Onyeobi TUS (1999) Potential of marble dust to stabilise red tropical soils for road construction. Eng Geol 53:371–380. https://doi.org/10.1016/S0013-7952(99)00036-8

7. Nagaraj HB, Rajesh A, Sravan MV (2016) Influence of soil gradation, proportion and combination of admixtures on the properties and durability of CSEBs
8. Venkatarama Reddy BV, Latha MS (2014) Influence of soil grading on the characteristics of cement stabilised soil compacts. Mater Struct Constr 47:1633–1645. https://doi.org/10.1617/s11527-013-0142-1
9. Wang D, Abriak NE, Zentar R, Chen W (2013) Effect of lime treatment on geotechnical properties of Dunkirk sediments in France. Road Mater Pavement Des 14:485–503. https://doi.org/10.1080/14680629.2012.755935
10. González-López JR, Juárez-Alvarado CA, Ayub-Francis B, Mendoza-Rangel JM (2018) Compaction effect on the compressive strength and durability of stabilized earth blocks. Constr Build Mater 163:179–188. https://doi.org/10.1016/j.conbuildmat.2017.12.074
11. Chiu CF, Zhu W, Zhang CL (2009) Yielding and shear behaviour of cement-treated dredged materials. Eng Geol 103:1–12. https://doi.org/10.1016/j.enggeo.2008.07.007
12. Preetham HK, Nayak S (2019) Geotechnical investigations on marine clay stabilized using granulated blast furnace slag and cement. Int J Geosynth Gr Eng 5:1–12. https://doi.org/10.1007/s40891-019-0179-5
13. Sekhar DC, Nayak S, Preetham HK (2017) Influence of granulated blast furnace slag and cement on the strength properties of lithomargic clay. Indian Geotech J 47:384–392. https://doi.org/10.1007/s40098-017-0228-8
14. Patra RK, Mukharjee BB (2017) Influence of incorporation of granulated blast furnace slag as replacement of fine aggregate on properties of concrete. J Clean Prod 165:468–476. https://doi.org/10.1016/j.jclepro.2017.07.125
15. Yi Y, Li C, Liu S (2015) Alkali-activated ground-granulated blast furnace slag for stabilization of marine soft clay. J Mater Civ Eng 27:1–7. https://doi.org/10.1061/(ASCE)MT.1943-5533.0001100
16. Choobbasti AJ, Kutanaei SS (2017) Microstructure characteristics of cement-stabilized sandy soil using nanosilica. J Rock Mech Geotech Eng 9:981–988. https://doi.org/10.1016/j.jrmge.2017.03.015
17. Nayak S, Sarvade PG (2012) Effect of cement and quarry dust on shear strength and hydraulic characteristics of lithomargic clay. Geotech Geol Eng 30:419–430. https://doi.org/10.1007/s10706-011-9477-y
18. Sekhar CD, Nayak S (2017) SEM and XRD investigations on lithomargic clay stabilized using granulated blast furnace slag and cement. Int J Geotech Eng 6362:1–15. https://doi.org/10.1080/19386362.2017.1380355
19. Sekhar CD, Nayak S (2018) Utilization of granulated blast furnace slag and cement in the manufacture of compressed stabilized earth blocks. Constr Build Mater 166:531–536. https://doi.org/10.1016/j.conbuildmat.2018.01.125
20. Yadu L, Tripathi RK (2013) Effects of granulated blast furnace slag in the engineering behaviour of stabilized soft soil. Procedia Eng 51:125–131. https://doi.org/10.1016/j.proeng.2013.01.019

Strength and Impact of Rice Husk Ash on Expansive Soil by Using Soil Stabilization

Sonoo Kumar, R. P. Singh, and S. K. Paswan

Abstract There are about 20% black cotton soil areas in India. It is one of the most problematic soils with a high tendency to shrink or swell due to water content changes. When it comes into contact with water, the volume of the soil increases and decreases when water evaporates or loses its touch. Because of the unpredictable nature of this soil, it is not easy to construct or use such land properly without proper treatment because it provides different settlements that pose risks to constructed structures. The rice husk is available in abundance, as India has a large rice production. The rice husk is produced from Rice Mills as a by-product. This husk contains some of the volatile matter and is converted to ash, which is produced during the firing of rice husk. The purpose of the present research is to study the impact of rice husk ash on swelling and mechanical characteristics of the black cotton soil. With increase in the contents of RHA into soil up to 20% by dry weight of soils—RHA composite mix starting from 5 to 20% at a rate of increase as 5%, RHA and expansive soil mixtures were prepared in adequate numbers. After three days of curing, differential free swell tests, Atterberg limit tests, shrinkage factor tests, specific gravity tests, light compaction tests, soaked CBR were carried out. The addition of RHA reduces differential free swell, plasticity index, flow index, shrinkage, maximum dry density, angle of internal friction. As the amount of RHA increases CBR, liquid limit, plastic limit, void ratio, porosity, peak deviator stress values of soil increases. In cohesion intercept variation with RHA content, the typical pattern is observed. Cohesion intercept begins to grow with the rise in RHA content, but starts decreasing after 10 percent RHA content. Based on this experimental analysis, it is found that RHA not only reduces the swelling and shrinking behavior of expansive soil but also significantly increases its strength.

S. Kumar (✉) · R. P. Singh · S. K. Paswan
Civil Engineering Department, NIT, Jamshedpur 831014, India

R. P. Singh
e-mail: rpsingh.ce@nitjsr.ac.in

S. K. Paswan
e-mail: skpaswan.ce@nitjsr.ac.in

A. K. Choudhary et al. (eds.), *Advances in Geo-Science and Geo-Structures*,
Lecture Notes in Civil Engineering 154,
https://doi.org/10.1007/978-981-16-1993-9_9

Keywords Expansive soils · Soil stabilization · Rice husk ash · Agricultural waste · Waste recycled product

1 Introduction

Soil supports all the structures through suitably designed foundation systems. However, the response of all soil types on loading differs in many ways, and some soils have characteristics to damage the structures on it. Expansive soil is such soil which have a typical swell-shrink behavior with water content. This poses challenge for the structures built on it.

Paddy is one of the main kharif crops in India and its annual production is around 100 million tons per year. The rice husk on burning can give the ash about 20% of it by weight. Since the rice husk ash (RHA) is available in abundance, it has been attempted to find its one of the good use in improving or controlling the swell-shrink characteristics of the black cotton soil.

1.1 Soil Stabilization

Soil stabilization is the changes brought into the soils to increase their physical properties. A method of soil stabilization includes mechanical methods and mixing different types of admixtures. The admixtures may be lime, cement, chemicals, fly ash, rice husk ash, bitumen, thermal and electrical methods, geo-textile and fabrics, recycled and waste products, etc. Soil stabilization is required when the clayey soil with high clay content is encountered as these soils swell on increase in moisture content. Soil stabilization has been investigated with industrial and agricultural wastes. India produces about 350 million tons of agricultural wastes every year.

Joe, MA and Rajesh AM investigated on soil stabilization using copper slag and lime as admixtures. It reports notable increase in the unconfined compressive strength of sub-base soil. Further inference of the works is the improvement in compaction parameters; the maximum dry density is increased at optimum moisture content [1].

Pandey PK and Jawaid, SM did a series of laboratory experiments with the mix of soil, red mud and fly ash. It has been concluded that while keeping the fly ash as 3% and varying the red mud in the range of 3–36% by weight of dry soil, CBR of the mix kept on increasing up to 30% red mud and the maximum value of CBR was found to be 7.354% [2].

GGBS (Ground Granulate Blast Furnace Slag) is a by-product in the blast furnace during extraction of iron or steel from iron ores. CBR of a clayey sub-base can be improved with GGBS and fly ash. Dayalan J did experiments on soil mix with fly ash and GGBS. It was inferred that the optimum value of fly ash and GGBS is 15% and 20%, respectively, with maximum CBR value of 8.6% [3].

The study on expansive soil stabilization with bagasse ash and lime by Wubshet M and Tadesse S states that the plasticity index of the mix of soil, bagasse ash (15% of soil) and lime (3% of soil) decreases substantially and maximum dry density. This implies that the swell-shrink in volume is also reduced with change in moisture content. The CBR of the soil increases [4].

Ali R et al. report that the addition of 4, 8 and 12 marble dust along with bagasse ash to expansive soil decreases the index properties. Notable increase in dry density at OMC of the expansive soil has been observed with addition of bagasse ash and marble dust [5].

Murari A et al. mention that the bagasse ash mixed soil shows decreased atterberg limits values. This change is due to cation exchange reaction as well as due to flocculation and aggregation of bagasse ash. It is also reported that the optimum moisture content of the mix is increased and the maximum dry density (MDD) is reduced [6].

Oviya R and Manikandan R study report improvement in the expansive soil due to addition of rice husk ash (RHA) and lime. The RHA mixed soil has reduced specific gravity, increased Atterberg Limits, reduced MDD with higher optimum moisture (OMC) content. When the expansive soil is treated with RHA and lime [Ca(OH)2], OMC increases, shrinkage limit decreases and swelling is reduced. CBR and UCS value of the mix are improved [7].

Shinde S. S. and Patil G. K. treated the weak subgrade soil with rice husk ash (RHA) and ground nut shell ash (GNSA) separately with different percentages. CBR value increased in both the cases, i.e., with RHA the increase was from 2.39% to 5.09% and with GNSA increase was from 2.39% to 4.30%. In both the cases, maximum dry density was found to increase, whereas the optimum moisture content was decreased [8].

2 Materials and Methodology

2.1 Black Cotton Soil

For this study, black cotton soil has been taken from the site near Sasaram in Bihar. The engineering properties of soil and rice husk ash used in the investigation are presented in Tables 1 and 2, respectively.

2.2 Rice Husk Ash

Rice husk ash (RHA) is an agricultural by-product and available in abundance. Particularly, in the rice mills, huge amount of husk is produced and used as fuel to generate electricity.

Table 1 Black cotton soil of expansive nature

Properties	Values
Color	Brownish black
Specific gravity	2.65
Differential free swell (in %)	57
Liquid limit (in %)	42.75
Plastic limit (in %)	13.3
Plasticity index (in %)	28.75
Classification (as per IS:1498–1970)	CI
OMC (in %)	19.5
MDD(in kN/m^2)	18.795
C_{UU}(in kN/m^2)	42.25
Φ_{UU}(in degrees)	18

Table 2 The basic geotechnical properties of RHA

Properties	Values
Color	Gray
Specific gravity	1.580
Differential free swell [DFS(%)]	0
Liquid limit (%), Plastic limit (%)	Non-plastic
Plasticity index (%)	Non-plastic

Table 3 Soil and RHA percentage in composite samples used for tests

Mixture designation	Materials in %	
	Expansive soil	Rice husk ash
ES	100	0
95% ES + 5% RHA	95	5.5
90% ES + 10%RHA	90	10.5
85% ES + 15% RHA	85	15.5
80% ES + 20%RHA	80	20.5

The black cotton soil and RHA from rice factory Sultanganj, Bhagalpur, Bihar were dried. Thus, prepared samples of soil and RHA were mixed. The amounts of RHA, in the soil-RHA composite samples, were varied from 5% to 20% at an increment rate of 5% by weight of dry soil sample. These mixes are listed in Table 3.

3 Results

Variation of specific gravity with change in RHA content is shown in Fig. 1. From the study, it is observed that the specific gravity of raw soil is 2.65 and that with RHA is 1.580, respectively. In the composite sample with an increase in the content of RHA, specific gravity decreases.

Fig. 1 Variation in specific gravity with RHA content

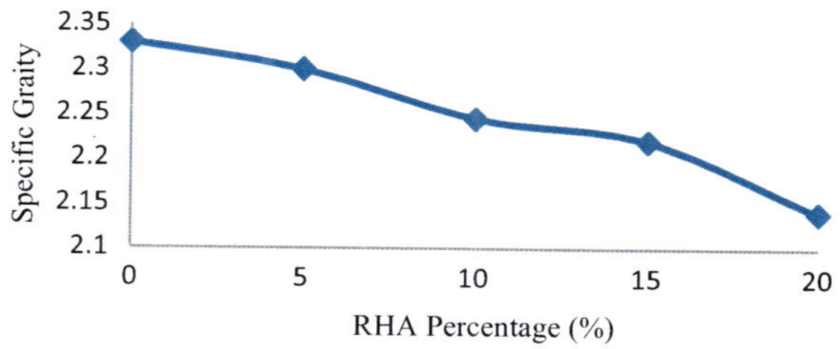

The grain size distribution curve is shown in Figs. 2 and 3 for the raw soil and RHA. This shows that the soil and RHA have considerable amount of clay particles and the RHA is mostly comprised of silica.

As a result of the RHA rise, w_l is increase was lower than w_p, with I_p decreases, with the RHA content increasing variation of I_t with an increase in RHA content is shown in Fig. 4. As the RHA content increases the liquid limit, the index of plasticity reduces the plasticity of the soil. Figure 4 shows the variation of I_p with w_L. It is observed that with an increase in w_L, the value of I_p decreases. The relationship between I_p and w_L is nearly linear.

Figure 5 shows the variation of MDD with RHA content in the composite sample. It is observed that with an increase in RHA content MDD value decreases.

Fig. 2 Grain size distribution curves of soil

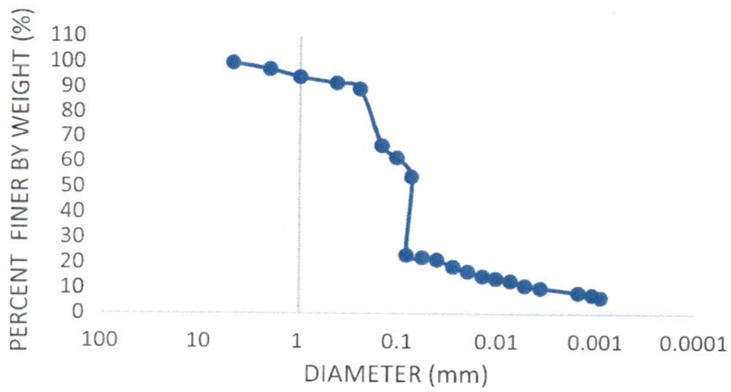

Fig. 3 Grain size distribution curves of RHA

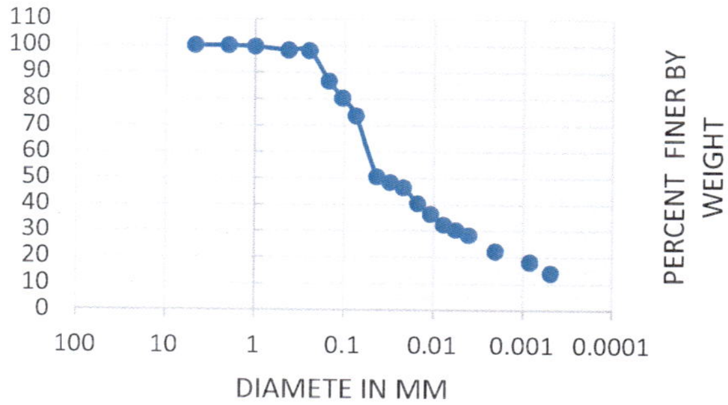

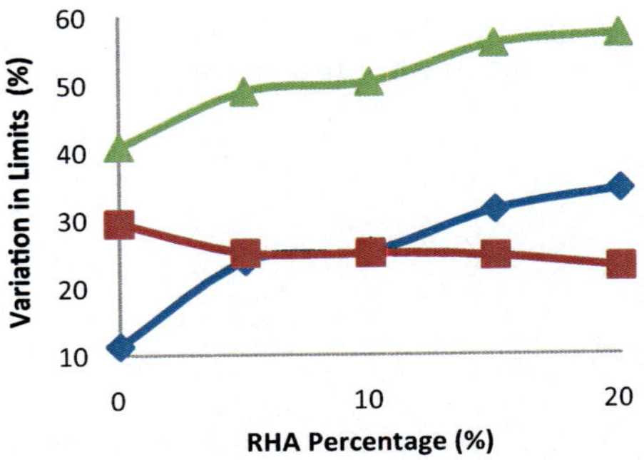

Fig. 4 Variation of liquid limit, plastic limit and plasticity index simultaneously

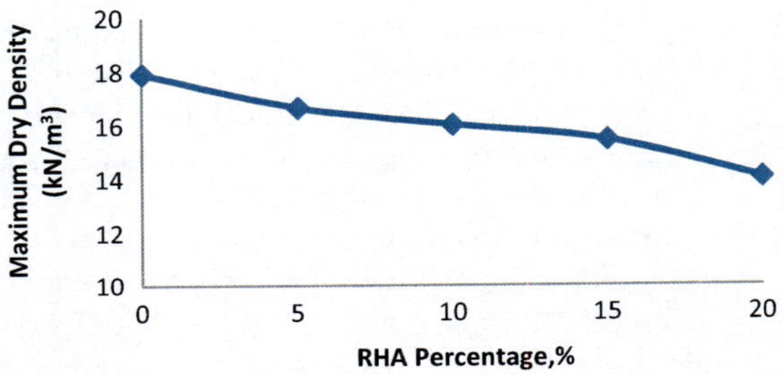

Fig. 5 Effect of RHA on compaction curves

CBR test results are shown in Fig. 6 shows the load—penetration curve of unstabilized and stabilized soil with different percentages of RHA. Figure 7 shows CBR value increases with increase in RHA content. Initially, the value of CBR found to be 1.2 with 0% RHA content which rises further to an appreciable value of 7.68 at 20% RHA content.

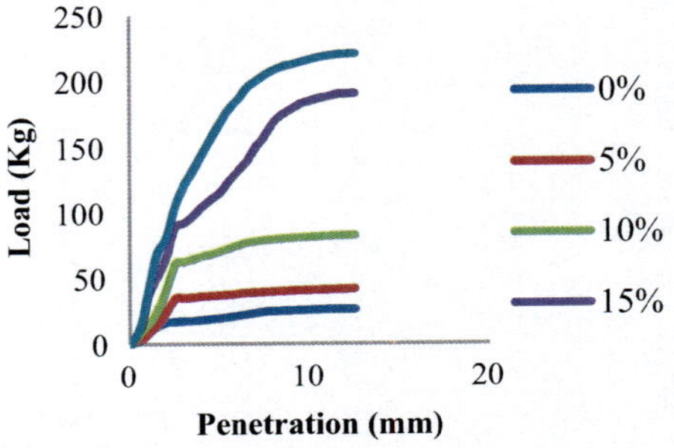

Fig. 6 Effect of RHA on load penetration curves

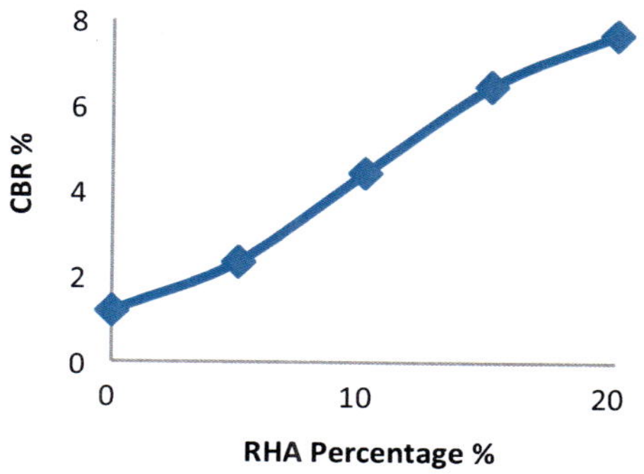

Fig. 7 Variation of CBR with RHA content

4 Conclusions

Based on the study carried out following conclusions can be made.

- Specific gravity of composite sample decreased, with increase in RHA content.
- Liquid limit (w_L) and Plastic limit (w_p) of soil increase with increase in RHA content. Increase in w_p was at high rate which results in decrease in plasticity index (I_p) value with increase in percentage RHA added to soil. Additions of RHA also change the value of I_p and flow index (I_f).
- The MDD value decreases and OMC value increases with increase in RHA content in the composite sample.
- Addition of RHA increases the void ratio and porosity due to decrease in MDD of composite samples.
- RHA addition shows as improvement in penetration resistance of compacted soil samples. With increase in RHA content CBR value of sub grade soil has increased significantly.

This is finally concluded that the wastes-like rice husk ash can be used effectively to stabilize the black cotton soil to a considerable extent.

References

1. Joe MA, Rajesh AM (2015) Soil stabilization using industrial waste and lime. Int J Sci Res Eng Technol 4(7):799–805
2. Panday PK, Jawaid SM (2015) Soil improvement using red mud and fly ash. Glob J Eng Sci Res 1:7–10
3. Dayalan J "Comparative study on stabilization of soil with ground granulated blast furnace slag (GGBS) and fly ash". Int Res J Engi and Technol 3:2198–2204
4. Wubshet M, Tadesse S (2014) Stabilization of expansive soil using bagasse ash and lime. J Eur Econ Assoc 32:21–26
5. Ali R et al (2014) Expansive soil stabilization using marble dust and bagasse ash. Int J Sci Res 3:2812–2816

6. Murari A et al (2015) "Stabilization of local soil with bagasse ash". SSRG Intl J Civil Eng 37–39
7. Oviya R, Manikandan R (2016) An experimental investigation on stabilizing the soil using rice husk ash with lime as admixture. Int J Inform Future Res 3:3511–3519
8. Shinde SS, Patil GK (2016) Study on utilization of agricultural waste as soil stabilizer. Int J Latest Trends Eng Technol 7(1):227–230

Improvement of Geotechnical Properties of Fine-Grained Soil by Using Lime and Guwahati Municipal Solid Waste Ash

Rahul Chauhan and Shailen Deka

Abstract Municipal solid waste (MSW) is a significant environmental problem that can have a negative ecological effect which may lead to future outbreaks of disease and epidemics. Utilization of incinerated municipal solid waste ash with lime for stabilization of soil is the main aspect of this study. Through this study, the geotechnical properties of fine-grained soil blended with different percentage of lime (1, 2 and 3%) and municipal solid waste ash (2, 4, 6, 8 and 10%) have been analyzed. The plasticity behavior of the soil has been studied by curing the samples for 0, 3, 7, 14 and 28 days. The strength behavior of the soil has been studied by curing the samples for 0, 3, 7 and 14 days. It is observed that the combined effect of lime, MSW ash and curing greatly influences the plasticity and strength behavior of the soil. The admixtures influenced the parameters of compaction by lowering the soil's maximum dry density and raising the optimum moisture content of the soil.

Keywords Lime · Municipal solid waste ash · Curing strength · Compaction · Plasticity · Soil stabilization

1 Introduction

In order to hold a structure, the foundation has to be strong enough to support the whole structure. To make the foundation effective, the soil around the foundation plays an extremely important key role [1]. Soil treatment and stabilization is a technique which is used worldwide to enhance soft soil. There are two main ways of improving soft soil: a) modification and b) stabilizing. According to Verma [2] ancient Chinese, Roman and Inca civilizations adopted different techniques to increase soil strength, etc., and few of these techniques were so successful that their structures and roads still persist. As per Sai [3], soil stabilization is a way to improve soil's geotechnical characteristics such as shear strength, shrinkage and bearing capacity of soil.

R. Chauhan (✉) · S. Deka
Department of Civil Engineering, Tezpur University, Tezpur, Assam, India

A. K. Choudhary et al. (eds.), *Advances in Geo-Science and Geo-Structures*, Lecture Notes in Civil Engineering 154,
https://doi.org/10.1007/978-981-16-1993-9_10

Lime is commonly used in soil stabilization. When lime is added in clayey soil, immediate reaction (ion exchange) and pozzolanic reactions (long-term reaction) take place it is the most commonly used stabilization method when the project require enhancement of local soil [4, 5]. As per Bell [4] for change or modification of soil, lime is usually added by weight between 1 and 3% lime.

India is a densely populated country with around 130 billion people, which generates millions of tonnes of municipal solid waste annually and increasing exponentially. Municipal solid waste management is a serious environmental issue which can have a detrimental ecological impact and can lead to potential disease outbreaks and epidemics. Hence, dumping of this generated MSW waste is a big challenging problem for the government, due to lack of landfill and incineration.

Guwahati is the fastest-growing metropolitan area in Assam with 23.89% of the state's total urban population. The Guwahati city is generating 317 TPD of municipal solid wastes. According to Taha et al. [6], incineration creates by-products known as bottom ash and fly ash which can be used in building activities. Show et al. [7] claimed that chemical composition and physical properties, MSWA is susceptible to pozzolanic reaction. According to Muhunthan et al. [8], incinerated ash, physical and chemical property depend on type and source of municipal solid waste. Many researchers utilized the incinerated ash as a road construction aggregate or as a filler material in embankments, reported Sherwood et al. [9]. Many researchers have used MSW ash to solve various geotechnical- and geo-environmental-related problems reported Kamon et al. [10]. The bottom ash produced after combustion of municipal solid waste has been found to have pozzolanic qualities that would be effective in enhancement of soil reported Berg and Neal [11] and Lin et al. [12]. As Per Kumar and Mittal [1], municipal solid waste (MSW) has cementitious property and so can be used as soil stabilizer. The use of MSWA with lime or cement will have a better pozzolanic reaction with age. Hence, utilization of incinerated municipal solid waste ash with lime for stabilization of soil is the main aspect of this study.

1.1 Objective of the Work

The main objectives of the present work are (a) To determine change in liquid limit, plastic limit and plasticity index upon treatment with lime and municipal solid waste ash for curing period of 0, 3, 7, 14 and 28 days. (b) To compare compaction behavior of untreated soil with that of treated soil for all mixing combinations. (c) To determine the change in strength of untreated soil upon treatment with the above admixtures for all mixing combinations by conducting unconfined compression test for curing periods of 0, 3, 7 and 14 days. The mix design combination of lime and municipal solid waste is given in Table 1.

Table 1 Mixing combination of lime and municipal solid waste ash

Mixing combination	Mixing combination	Mixing combination
1% Lime + 2% MSWA	2% Lime + 2% MSWA	3% Lime + 2% MSWA
1% Lime + 4% MSWA	2% Lime + 4% MSWA	3% Lime + 4% MSWA
1% Lime + 6% MSWA	2% Lime + 6% MSWA	3% Lime + 6% MSWA
1% Lime + 8% MSWA	2% Lime + 8% MSWA	3% Lime + 8% MSWA
1% Lime + 10% MSWA	2% Lime + 10% MSWA	3% Lime + 10% MSWA

1.2 Literature Review

Kumar [1] studied the behavior of compaction and strength properties of soil with addition of MSW ash. The MSW ash was added in percentage of 0, 5, 10, 15 and 20% by dry weight of soil. The MSW ash-stabilized soil shows decline in maximum dry density, when the percentage of MSW ash content increases. Addition of 20% MSW ash increases the optimum moisture content and the unconfined compression value and reduces the maximum dry density of soil.

Mohamedzein and Al-Aghbari [13] investigated the use of MSW ash to improve the dune sand, by using 10 to 80% of MSW ash by weight of dry soil. The soil and MSW ash mixture was allowed to cure for 7–90 days. The findings revealed that the maximum dry density stays about steady to 30% and then declines as the ash content is increased. With ash content up to 30%, unconfined compressive strength increases significantly and then decreases with increase in ash content.

Singh and Kumar [14] studied the compaction behavior and strength characteristics by inclusion of cement and MSW ash. The specimens were cured for 7, 14 and 28 d. The test result shows that the maximum dry density reduces and optimum moisture content rises with inclusion of cement and MSW ash. The UCS increases with increase in curing period. The strength increases at initial stage of curing, i.e., 7 days of curing.

1.3 Need of Present Work

The laboratory investigation for Atterberg limit (liquid limit, plastic limit and plasticity index) is less performed for all the mixing combination and for each curing period. The OMC and MDD for the treated soil are generally taken same as that of untreated soil, and hence, for each mixing combination, the value of OMC and MDD is needed to be determined. The MSWA admixture used in much research work was stabilized alone with the soil. The overall cementitious properties of MSWA are not much if stabilized alone with the soil, and hence, for better pozzolanic reactions, lime or cement can be used.

2 Material and Methods

2.1 Materials

The research work was carried out with locally available soil, lime and municipal solid waste ash. The summary of the materials is briefly discussed in detail.

Soil.

The soil samples used in the analysis were collected from Krishi Vigyan Kendra, Napaam, located about 10 km from the district headquarter Tezpur (Assam). The clayey soil was extracted at a depth of about 1 m. Laboratory tests have been carried out to classify the soil. Classification of the soil was done as per unified soil classification system (USCS). The physical characteristics of the soil are given below in Table 2.

Lime.

The lime used in the analysis was collected from Tezpur market (Assam). The specific gravity of lime was about 2.8. The minimum lime percentage that achieves a pH of 12.4 at 25°C is the effective lime percentage for soil stabilization, and it is recommended as the minimum lime content or initial lime consumption reported Gandhi et al. [15]. According to Dash and Hussain [5], pozzolanic reactions occurs when the pH of the soil is greater or equal to 12 and thus forms cementitious products

Table 2 Physical characteristics of soil

Parameter	Values
Specific gravity	2.65
Optimum moisture content (%)	22.3
Maximum dry density (g/cc)	1.7
Liquid limit (%)	37.4
Plastic limit (%)	21.6
Plasticity index (%)	15.9
Classification of the soil as per (USCS)	CI
% Gravel	0
% Coarse grained	17.7
% Fine grained	82.3
Unconfined compressive strength (kN/m2)	132.69

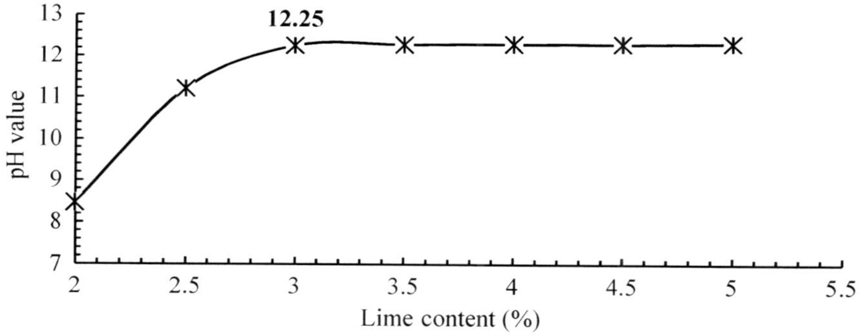

Fig. 1 pH values of lime-mixed soil

like C-S–H gel. The peak pH value achieved in this study was 12.25% at 3% lime content as shown in Fig. 1.

Municipal solid waste ash.

The incinerated ash was collected from Guwahati Municipal Corporation (GMC), Guwahati. The physical characteristics of the solid waste include 61.45% organic waste, 4.50% plastic, 20.38% paper, 9.14% rubber and leather, 2.90% glass and 1.63% metals. Specific gravity of MSW ash is 2.3. The energy dispersive X-ray was used to determine the elemental composition of the soil. Figure 2 shows an EDX spectrum of the soil. Table 3 lists the elements observed in the soil.

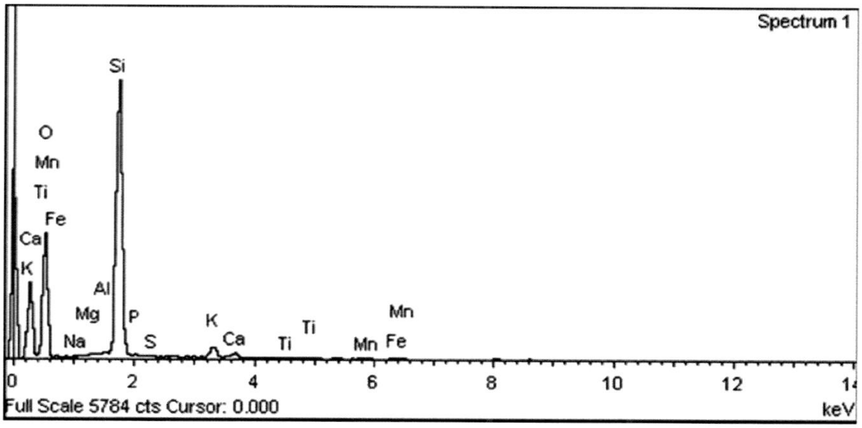

Fig. 2 Energy dispersive X-ray (EDX) spectrum of MSWA

Table 3 Elemental composition of MSWA as observed from EDX

Element	Weight%
O	56.19
Na	0.08
Mg	0.35
Al	0.31
Si	38.35
P	0.44
S	0.12
K	1.69
Ca	0.45
Mn	0.21
Fe	1.30

2.2 *Methods*

The soil was prepared by scraping the soil followed by drying the soil and breaking the lumps and by sieving through the specified sieve size for different tests. Trial percentage for MSWA (2, 4, 6, 8 and 10%) and lime (1, 2 and 3%) was selected. The prepared soil was then mixed with different combination of lime and MSWA to study their effect on the various index properties, compaction characteristics and unconfined compressive strength test.

Liquid limit test was performed as per IS 2720–5, (1985). Plastic test was performed as per IS 2720–5, (1985). Specific gravity test was performed as per IS-2720-part-3, (1980). The grain size distribution of soil was performed as per IS-2720-part4, (1985). The hydrometer test was performed to determine the particle size distribution of the fine-grained soil passing a sieve of 75μ, and the test was performed as per IS- 2720-part4, (1985). The proctor compaction test is done as per IS- 2720-Part 7, (1980). The unconfined compression test was performed as per IS 2720-part 10, (1991). Unconfined compressive strength testing was performed for both the untreated clay sample and the treated clay sample. The curing periods of the unconfined compressive strength sample were 0, 3, 7 and 14 days.

3 Results and Discussion

3.1 *Compaction Behavior*

The proctor compaction test is done as per IS- 2720-Part 7, (1980). The cause behind OMC's increase may be due to the pozzolanic reaction between the mixes that can enhance or improve affinity toward water or due to activation of pozzolanic reaction

Table 4 OMC and MDD data for treated soil

Mixing combination	OMC %	MDD (g/cc)
Untreated	22.33	1.686
1% Lime + 2% Municipal solid waste ash	23.8	1.565
1% Lime + 4% Municipal solid waste ash	24.15	1.551
1% Lime + 6% Municipal solid waste ash	24.5	1.538
1% Lime + 8% Municipal solid waste ash	25.5	1.51
1% Lime + 10% Municipal solid waste ash	26.5	1.482
2% Lime + 2% Municipal solid waste ash	25	1.492
2% Lime + 4% Municipal solid waste ash	26.295	1474
2% Lime + 6% Municipal solid waste ash	27.59	1.437
2% Lime + 8% Municipal solid waste ash	28.295	1.437
2% Lime + 10% Municipal solid waste ash	29	1.418
3% Lime + 2% Municipal solid waste ash	26.5	1.465
3% Lime + 4% Municipal solid waste ash	27.1	1.447
3% Lime + 6% Municipal solid waste ash	27.5	1.43
3% Lime + 8% Municipal solid waste ash	28.7	1.425
3% Lime + 10% Municipal solid waste ash	29.9	1.42

of soil and lime, which needs more water reported Choobbasti et al. [16]. The mix with 3% lime and 10% MSWA delivered the maximum OMC. The OMC in mix 3% lime and 10% MSWA increased by 33.9% compared to the OMC of untreated soil.

The mix 3% lime and 10% MSWA has been observed to have a maximum decline in MDD. The MDD in mix 3% lime and 10% MSWA is decreased by 15.77% compared to the MDD of untreated soil. The reduction in dry density could be due to an immediate formation of cementitious products which reduce compactibility and hence the density of the treated soil reported Lees et al. [17]. The results of the treated and untreated OMC and MDD are shown in Table 4.

3.2 Plasticity Behavior

The soil's plasticity behavior generally depends on the amount, variety, structure and fabric of the soil. This type of behavior can be evaluated using index properties, namely liquid limit (LL), plastic limit (PL) and plasticity index (PI).

Liquid limit.

The change in LL due to addition of MSW ash and lime has been presented in Figs. 3, 4, 5. It can be observed that there is a steep reduction in LL with the addition of lime and MSW ash.

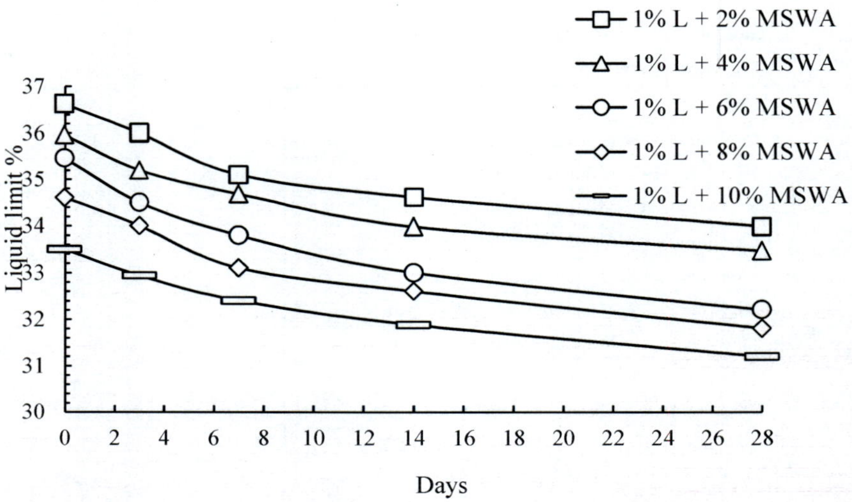

Fig. 3 Liquid limit of treated soil with (1% lime and 2, 4, 6, 8 and 10% MSWA)

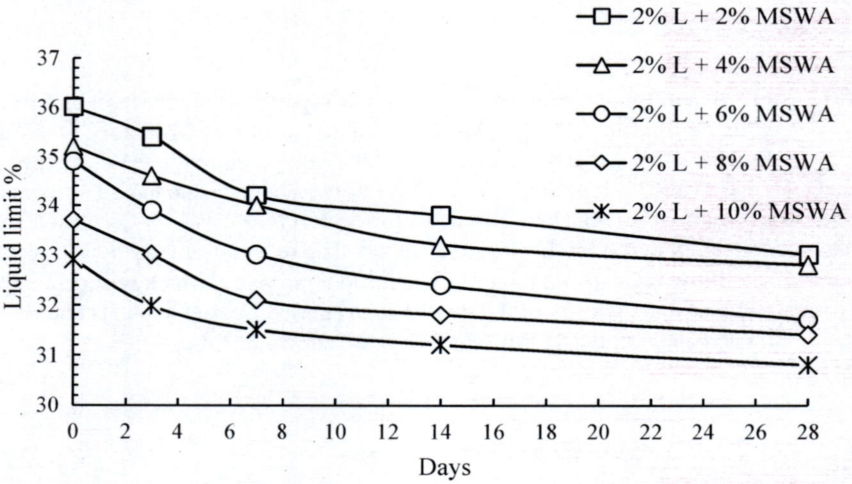

Fig. 4 Liquid limit of treated soil with (2% lime and 2, 4, 6, 8 and 10% MSWA)

This reduction in LL itself indicates an improvement in the properties of the soil. The explanation for one such reduction can be understood by cation exchange, cation ions present in lime get replaced or substituted by the cation present in the clay surface, and this exchange process of cation leads to reduction ion diffuse double layer [4, 5]. The reduction in DDL can lead to flocculation of soil particles which in turn reduce LL [18].

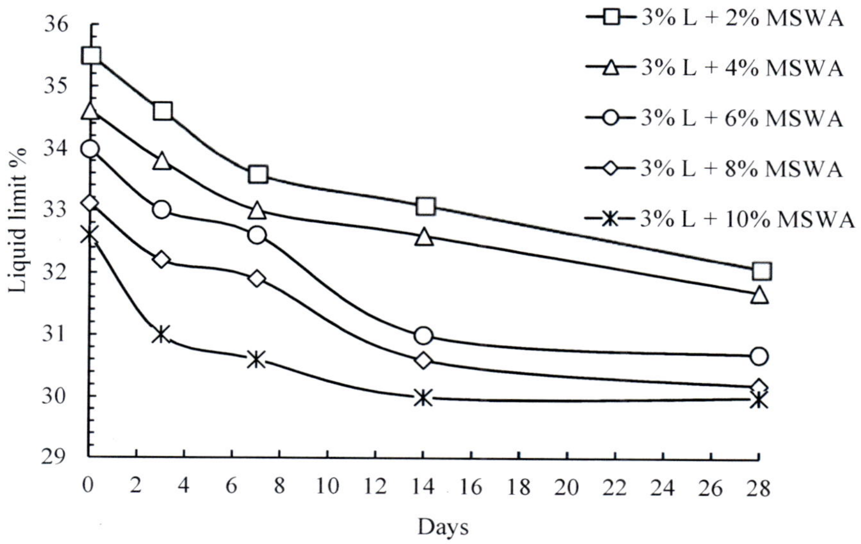

Fig. 5 Liquid limit of treated soil with (3% lime and 2, 4, 6, 8 and 10% MSWA)

Plastic limit.

PL for all mixed combinations is found to have risen, due to lime treatment and municipal solid waste ash treatment as shown in Figs. 6, 7 and 8. The reduction of DDL leads to flocculation of soil particle, and hence, clay particles aggregate to form larger particles which increase the shear resistance as well the plastic limit soil reported Bell [4]; Dash and Hussain [5].

Plasticity index.

Plasticity index indicates the soil workability and ease to deform it. The PI of all mix combination is found to be significantly decreasing as shown in Figs. 9, 10 and 11. The application of lime and MSW ash might have altered the soil structure by flocculating the clay particles that could have substituted the finer clay particles with some coarser particles, thereby enhancing their workability and reducing PI. It can be seen that, curing effect a lot on soil, lime and MSW ash mix.

3.3 *Strength Behavior*

Unconfined compressive strength test is the most reliable method to determine the compressive strength of cohesive soil, as testing and sample preparation is simple as compared to other test. The pozzolanic reaction of lime and MSW ash-stabilized

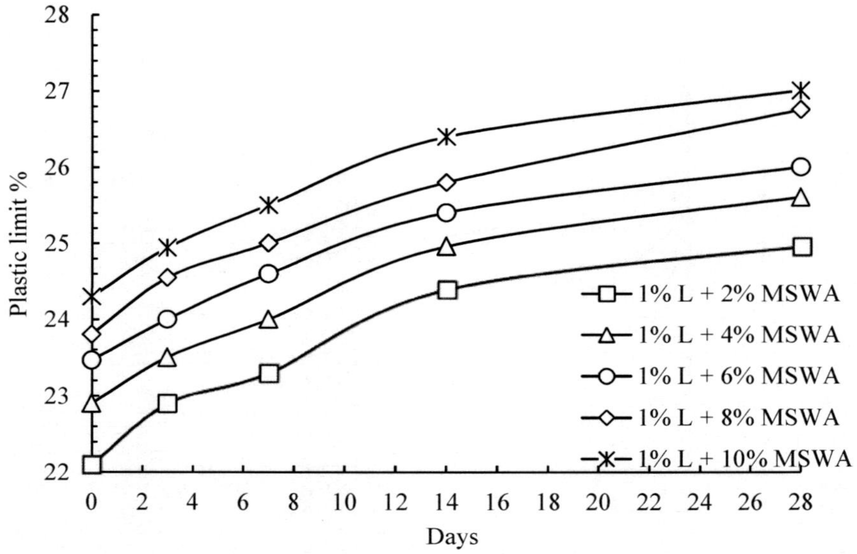

Fig. 6 Plastic limit of treated soil with (1% lime and 2, 4, 6, 8 and 10% MSWA)

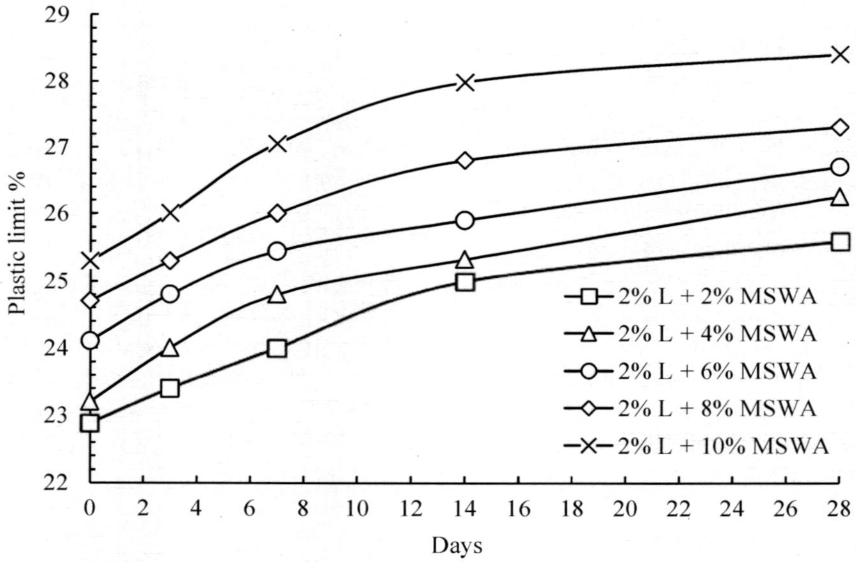

Fig. 7 Plastic limit of treated soil with (2% lime and 2, 4, 6, 8 and 10% MSWA)

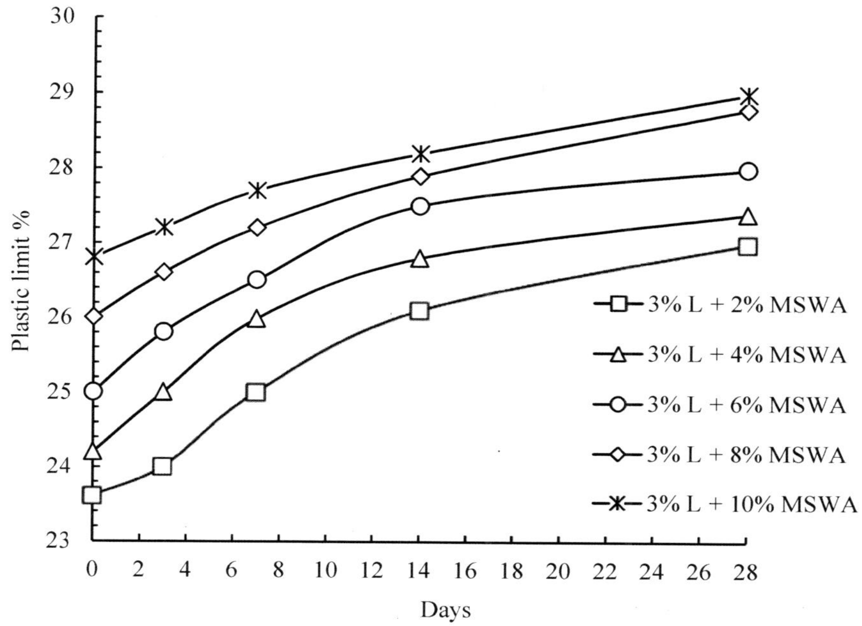

Fig. 8 Plastic limit of treated soil with (3% lime and 2, 4, 6, 8 and 10% MSWA

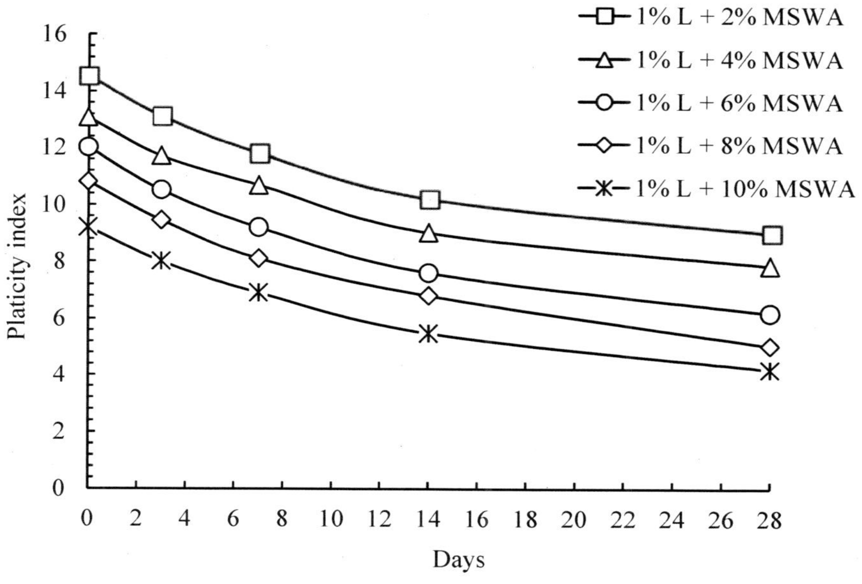

Fig. 9 Plasticity index of treated soil with (1% lime and 2, 4, 6, 8 and 10% MSWA)

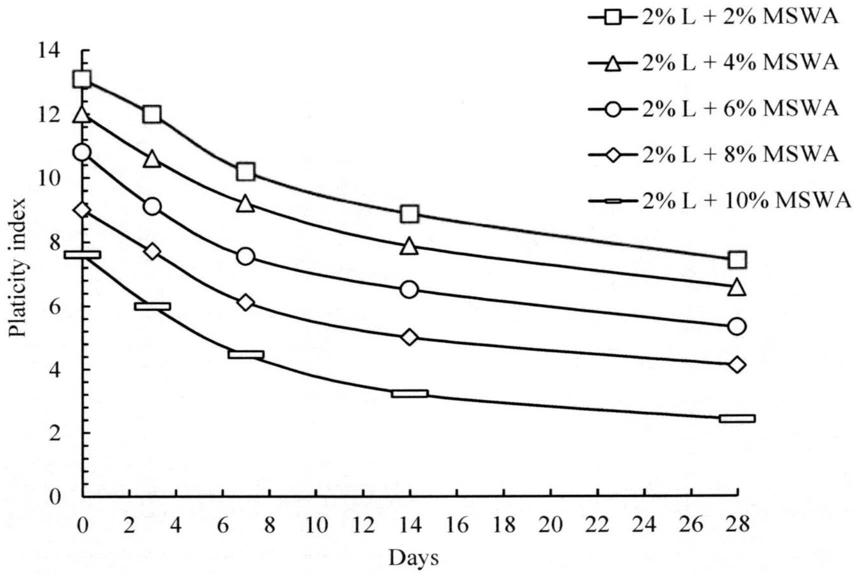

Fig. 10 Plasticity index of treated soil with (2% lime and 2, 4, 6, 8 and 10% MSWA)

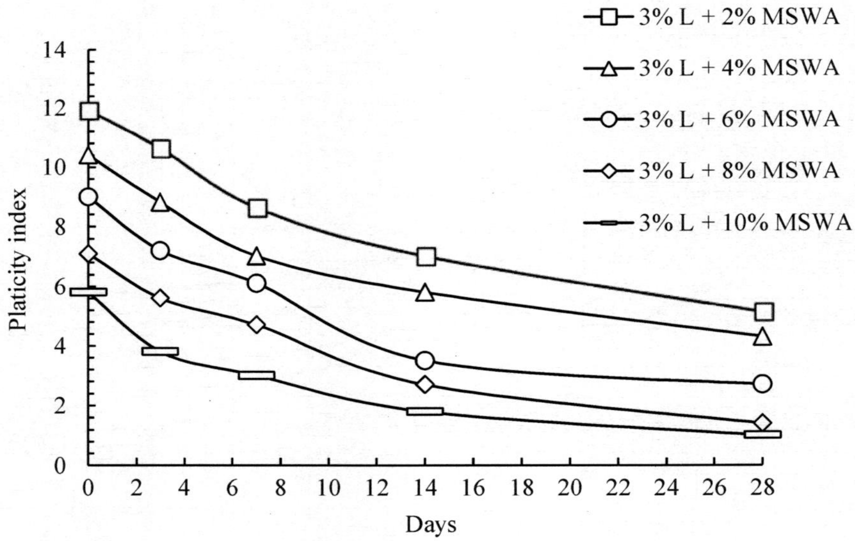

Fig. 11 Plasticity index of treated soil with (3% lime and 2, 4, 6, 8 and 10% MSWA)

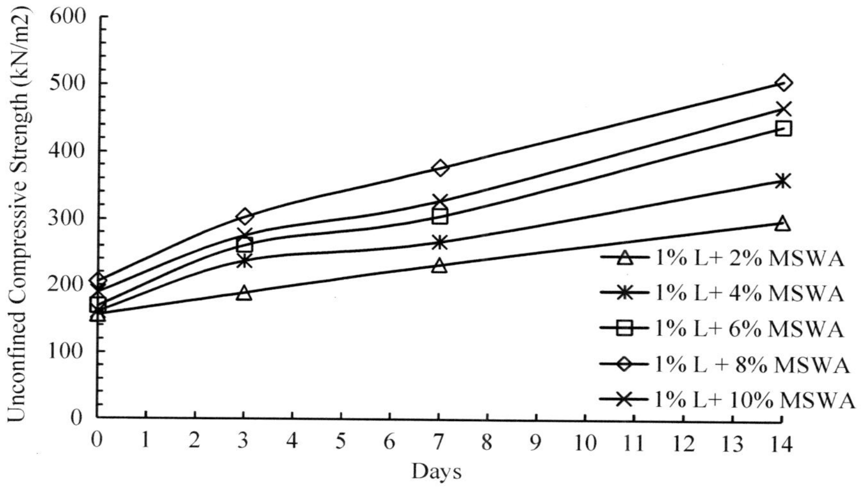

Fig. 12 UCS versus curing days

soil occurs over time, and hence, the UCS tests were performed on samples cured for varying curing times, i.e., 0, 3, 7 and 14 days.

UCS sample (1% lime and varying percentage of MSW ash).

The soil is treated with 1% lime, and the percentage of MSW ash varies (2, 4, 6, 8 and 10%). In 0 day curing, strength increases to 205.21 kPa at 8% MSW ash and then decreases to 188.95 kPa at 10% MSW ash. This rise is approximately 54.65% relative to untreated soil. At 14 day curing, strength increases to 509.14 kPa at 8% MSW ash and then reduces 470.71 kPa at 10% MSW ash. Compared to untreated soil, this rise is about 283.70%. The UCS versus curing days for 1% lime and varying percentage of MSW ash are shown in Fig. 12.

UCS sample (2% lime and varying percentage of MSW ash).

The soil is treated with 2% lime, and the percentage of MSW ash varies (2, 4, 6, 8 and 10%). In 0 day curing, strength increases to 243.32 kPa at 8% MSW ash and then decreases to 196.92 kPa at 10% MSW ash. This rise is approximately 83.37% relative to untreated soil. At 14 day curing, strength increases to 768.51 kPa at 8% MSW ash and then reduces 691.66 kPa at 10% MSW ash. Compared to untreated soil, this rise is about 479.17%. The UCS versus curing days for 1% lime and varying percentage of MSW ash is shown in Fig. 13.

UCS sample (3% lime and varying percentage of MSW ash).

The soil is treated with 3% lime and the percentage of MSW ash varies (2, 4, 6, 8 and 10%). In 0 days curing, strength increases to 302.89 kPa at 8% MSW ash and then decreases to 260.75 kPa at 10% MSW ash. This rise is approximately 128.26%

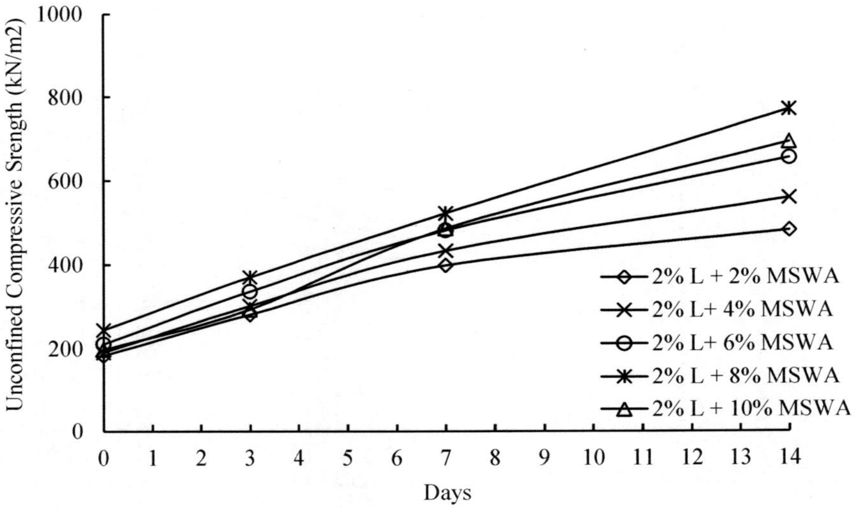

Fig. 13 UCS versus curing days

relative to untreated soil. At 14 day curing, strength increases to 942.09 kPa at 8% MSW ash and then reduces 883.79 kPa at 10% MSW ash. Compared to untreated soil, this rise is about 610%. At 3% lime content, pH of soil–lime mix becomes 12.25 which starts pozzolanic reaction producing cementious material which binds the particles with time. Hence, 3% lime content enhances the strength to some extent. The UCS versus curing days for 1% lime and varying percentage of MSW ash is shown in Fig. 14.

4 Conclusions

Different test have been performed to determine the engineering properties of the fine-grained soil. The soil was mixed with varying percentage of lime (1, 2 and 3%) and MSWA (2, 4, 6, 8 and 10%), and the sample was cured for (0, 3, 7 and 14 days) before performing the experiment. The below conclusions have been drawn on the basis of the data obtained.

1. The optimum moisture content of the soil increases exponentially with increase in lime and MSW ash percentage. The maximum dry density of the soil decreases simultaneously with increase in lime and MSW ash. The OMC in treated soil is increased by 33.9% compared to the OMC of untreated soil. The MDD in treated soil is decreased by 15.77% compared to the OMC of untreated soil.

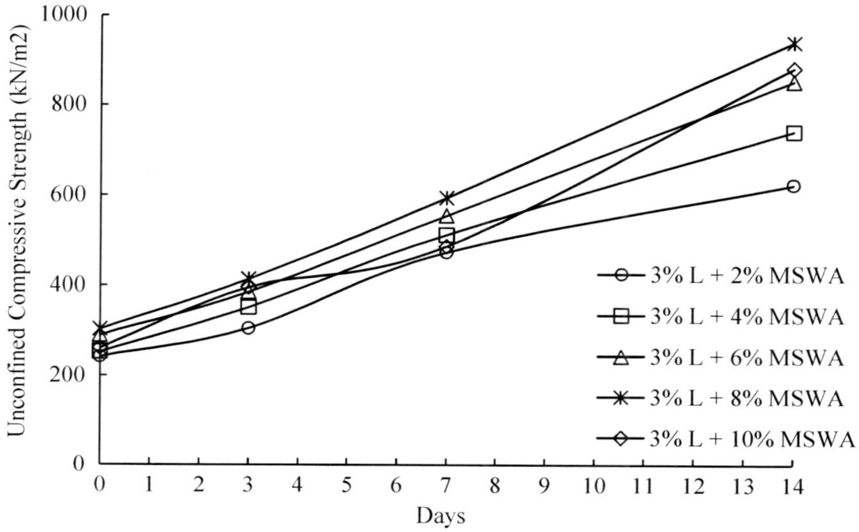

Fig. 14 UCS versus curing days

2. It can be observed that there is a steep reduction in LL with the addition of lime and MSW ash and with curing. The reduction is due to cation exchange process, flocculation of soil particle or change in soil particles.
3. Plastic limit increases with addition of lime, MSWA and curing. This rise in plastic limit may be due to decrease in DDL thickness, which leads to alternation of particles, and thus, clay particles undergo to from aggregates (silt) due to conglomeration. The formation of silt increases the shear resistance and hence increasing plastic limit.
4. The decrease in LL and increase in PL lead to reduction of plasticity index. Plasticity index indicates the soil workability and ease to deform it. The application of lime and MSW ash might have altered the soil structure by flocculating the clay particles that could have substituted the finer clay particles with some coarser particles, thereby enhancing their workability and reducing PI.
5. The strength behavior increases with curing period, lime content and MSW ash. It is observed that 3% lime and 8% MSW ash mix is the optimum mix at which the strength of the soil increased with curing. The increase in strength may be due to increase in lime content, as higher percentage lime allows good pozzolanic reaction with age. Although the strength of the soil decreases with curing at 3% lime and 10% MSW ash, more pozzolanic reaction makes the soil more porous.

References

1. Kumar A, Mittal A (2019) Utilization of municipal solid waste ash for stabilization of soil. Environ Geotechnol 31:133–139. https://doi.org/10.1007/978-981-13-7010-6_12
2. Verma H (2020) Soil stabilization techniques. Int J Creative Res Thoughts 8(6):252–256
3. Sai H (2018) Soil stabilization using lime. Int J Res Appl Sci Eng Technol 6(1):1096–1100. https://doi.org/10.22214/ijraset.2018.1167
4. Bell FG (1996) Lime stabilization of clay minerals and soils. Eng Geol 42(4):223–237. https://doi.org/10.1016/0013-7952(96)00028-2
5. Dash SK, Hussain M (2012) Lime stabilization of soils: reappraisal. J Mater Civ Eng 24(6):707–714. https://doi.org/10.1061/(ASCE)MT.1943-5533.0000431
6. Taha RA, Mohamedzein YEA, Al-Aghbari MY (2006) Stabilization of desert sands using municipal solid waste incinerator ash. J Geotech Geol Eng 24:1767–1780. https://doi.org/10.1007/s10706-006-6806-7
7. Show KY, Tay JH, Goh ATC (2003) Reuse of incinerator fly-ash in soft soil stabilization. J Mater Civ Eng 15(4):335–343. https://doi.org/10.1061/(ASCE)/0899-1561(2003)15:4-335
8. Muhunthan B, Taha R, Said J (2004) Geotechnical engineering properties of incinerator ash mix. J Air Waste Manage Assoc 54(8):985–991. https://doi.org/10.1080/10473289.2004.10470959
9. Sherwood PT, Ryley MD (1986) The use of pulverized fuel ash in road construction. RRL Rep, London
10. Kamon M, Katsumi T, Sano Y (2000) MSW fly ash stabilized with coal ash for geotechnical application. J Hazardous Mater 76(2):265–283. https://doi.org/10.1016/S0304-3894(00)00203-X
11. Berg ER, Neal JA (1998) Municipal solid waste bottom ash as portland cement concrete ingredient. J Mater Civ Eng 10(3):168–173. https://doi.org/10.1061/(ASCE)0899-1561(1998)10:3(168)
12. Lin KL, Wang KS, Tzeng BY, Lin CY (2003) The reuse of municipal solid waste incinerator fly ash slag as a cement substitute. J Res Conserv Recycling 39(4):315–324. https://doi.org/10.1016/S0921-3449(02)00172-6
13. Mohamedzein YEA, Al-Aghbari MY (2012) The use of municipal solid waste incinerator ash to stabilize dune sands. J Geotech Geol Eng 30(6):1335–1344. https://doi.org/10.1007/s10706-012-9548-8
14. Singh D, Kumar A (2017) Geo-environmental application of municipal solid waste incinerator ash stabilized with cement. J Rock Mech Geotech Eng 9(2):370–375. https://doi.org/10.1016/j.jrmge.2016.11.008
15. Gandhi SR, Robinson RG, Bhuvaneshwari S (2014) Behavior of lime treated cured expansive soil composites. Indian Geotech J 44:278–293. https://doi.org/10.1007/s40098-013-0081-3
16. Choobbasti AJ, Ghodrat H, Vahdatirad MJ, Firouzian S (2010) Influence of using rice husk ash in soil stabilization method with lime. Frontier Earth Sci China 4(4):471–480. https://doi.org/10.1007/s11707-010-0138-x
17. Lees G, Abdelkater MO, Hamdani SK (1982) Effect of the clay fraction on some mechanical properties of lime-soil mixtures. The Highway Engineer 29(11):2–9
18. Muhmed A, Wanatowski D (2013) Effect of lime stabilization on the strength and microstructure of clay. IOSR J Mech Civ Eng 6(3):87–94. https://doi.org/10.9790/1684-638794

Numerical Modeling of Soil-Nailed Slope Using Drucker–Prager Model

Sagar Jaiswal, Ananya Srivastava, and Vinay Bhushan Chauhan

Abstract Soil nailing is an efficient in-situ ground improvement technique that is being used extensively in the field to stabilize the unstable artificial or natural soil slopes. In this paper, a parametric study is done by varying the soil nailing parameters, namely the length (L) and the inclination of soil nails (α), with respect to the horizontal for a steep soil slope ($\beta = 40°$). A finite element method-based program (Optum G2) has been used to analyze the soil slope using the shear strength reduction method. The Drucker–Prager failure criterion is considered in this study to model the soil slope. A soil slope of height (H) 6 m is considered, and the nail inclination (α) is varied from 0° to 50° in 10° intervals from the horizontal to get the best orientation of nail arrangement and study their effects on the performance of the overall slope based on the calculated factor of safety of the soil slope. Findings of the present study indicate that the length and the inclination of the soil nail are the major governing parameters to control the stability of the slope and based on above-mentioned outcomes, optimum length, and the orientation of soil nail are recommended and the transition of possible modes of failure of the slope are discussed for reinforced and unreinforced slope.

Keywords Slope stability · Soil nail · Drucker–Prager criterion · Strength reduction · Factor of safety

1 Introduction

Analysis of soil slope stability has gained attention in the recent past due to an increase in construction near the slopes. The soil slopes are susceptible to collapse due to nearby excavation activities, rainfall infiltration, or natural calamities and ground surface failure owing to detachment of the soil layers. The extent and possibility of failure of soil slopes mainly depend upon the shear strength of soil, soil slope inclination, erosion due to precipitation, and other man-made activities near the site.

S. Jaiswal · A. Srivastava · V. B. Chauhan (✉)
Department of Civil Engineering, Madan Mohan Malaviya University of Technology, Gorakhpur 273010, India

A. K. Choudhary et al. (eds.), *Advances in Geo-Science and Geo-Structures*, Lecture Notes in Civil Engineering 154,
https://doi.org/10.1007/978-981-16-1993-9_11

Soil nailing is an in-situ ground improvement technique in which reinforcement of ground is done with the nails, which are generally made up of hollow steel tubes or high yielding strength deformed (HYSD) bars. In this technique, the reinforcement is generally needed to be installed horizontally or slightly inclined parallel to the direction of tensile strain to develop the maximum tensile force in the soil nails, which is passive structural inclusion to stabilize the unstable ground. This technique is often used as a permanent or temporary soil slope/earth-retention system for steep slopes. Usually, soil nails are installed in predrilled holes and grouted under pressure with a joint hose, and thus, the mortars transfer the load to the soil nails through friction which leads to soil stability [1].

There are many advantages associated with this ground improvement technique, which can be roughly divided into three major categories, i.e., cost, design, and performance [2]. The construction of the soil-nailed slope is relatively quick, inexpensive, and sustainable compared to many other practices to stabilize the slope because it is less disruptive to traffic. Soil nails are flexible in terms of performance, and their deflections are usually within the tolerable limit.

It has been previously reported that the force generated by soil nails can reduce the principal strain in soil by decreasing the displacement of soil layers and subsequently improve the slope stability by creating a bond between the surrounding soil or friction between the adjacent soil that is reinforced between two soil nails [3]. The slope reinforced with soil nail can be divided into two zones, i.e., active zone and passive zone. At the time of slope failure, an axial shift in soil nails is observed due to the deformation of the active zone. The displacement of the reinforced bars tends to develop pulling force in the soil nails present in the passive region, which in turn resist the deformation in the active region. The length of soil nails is considered in such a manner that they are embedded sufficiently in the passive zone to resist the pull-out generated in the soil nail due to the friction produced between the surrounding soil and the nails.

Several methods of soil nailing are currently available for slope stabilization, such as the German method [4], the French method [5], the Davis method [6], and the finite element method [7], of which the first three methods are based on limit equilibrium approach, whereas the latter one is based on the limit analysis. However, an upper bound and lower bound solutions give an exact idea of the range of the factor of safety which can be certainly used to design the slope with confidence, which is limited in the available literature.

Given above, in the present work, slope stability analysis has been carried out using a finite element program taking into account the Drucker–Prager criterion for the soil. Also, the effect of soil nailing technique has been analyzed by varying the length of soil nails ranging from 2 m to 8 m having an interval of 2 m to get the optimum length by considering the spacing between soil nail, $s = 1$ m, for each simulation. Moreover, the effect of the soil nail inclination with the horizontal (α) on the stability of the slope is studied by varying α as shown in Fig. 1. The upper bound and lower bound strength reduction analysis is performed to examine the factor of safety (FOS) of a slope with a height, $H = 6$ m, and a slope angle (β) of 40°.

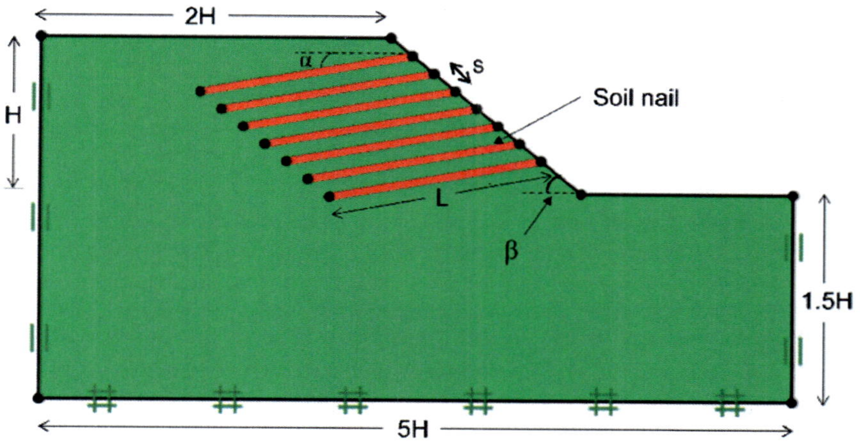

Fig. 1 Schematic diagram of slope reinforced with soil nails

2 Numerical Modeling

A slope stability analysis has been carried out for a soil slope with height, $H = 6$ m and a slope angle of $\beta = 40°$ from the horizontal, using two-dimensional finite element method-based numerical tool Optum G2 [8]. The soil slope is reinforced with soil nails at six different inclinations from the horizontal, i.e., 0°, 10°, 20°, 30°, 40°, and 50°, respectively (as shown in Fig. 1), and their effects on the factor of safety (FOS) have been studied using shear strength reduction method. The Drucker–Prager model is chosen to represent the soil behavior in the numerical model for performing finite element analysis, and the plane strain condition with associated flow rule is considered in the present study. The boundary conditions considered in this study are roller support at the vertical ends of the boundary and fixed support at the bottom. A schematic representation of the geometric configuration of soil slope reinforced with soil nails is shown in Fig. 1.

The strength reduction in Optum G2 is analyzed by calculating a strength reduction factor in order to reduce the material parameters to attain the state of incipient collapse [9]. The strength parameters of the Drucker–Prager model are cohesion, k (kPa), and friction coefficient, M, and the factor of safety for the considered failure criterion is defined in Eq. 1 [9]:

$$\text{FOS} = \frac{k}{k_{\text{red}}} = \frac{M}{M_{\text{red}}} \tag{1}$$

The accuracy of a numerical model analyzed with finite elements simulations depends upon the selection of the total number of elements for a generated mesh as noted from the previous findings [10]. For a high precision of FOS, it is necessary to take a large number of elements, whereas taking a large number of elements for a given mesh may elapse elongated computational time than usual. On account of the above, a sensitivity analysis is performed to find out an optimum number of elements for the mesh considered in this study. The number of elements has been increased from 1000 to 6000 at an interval of 1000 each, and FOS is observed in both the lower bound (LB) and upper bound (UB) analyses, and it is observed that 4000 elements are enough for the mesh considered in this analysis (Fig. 2). The typical properties considered in this study are shown in Table 1.

In the present study, a soil slope with angle $\beta = 40°$ is considered and its FOS has been determined by strength reduction analysis. Furthermore, the soil slope is reinforced with soil nails at various inclinations from the horizontal, and again, the FOS has been determined. A parametric study is also performed to ascertain the FOS of the soil slope, the length of soil nails is varied from 2 to 8 m with an increment of 2 m each, and the best length is determined to carry out further study. The analysis

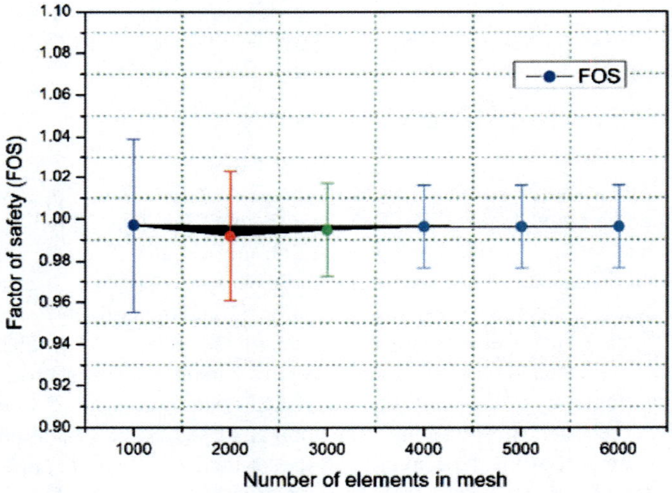

Fig. 2 Sensitivity analysis for the considered mesh in the present study

Table 1 Material properties used in this study

Properties	Soil	Soil nail
Lateral strength, $F_{L,\,max}$ (kN/m) [a]	–	42
Axial strength, $F_{A,\,max}$ (kN/m) [a]	–	415
Young's modulus, E (MPa) [b, a]	30	2.6×10^4
Cohesion, k (kPa) [b]	12.33	–
Friction coefficient, M [b]	0.613	–
Poisson's ratio, ν [b]	0.25	–
Dry unit weight, γ_{dry} (kN/m^3) [b]	22	–

[a] Rawat and Gupta [11], [b] Abderrahmane and Abdelmadjid [12]

of failure plane obtained through numerical simulations for the unreinforced slope is performed, and the improvement in FOS is calculated after the reinforcement of the slope.

3 Results and Discussion

The stability of the unreinforced slope is examined, and the FOS is found to be 1.02 which is well compared with the results reported by Abderrahmane and Berga [12]. Figure 3 indicates the failure plane obtained without reinforcing the soil slope for $\beta = 40°$ with horizontal. The current model is therefore validated, and it shows that the current model is suitable for carrying out further analysis. Furthermore, parametric studies were conducted using the parameters mentioned earlier.

Figure 4 shows the variation of the length of soil nail with FOS obtained from the numerical analysis for $\alpha = 10°$ with horizontal. The obtained variation indicates that the smaller length of soil nails could not contribute toward the enhancement of the overall factor of safety of the slope system. However, the FOS of the slope tends to increase with the increase in the length of soil nail (L), and it is found to be 1.55 (>1.5) for $L = 8$ m.

It is observed that for the nail length (L) 2 m, 4 m, and 6 m, the reinforcement length could not reach to the passive region of the failure zone, due to which the utilization of tensile strength of the nails could not be realized by the soil slope and the FOS thus obtained is smaller than the criteria recommended by the FHWA [2]. The sliding circle formed by the respective reinforcement (2 m, 4 m, and 6 m) was also relatively smaller, which does not support the backfill. Global failure is observed in the case of failure plane obtained when the soil nail length does not reach to the passive region of the reinforced slope. It should be noted that the nail length of 8 m brings the FOS of the considered slope to the recommended minimum value of FOS, which is 1.5. Thus, a nail length of 8 m is found to be most suitable for the provision

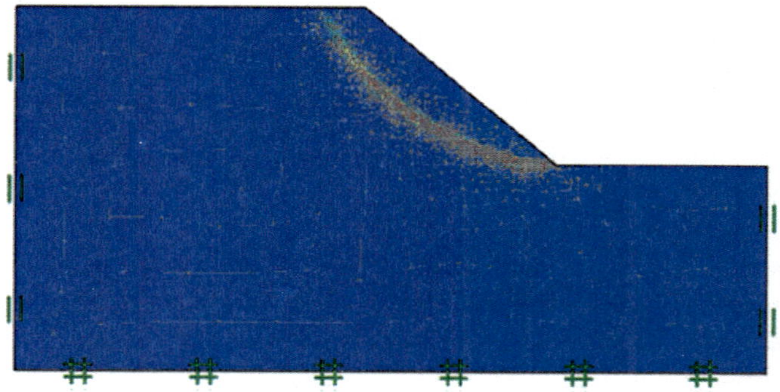

Fig. 3 Failure surface of unreinforced slope considered in the present study

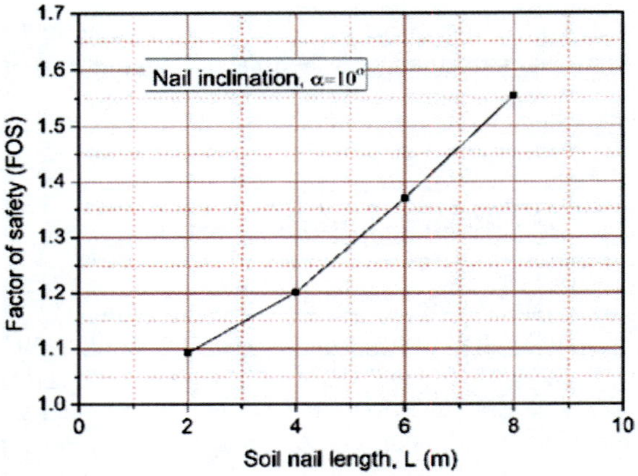

Fig. 4 Variation of FOS with reinforcing soil nail length, L (m)

of the slope under consideration. Lin et al. [13] conducted a parametric study on the length of soil nails and reported similar results.

Furthermore, analysis is performed while keeping the nail length and spacing invariable, and the factor of safety is determined with the variation of nail inclination, $\alpha = 0°$ to $50°$ with horizontal, to determine the best orientation of nail to maximize the FOS of the reinforced slope.

The variation of factor of safety and soil nail inclination (α) with horizontal is shown in Fig. 5, which indicates that initially the FOS increases with an increase in nail inclination (α) up to 30°, but any further increase in α decreases the FOS rapidly. This can be attributed to the fact that a too steep nail inclination (α) can cancel out the advantages since the direction of the nail is not optimal for stability. Nails that are steeper than 30° may be in compression, which can be the possible reason for a reduction in FOS.

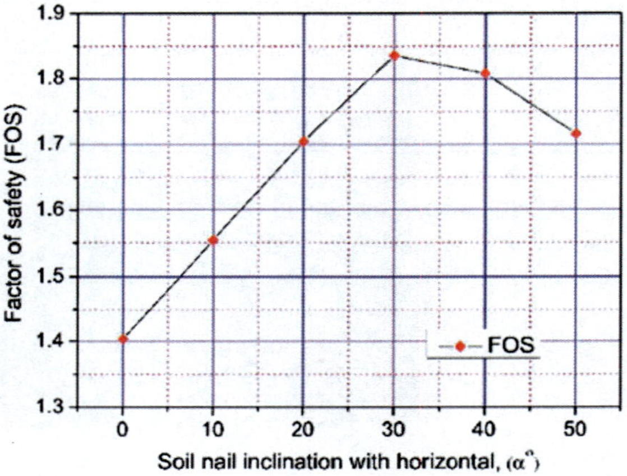

Fig. 5 Variation of FOS with nail inclination (α) at soil slope, $\beta = 40°$, and nail length, $L = 8$ m

It is also observed that if the nail orientation is kept parallel to the horizontal, the FOS only rises to 1.4 ± 0.01 (<1.5) (where the first term signifies the average of LB and UB values of FOS, whereas the latter one denotes the difference between the average value of FOS and the UB and LB values of FOS), which is not the recommended value for design. For this reason, a nail inclination of less than 10° should be avoided to prevent creating voids from forming, when grouting is to be carried out after reinforcement since voids can reduce the pull-out resistance and reduce corrosion protection. As α increased to 10°, 20°, and 30°, the FOS increased significantly to 1.55 ± 0.01, 1.70 ± 0.02, and 1.83 ± 0.03, respectively. The maximum FOS was achieved at 30° inclination of the nail with horizontal. It is evident from Fig. 5, as α increases to 40° and 50°, a reduction of 1.7% and 7% in FOS is observed.

The potential failure curve obtained in the above study is also analyzed for the different inclinations of the soil nails for the same soil slope as shown in Fig. 6.

Figure 6a shows potential failure curve of soil slope reinforced with nails that are kept parallel to the horizontal (i.e., $\alpha = 0°$) and noted that the slip surface form during failure of a soil slope is larger in this case than the unreinforced case (Fig. 3). It is evident that as α increases from 0°, the slip surface tends to enlarge (as shown in Fig. 6b and c). However, the slip surface formed by the soil nail inclined at 30° (Fig. 6d) is the largest than all the cases considered in this study, and this can be

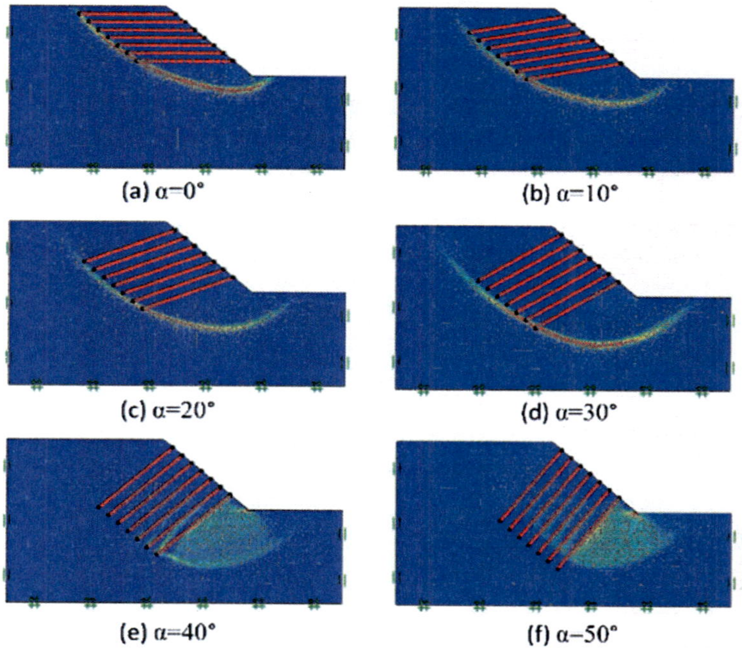

Fig. 6 Potential failure curve for soil slope ($\beta = 40°$) reinforced with soil nail at inclination **a** $\alpha = 0°$; **b** $\alpha = 10°$; **c** $\alpha = 20°$; **d** $\alpha = 30°$; **e** $\alpha = 40°$; and **f** $\alpha = 50°$

attributed that maximum nail length is utilized to resist the shearing action. It can be also seen that no nail breakage has occurred, and this means that the load transfer mechanism is driven by the pull-out capacity of the soil nail.

A complex failure pattern is observed at $\alpha = 40°$ and 50° (Fig. 6e and f). The soil mass tends to move with the mobilization of the toe of soil slope and also the upper tension bearing member does not seem to take any load on it, which is why the failure surface originates at the bottom-most nail and a global failure is observed in both the cases. This could be the reason for the reduction in the factor of safety of reinforced slope when the soil inclination is increased at 40° and 50°, respectively.

4 Conclusion

A parametric study is carried out on a 6 m high slope to determine the behavior of soil slope when reinforced with soil nails at various nail length (L) and inclination with the horizontal (α), using Drucker–Prager criterion for the soil under plane strain condition by a finite element numerical tool. A parametric study is done by varying the length (L) of soil nails from 2 m to 8 m in 2 m intervals and the inclination of soil nails (α) from 0° to 50° in 10° intervals, with respect to the horizontal for a steep soil slope ($\beta = 40°$). The significant conclusions obtained in this study are as follows:

1. The factor of safety increases with an increase in nail length and the maximum value is achieved at nail length, $L = 8$ m and $s = 1$ m, for a given slope 6 m high and $\beta = 40°$. This may be attributed to the fact that the slope surface holds an elongated path in order to form a deeper sliding surface with the increasing nail length, which in turn leads to an increase in factor of safety.
2. It is noted that as α increases from 0°, the factor of safety (FOS) increases up to 30°, but as soon as it reaches 40°, the FOS slightly decreased, and for $\alpha > \beta$, the value of FOS is decreased by 7%, which shows that soil nails are in compression when the inclination is too steep that is why too steep nails (>30°) must be avoided while reinforcing the soil slope with soil nails. The failure plane observed in too steep nail inclination (>30°) is not optimal for stability and can cancel out the advantages of soil nailing.
3. It is observed that the maximum factor of safety is obtained at $L = 8$ m and $\alpha =$ 30°, i.e., 1.83 ± 0.026, for the slope $\beta = 40°$; also, the result demonstrates that for a 6 m high slope, the factor of safety to be greater than the recommended value (i.e., 1.5), the L/H ratio must be greater than 1.

References

1. Taenaka S, Ishihama Y, Terada Y, Terada H (2015) Investigation of the bond strength on the surface of a mini-steel pipe with dimples. Jpn Geotech Soc Spec Publ 1(5):17–22. https://doi.

org/10.3208/jgssp.JPN-27
2. Lazarte CA, Robinson H, Gómez JE, Baxter A, Cadden A, Berg R (2015) Geotechnical engineering circular No. 7 soil nail walls—Reference manual. US Department of Transportation Publication No. FHWA-NHI-14–007, Federal Highway Administration, FHWA, Washington, DC
3. McGown A, Andrawes KZ, Al- MM (1978) Effect of inclusion properties on the behaviour of sand. Geotechnique 28(3):327–346. https://doi.org/10.1680/geot.1978.28.3.327
4. Stocker ME, Korcber GW, Gassler G, Gudehus G (1979) Soil nailing. Int Conf Soil Reinforcement, 469–474
5. Schlosser F (1982) Behaviour and design of soil nailing. Int Symp Asia Institute of Technology, 399–419
6. Mitchell JK, Villet WC (1987) Reinforcement of earth slopes and embankments. NCHRP report, New York (290)
7. Griffiths DV, Lane PA (1999) Slope stability analysis by finite elements. Geotechnique 49(3):387–403. https://doi.org/10.1680/geot.1999.49.3.387
8. Optum G2 (2020) Finite Element Program for Geotechnical Analysis, Optum Computational Engineering, www.optumce.com
9. Pandey A, Jaiswal S, Chauhan VB (2021) Numerical Studies on the Behavior of Slope Reinforced with Soil Nails. In: Sitharam TG, Jakka R, Govindaraju L (eds) Local Site Effects and Ground Failures. Lecture Notes in Civil Engineering, 117:217–228, Springer, Singapore. https://doi.org/10.1007/978-981-15-9984-2_19
10. Pandey A, Chauhan VB (2020) Evaluation of pull-out capacity of helical Anchors in clay using finite element analysis. Proc Geo-Congress (2020) Modeling, geomaterials, and site characterization, GSP 317, ASCE, 350–359. https://doi.org/10.1061/9780784482803.007
11. Rawat S, Gupta AK (2016) Analysis of a nailed soil slope using limit equilibrium and finite element methods. Int J Geosynth Ground Eng 2(4):34. https://doi.org/10.1007/s40891-016-0076-0
12. Abderrahmane TH, Abdelmadjid B (2016) Analyzing of slope stability by difference model of behavior. Asian Eng Review 3(1):1–9. https://doi.org/10.20448/journal.508/2016.3.1/508.1.1.9
13. Lin H, Xiong W, Cao P (2013) Stability of soil nailed slope using strength reduction method. European J Environ Civil Eng 17(9):872–885. https://doi.org/10.1080/19648189.2013.828658

Winkler's Based Parametric Analysis of Unplugged Short Pipe Pile

Aparna Verma and S. M. Ali Jawaid

Abstract The North India's lowlands consist of soft alluvial deposits, which are saturated throughout the year. The construction of the conventional shallow foundation is very difficult in this area due to high groundwater table. A short open-ended steel pipe pile with soil plug is being proposed for this region. By using shear stress–displacement response for alluvial deposits, the proposed foundation is analyzed. Parametric studies are carried out to study the effects of length to diameter ratio (L/d_o), diameter ratio (d/d_o), and angle of shearing resistance of core plug (φ_{sp}) on the load sharing and the settlement of the proposed foundation.

Keywords Lowlands · Short pipe piles · Soil plug · Load-carrying capacity · Soil–pile interaction

1 Introduction

The alluvial lowlands deposits are loose or soft in nature having low strength and high compressibility [1]. Due to regular flood and the presence of a high-water table in lowlands [2], the construction of the conventional shallow foundation is very difficult. Thus, for this region, a new unplugged short open-ended pipe pile foundation is proposed. In unplugged condition, the soil elevation in the inner and outer sides of the pipe pile remains the same. This occurs in soft soil [3]. The purpose of this study is to understand the variation of normalize settlement and load carried by steining with frictional angle (φ). The effect of variation of length to diameter ratio (L/d_o) and diameter ratio (d/d_o) on the normalized load carried by steining is also studied.

A. Verma (✉) · S. M. A. Jawaid
Madan Mohan Malaviya University of Technology, Gorakhpur, Uttar Pradesh, India

A. K. Choudhary et al. (eds.), *Advances in Geo-Science and Geo-Structures*,
Lecture Notes in Civil Engineering 154,
https://doi.org/10.1007/978-981-16-1993-9_12

2 Analysis of Proposed Foundation

The applied load (Q) is shared at the top (Fig. 1) by pile shaft (Q_{st}) and the soil plug (Q_{sp}). The relative proportions of loads transferred to the shaft (Q_{st}), and the soil plug (Q_{sp}) are controlled by the stiffness, geometry, and interfacial shear stresses. Vertical equilibrium force for unplugged pile along soil plug inner side is (Fig. 2a) expressed below

$$Q = Q_{\mathrm{st}} + Q_{\mathrm{sp}} \tag{1}$$

$$Q = Q_{\mathrm{st},s} + Q_{\mathrm{st},b} + Q_{\mathrm{sp},L} \tag{2}$$

where.

$Q_{\mathrm{st},s}$ = loads shared by outer pile shaft.

$Q_{\mathrm{st},b}$ = loads shared by outer steining base.

$Q_{\mathrm{sp},L}$ = load transferred by the base of soil plug.

Expressing the forces Q, in the terms of stresses (Fig. 2b and c).

$$Q = 4\left(\frac{L}{d_o}\right)\tau + \left[1 - \left(\frac{d^2}{d_o^2}\right)\right]q_{\mathrm{st},b} + \left(\frac{d^2}{d_o^2}\right)q_{\mathrm{sp},L} \tag{3}$$

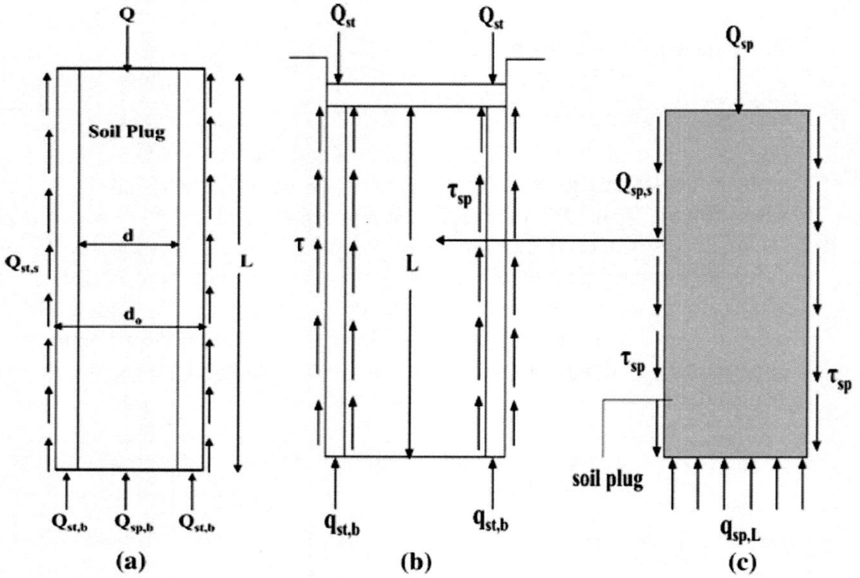

Fig. 1 Forces and stresses on **a** pile cap, **b** pile steining and **c** soil plug

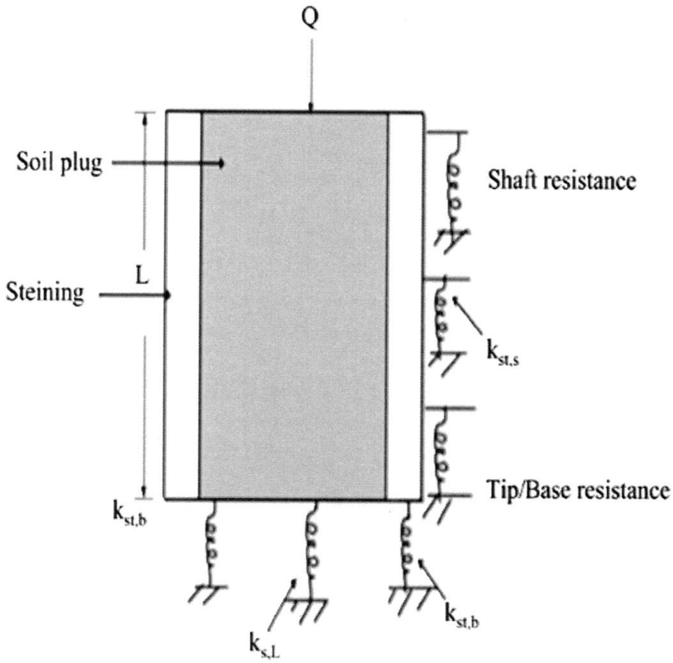

Fig. 2 Winkler representation of soil response to the load on the unplugged open-ended foundation

where.

q = average stresses on the foundation.

τ = average stresses on the outer surface of steining and,

$q_{st,b}$ = average stresses on the steining base.

$q_{sp,L}$ = average stresses on the soil plug, respectively.

The soil plug gets loaded from top cap Q_{sp}, and also from down-drag stress τ_{sp}. The vertical equilibrium of an element of core plug, neglecting its weight, is

$$(\sigma_z + \Delta\sigma_z)\left(\frac{\pi}{4}\right)d^2 - \sigma_z\left(\frac{\pi}{4}\right)d^2 - \tau_{sp}(\pi d) = 0 \tag{4}$$

On simplifying,

$$\left(\frac{d\sigma_z}{dz}\right) - \left(\frac{4}{d}\right)\tau_{sp} = 0 \tag{5}$$

where σ_z is the normal stress on the top planes and $(\sigma_z + \Delta\sigma_z)$ is the normal stress on the bottom planes and τ_{sp} is the shear stress is the soil plug.

Assuming full mobilization of shaft resistance between soil plug and inside surface of the pipe shaft, i.e., $\tau_{sp} = K\sigma_z \tan\delta$, where K is the coefficient of lateral earth pressure and δ is the wall friction angle. Putting the value of τ_{sp} in Eq. 8 and finding the solution as given below,

$$\sigma_z = c_o \exp(\text{tz}) \tag{6}$$

where $t = \left(\frac{4}{d}\right)K \tan\delta$ and c_o is a constant.

At the top of the unplugged pipe pile ($z = 0$), $\sigma_z = q_{sp}$, and hence, one gets $c_o = q_{sp}$.

so,

$$\sigma_z = q_{sp} \exp(c_1 z) \tag{7}$$

The stress transferred by the core plug, $q_{sp,L}$, to the soil below, i.e., at $z = L$, becomes,

$$Q_{sp,L} = q_{sp} R_{sp} \tag{8}$$

where $R_{sp} = \exp(c_1 L)$.

Settlement W_{sL}, of the soil, below the soil plug, i.e., at $z = L$

$$W_{sL} = q_{sp,L}\left(\frac{d\left(1 - \upsilon_s^2\right)I_f}{E_s}\right) = \left[\frac{q_{sp,L}}{k_{sL}}\right] \tag{9}$$

where I_f is an influence factor and its value for a circular loaded area its constant value is 1.0 and k_{sL} is the modulus of subgrade reaction of soil below the soil plug $= \left[E_s/d\left(1 - \upsilon_s^2\right)\right]$.

The core plug is under $K \geq K_o$ condition its compression Δw_{sp} is calculated by integrating one-dimensional compression equation for an element as

$$\Delta W_{sp} = \int_0^L \varepsilon_z \text{dz} \tag{10}$$

$$\Delta W_{sp} = \int_0^L \left(\frac{\sigma_z}{D_{sp}}\right) \text{dz} \tag{11}$$

$$\Delta W_{sp} = \left(\frac{q_{sp}}{D_{sp}}\right)\left[\frac{d\left(R_{sp} - 1\right)}{t}\right] \tag{12}$$

where

$$R_{\text{sp}} = \exp\left(t\left(\frac{L/d_o}{d/d_o}\right)\right) \tag{13}$$

$t = 4K \tan\delta$ and D_{sp},

$$\text{the constrained modulus} = \left[\frac{E_{\text{sp}}(1 - v_{\text{sp}})}{(1 + v_{\text{sp}})(1 - 2v_{\text{sp}})}\right]$$

$$= \beta E_{\text{sp}} \tag{14}$$

and

$$\beta = \frac{(1 - v_{\text{sp}})}{[(1 + v_{\text{sp}})(1 - 2v_{\text{sp}})]} \tag{15}$$

The settlement at the top of soil plug, $W_{\text{sp},0}$, i.e., at $z = 0$, is a sum of compression of the core and the settlement of the soil below and is obtained as

$$W_{\text{sp},o} = W_{\text{sL}} + \Delta W_{\text{sp}} \tag{16}$$

$$= q_{\text{sp}} f_1 \tag{17}$$

where

$$f_1 = \left[\left\{\frac{R_{\text{sp}}}{k_{\text{sL}}}\right\} + \left\{d\left(\frac{R_{\text{sp}} - 1}{D_{\text{sp}} t}\right)\right\}\right] \tag{18}$$

The steining is assumed to be rigid and settles by W_{st}. Compatibility of displacements of the pile shaft and the top of the soil plug requires

$$W_{\text{st}} = W_{\text{sp},o} \tag{19}$$

Based on Scott [4], the shear stress on the outer surface of steining (τ), related to the displacement W_{st}, (Fig. 2).

$$\tau = k_{\text{st},s} W_{\text{st}} = k_{\text{st},s} W_{\text{sp},o} \tag{20}$$

and

$$q_{\text{st},b} = k_{\text{st},b} W_{\text{st}} = k_{\text{st},b} W_{\text{sp},o} \tag{21}$$

where

$$k_{\text{st},s} = \alpha_\tau k_{\text{st},b} \tag{22}$$

$$k_{\text{st},b} = \alpha_b k_{\text{sL}} \tag{23}$$

$$Q = 4\left(\frac{L}{d_o}\right)k_{\text{st},s}W_{\text{st}} + \left[1 - \left(\frac{d^2}{d_o^2}\right)\right]k_{\text{st},b}W_{\text{st}} + \left(\frac{d^2}{d_o^2}\right)q_{\text{sp},L} \tag{24}$$

or

$$q = W_{\text{st}} f_2 + q_{\text{sp}} f_3 \tag{25}$$

where

$$f_2 = 4\left(\frac{L}{d_o}\right)k_{\text{st},s} + \left[1 - \left(\frac{d^2}{d_o^2}\right)\right]k_{\text{st},b}$$

and

$$f_3 = \left(\frac{d}{d_o}\right)^2 R_{\text{sp}}$$

Substituting for the value of W_{st}, we get

$$W_{\text{st}} = q_{\text{sp}} f_1 \tag{26}$$

and

$$q = q_{\text{sp}}(f_1 f_2 + f_3) \tag{27}$$

or

$$Q = q_{\text{sp}} F \tag{28}$$

or the force transmitted to the core plug is

$$Q_{\text{sp}} = \frac{Q}{F} d_r^2 \tag{29}$$

where

$$d_r = d/d_o$$

and the fraction of the load transferred to the steining becomes

$$\frac{Q_{\text{st}}}{Q} = \left(1 - \frac{Q_{\text{sp}}}{Q}\right) \tag{30}$$

Normalizing, W_{st}, the settlement of pipe pile is

$$W_{st} = q_{sp} f_1 = q_{sp}\left(\frac{R_{sp}}{k_{sL}} + \frac{d(R_{sp}-1)}{D_{sp}t}\right) \quad (31)$$

Putting the value of Q and simplifying

$$\frac{W_{st} d_o E_s}{Q} = \left(\frac{1}{F}\right)\left(\frac{4(d/d_o)(1-\upsilon_s^2)}{\pi}\right)\left(R_{sp} + \frac{(R_{sp}-1)}{Rt}\right) \quad (32)$$

3 Result and Discussion

The results are obtained for a range of parameters to illustrate their influences on behavior of the short open-ended unplugged pipe pile. The important parameters considered in this research paper are following in Table 1.

Influence of E_{sp}/E_s on normalized steining load sharing (Q_{st}/Q) is depicted in Fig. 3. At (L/d_o) of 5, load shared by steining is 98.1%, 97.7%, and 97.6% at modular ratio $(E_{sp}/E_s) = 1$, 6, and 12, respectively, and variation between them is almost negligible. It is because as length-to-diameter ratio (L/d_o) increases, more load is taken by steining of short pipe piles than the pile base. A similar trend is observed at higher (L/d_o) ratio. Based on the above observation, it is found that load shared

Table 1 Values of important parameters considered in this analysis

Parameters	Assigned value/Range
Poisson's ratio of soil (υ_s)	0.3
Poisson's ratio of core plug (υ_{sp})	0.25
Modular ratio $\left(\frac{E_{sp}}{E_s}\right)$	1, 6, and 12
Relative steining base stiffness (α_b)	0.25
Relative steining–soil interface/base stiffness (α_t)	1.0
Diameter ratio $\left(\frac{d}{d_o}\right)$	0.95
Length to diameter ratio $\left(\frac{L}{d_o}\right)$	5 - 15
Angle of shearing resistance of core plug (φ_{sp})	10° - 40°
Coefficient of lateral pressure (K)	0.5

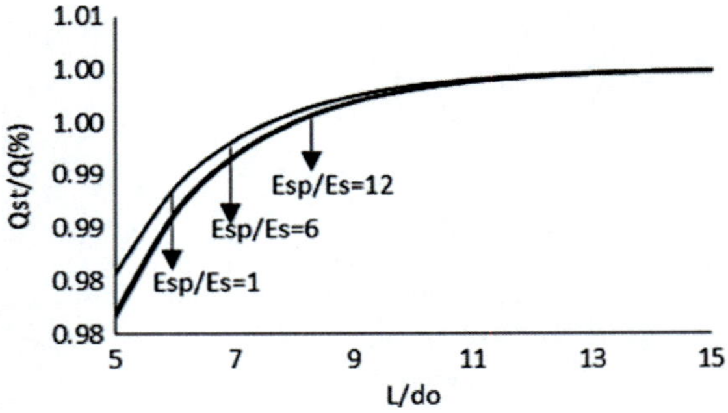

Fig. 3 Effect of modular ratio (E_{sp}/E_s) on per cent load carried by steining

by steining at (L/d_o) ratio between 5 and 15 is almost independent of modular ratio (E_{sp}/E_s).

The relationship between load carried by steining verses different lengths to dia ratio (L/d_o) at a different angle of friction 'φ' is plotted in Fig. 4. It is observed that load carried by steining increases by 1.02% with an increase of (L/d_o) ratio from 5 to 10 at $\varphi = 10°$. Beyond that, it is almost constant. However, the variation of load sharing by steining is almost constant at a higher angle of friction value. It simply means that as the angle of shear resistance increases, shear strength along with the shaft increases, and hence, more load is transferred to steining.

The variation of normalized settlement ($W_{st}E_s\ d_o/Q$) with (L/d_o) is shown in Fig. 5. It is evident from this graph that settlement is around 0.72 at (L/d_o) of 5, and around 0.3 at (L/d_o) ratio of 15, the variation is around 1.67%. It is due to the fact that at higher L/d_o ratio, the steining takes more load due to an increase in frictional resistance and thus decrease in the settlement.

The effect of angle of friction 'φ' on the settlement is shown in Fig. 6. At (L/d_o) of 5, the settlement is 0.72, but it reduces to 0.88 with the increase in 'φ' from 10° to 40°. Thus, the angle of internal friction has a marked effect on the settlement.

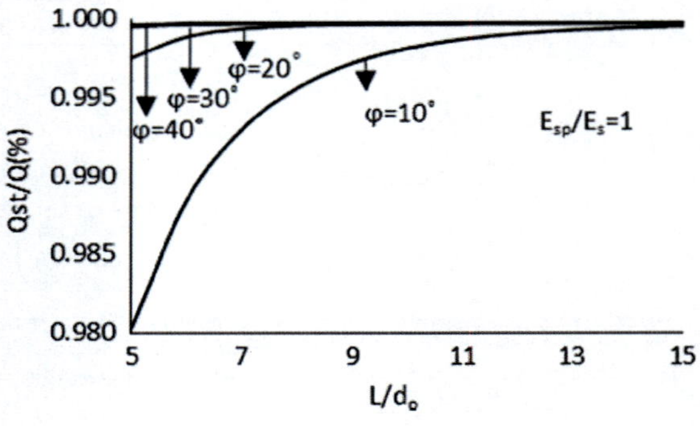

Fig. 4 Effect of angle of friction of soil on per cent load carried by steining

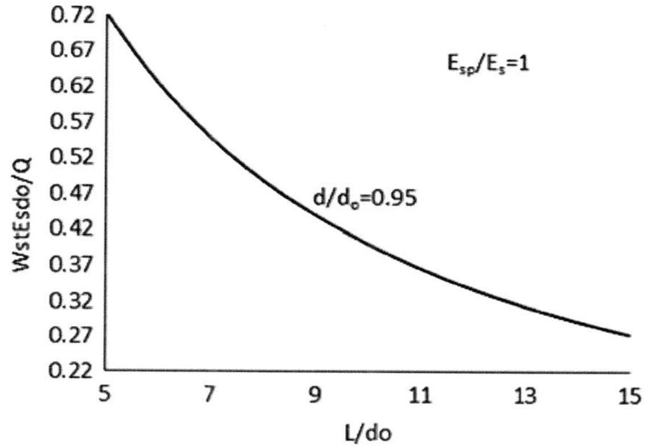

Fig. 5 Variation of the settlement of pile core at the base with (L/d_o) and (d/d_o)

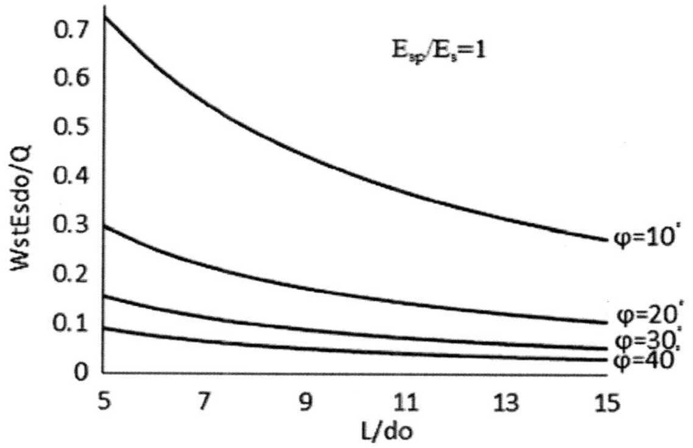

Fig. 6 Effect of angle of friction of soil on settlement

4 Conclusion

A simple theoretical approach using Winkler-type response, i.e., linear shear stress–displacement relationship, for short open-ended unplugged pipe pile is presented in this paper. The parametric study quantifies the effects of L/d_o, inner-to-outer diameter ratio, d/d_o, and modular ratio (E_{sp}/E_s) on load-carrying capacity. Based on this study, the following conclusions may be drawn:

1. The load carried steining by short open-ended unplugged pipe pile is almost independent of modular ratio (E_{sp}/E_s).
2. The load carried by steining of short pipe pile increases with an increase in frictional angle (φ).
3. The normalized settlement decreases with increase in L/d_o ratio because with the increase in L/d_o ratio, the friction resistance along with shaft increases.
4. The normalized settlement decreases with an increase in the frictional angle (φ).

References

1. Alamgir M (1996) Analysis of soft ground reinforced by columnar inclusions. Ph.D. thesis: Saga University, Saga, Japan
2. Miura N, Madhav MR (1994) Engineering of ground in lowlands. Proc of Symp of development in geotechnical engineering (from Harvard to New Delhi 1936–94) Bangkok, 6:51–62
3. Paikowsky SG, Whitman RV, Baligh MM (1989) A new look at the phenomenon of offshore pile plugging. J Marine Geotechnol 8:213–230
4. Scott RF (1981) Foundation analysis. Prentice-Hall New Jersey, 53

Study of Geotechnical Characteristics of Frost-affected Soil Stabilized with Cement and Wood Ash

Yamem Tamut, Ajanta Kalita, and S. K. Singh

Abstract The scope of present study is to assess the suitability of locally available waste products such as wood ash as a soil admixture to improve the engineering properties of frost-affected subgrade soil against the seasonal freezing–thawing effects. The proctor compaction analysis was carried out to evaluate the OMC and MDD of the frost-affected soil treated with wood ash and cement. Freezing–thawing effect on the hydraulic conductivity and strength was evaluated by falling head permeability method and UCS tests, respectively. The prepared samples were frozen at -18 °C for 24 h, and then, it was allowed to thaw at room temperature for 24 h to complete one cycle. Several freezing and thawing cycles were performed for each mixes. In this study, the frost-affected soil was treated with cement and wood ash with percentage varying from 2%, 4%, 6% for cement and 5%, 10%, 15%, 20% for wood ash, respectively. It was found that addition of wood ash increases the OMC and decreases the MDD, but when it was treated with cement, only OMC decreases, and MDD increases. After application of two cycles of freezing–thawing, the coefficient of permeability of soil was found to be zero or no voids with the mix proportions of S + 4% C + 20% WA. From the UCS tests, it was observed that parent soil, S + C and S + WA mixes show decrease in strength when subjected to freeze–thaw cycles compared to S + 4%C + 20%WA mix which shows higher UCS value after several freeze–thaw cycles.

Keywords Frost-affected soil · OMC and MDD · Permeability · UCS

Y. Tamut (✉) · A. Kalita · S. K. Singh
NERIST, Nirjuli, Arunachal Pradesh, India
e-mail: ymt@nerist.ac.in

A. Kalita
e-mail: ajanta@nerist.ac.in

S. K. Singh
e-mail: sks@nerist.ac.in

A. K. Choudhary et al. (eds.), *Advances in Geo-Science and Geo-Structures*, Lecture Notes in Civil Engineering 154,
https://doi.org/10.1007/978-981-16-1993-9_13

1 Introduction

It is a common delusion that pavement failures such as cracking, rutting and potholes develop merely as a result of heavy traffic flow. But as a matter of fact, traffic is usually one contributor to the underlying problems. Pavement failures are mainly contributed by weak subgrade soil, poor drainage system, water table depth [4] and adverse climatic conditions. In seasonal frozen areas, when the soils are exposed to freezing–thawing behavior, it has significant weakening effect on the compacted soil [6]. Freezing–thawing phenomena are not only harmful to the pavement structure but also it reduces the compressive strength, modulus of elasticity and failure strain of compacted fine-grained soil. Frost heave and boiling are the major factors that cause road damage in cold regions [7] and ultimately lead to permanent damage of the structure.

If the natural soil is weak or soft due to extreme weather conditions, a treatment is given to improve the weak soil to achieve required strength to be used as pavement subgrade layer. The most common way to improve bearing capacity of a subgrade soil can be by densification or compaction of the soil. The materials selected as admixtures for use in the improvement of subgrade must be adequately strong, and at the same time, it must be economical for use. Before conducting strength tests, the proper dosage of the admixture has to be optimized, and OMC and MDD of the soil stabilized mixture must be determined. In order to prevent the percolation of water into the subgrade after thawing of frost and capillary action of groundwater table, the admixture used in subgrade must be sufficiently impervious.

Numerous researchers have studied different methods of soil stabilization to derive practical measures against the detrimental effects of the freezing–thawing process. [1] studies the various additives to improve the frost susceptibility of soils, and durability performance of clayey subgrade soils stabilized with lime associated with freeze–thaw (F-T) cycles was studied by [8]. The samples stabilized with silica fume–lime, fly ash lime and red mud–cement additive mixtures show high freezing–thawing durability as compared to unstabilized samples [5]. Use of wood fly ash as soil stabilization for unbound pavement layers was studied by [9]. Very limited works have been studied on use of wood ash waste and cement as admixture for improving seasonal freezing–thawing effects of subgrades.

The present study is an attempt to evaluate the suitability of locally available waste products such as wood ash as a soil admixture. Its main focus is to improve the engineering properties of the frost-affected subgrade soil to resist against seasonal freezing–thawing effects. Further, effectiveness of the admixture used on permeability after application of freeze–thaw cycles shall be studied.

2 Materials

The frost-affected soil sample used in the study was collected from a high altitude region, namely Tawang, Arunachal Pradesh (India) at an elevation of 3794 m, coordinates Lat. N 27° 31.924' and Long. E 92° 03.306'. Figure 1 shows the grain size distribution curve, and the soil was characterized as silty sand (SM). The physical properties of the soil are shown in Table. 1, and soil sample passing through 4.75 mm IS sieve size is shown in Fig. 2.

The wood ash used in the present study is a residue left after combustion of wood in closed chamber (locally called Bukhari) for heating purpose. It was collected from a household in Tawang, Arunachal Pradesh. Many studies have been done on

Fig. 1 Grain size distribution

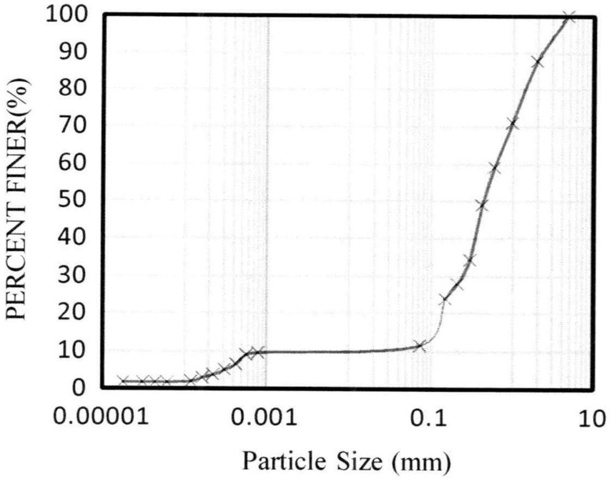

Table 1 Physical properties of the frost-affected soil

Sl.No	Properties	Value
1	Sieve analysis i. Gravel size (> 4.75 mm) ii. Sand size (0.075 mm-4.75 mm) iii. Silt (0.075 mm - 0.002 mm) iv. Clay (< 0.002 mm)	00.00% 86.99% 11.01% 2.00%
2	**Coefficient of uniformity (C_c)**	8
3	**Coefficient of curvature (C_u)**	1
4	**IS soil classification**	Silty Sand (SM)
5	**Specific gravity**	2.67
6	Liquid limit	27.55%
7	Plastic limit	NP (non-plastic)
8	Maximum dry density (MDD)	19.03 kN/m^3
9	Optimum water content (OMC)	11.40%

Fig. 2 Frost-affected soil sample

the chemical composition of wood ash, and major components of the wood ash were lime, calcite, portlandite and calcium silicate [3]. The wood ash considered in this study has been oven dried for 24 h and then allowed to pass through 75 micron size. Figure 3 and Fig. 4 show the picture of Bukhari and wood ash.

The cement used in this study was Dalmia DSP (Dhalai Expert) cement of 53 grades. Here, S, C and WA indicate soil, cement and wood ash, respectively. 20% WA means 20% by weight of wood ash (Tables 1 and 2).

Fig. 3 Bukhari

Fig. 4 Wood ash

Table 2 Physical properties of the wood ash

Sl. No	Properties	Value
1	Grain size distribution	
	i. Silt	88.53%
	ii. Clay	11.47%
2	**Specific gravity**	2.25

3 Results and Discussion

Standard proctor tests as per IS 2720, Part-1, 1983, were conducted on frost-affected soil with different mix proportions. A wide range of maximum dry densities (MDD) and optimum moisture contents (OMC) were determined on different mix proportions treated with wood ash, cement and wood ash–cement with the frost-affected soil. The variations of (MDD) and (OMC) with the parent soil are discussed in the following sections.

3.1 Study of Soil–cement Mixes Behavior

Figure 5 shows that OMC decreases with the addition of varying percentage of cement while MDD increases on addition of cement. This is due to the fact that cement is being denser material than the soil sample.

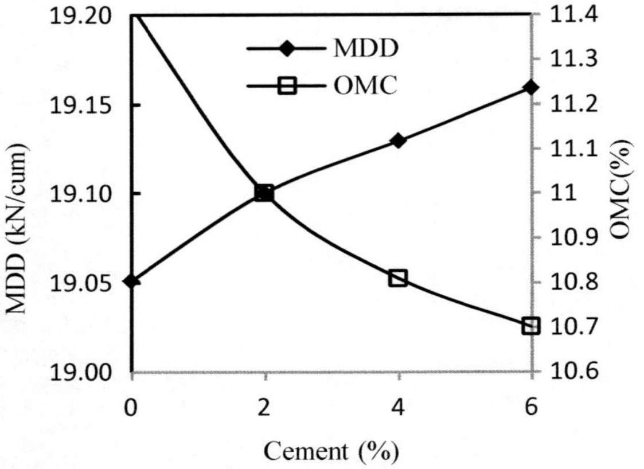

Fig. 5 MDD and OMC with cement

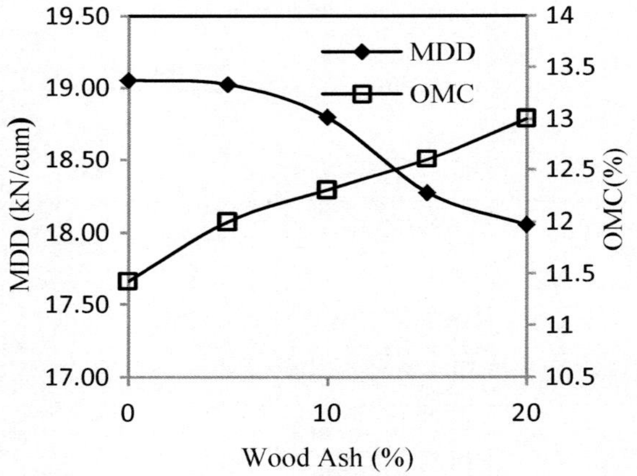

Fig. 6 MDD and OMC with wood ash

3.2 Study of Soil–wood Ash Mixes Behavior

Figure 6 shows that MDD decreases while OMC increases on addition of wood ash. This is owing to wood ash being lighter material than the soil sample. Though the OMC continues to increase, the wood ash percentage was limited to 20% in order to avoid too much fine particle which may weaken the strength of the soil.

3.3 Study of Soil–cement–wood Ash Mixes Behavior

As shown in Fig. 7a–c, OMC for soil increases with increase in cement percentage of 2%, 4%, 6% and wood ash percentage from 5 to 20%, whereas the MDD decreases.

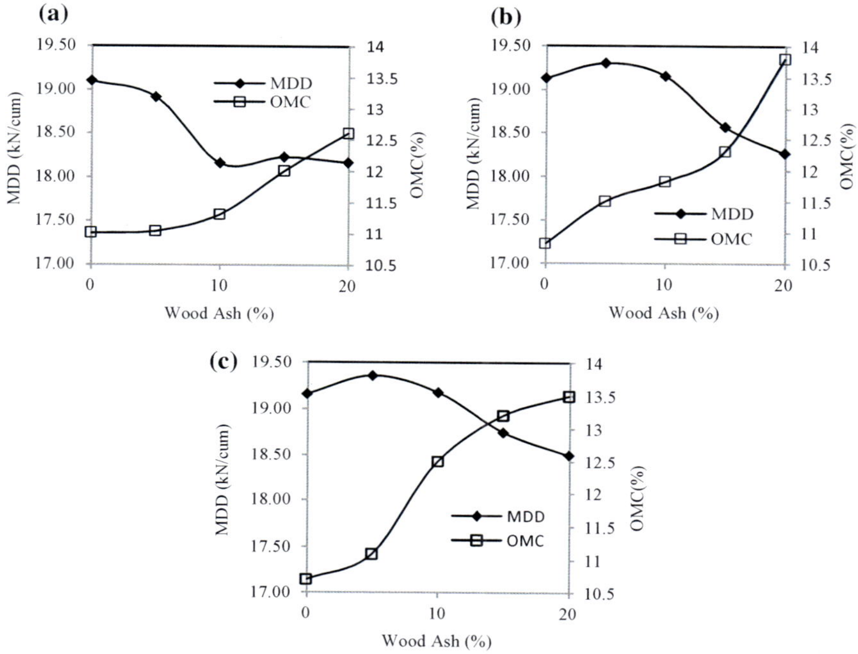

Fig. 7 a S + 2% C + diff % WA. **b** S + 4% C + diff % WA. **c** S + 6% C + diff % WA

This increase was more prominent for 4% cement with 20% wood ash-treated soil mixes.

3.4 *Effect of Freezing–thawing on Permeability*

A sufficiently impervious barrier is required to prevent the entry of water into the subgrade soil and capillary action due to groundwater. In order to assess the effectiveness of the admixtures as such materials, the samples were prepared at their MDD and OMC. And to verify the extreme conditions due to freezing–thawing effect, it was frozen in a deep freezer at −18 °C for 24 h and allowed to thaw at room temperature of about +20 °C for 24 h. Two cycles of freezing–thawing were repeated on every mix proportions. Various mix proportions were tried, and freeze–thaw cycles were performed on each trial until an impervious combination was found. To compare the effect of freeze–thaw and without freeze–thaw, curing was done on the same mix proportions for four days in desiccators (Fig. 8 and Fig. 9). Upon completion of two freeze–thaw cycles and four days curing period, falling head permeability tests (Fig. 10) were conducted to obtain coefficient of permeability.

Fig. 8 Freezing–thawing

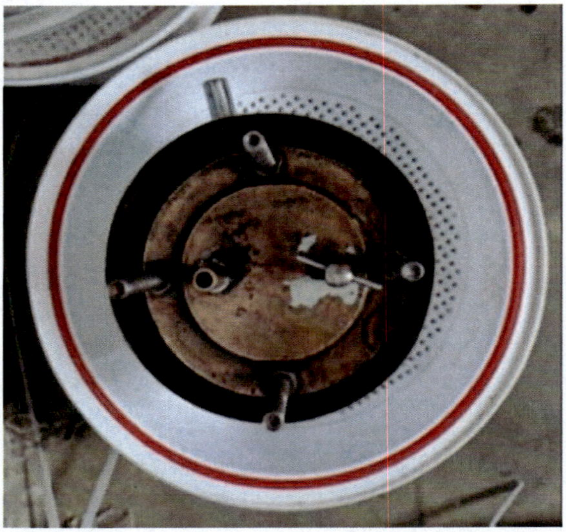

Fig. 9 Curing

From the experimental investigation, it was observed that when the soil was exposed to freeze–thaw cycles, permeability increases (Fig. 11). Addition of cement and wood ash alone did not improve much on the permeability (Figs. 12, 13 and 14). However, when the soil was treated with cement and wood ash, noticeable improvement in the permeability was observed (Figs. 15, 16, 17 and 18). It was found that mix proportion of S + 4%C + 20%WA and S + 6%C + 15% WA resulted into zero permeability or zero void ratio. This is valid for both freezing–thawing and curing.

Fig. 10 FH permeameter

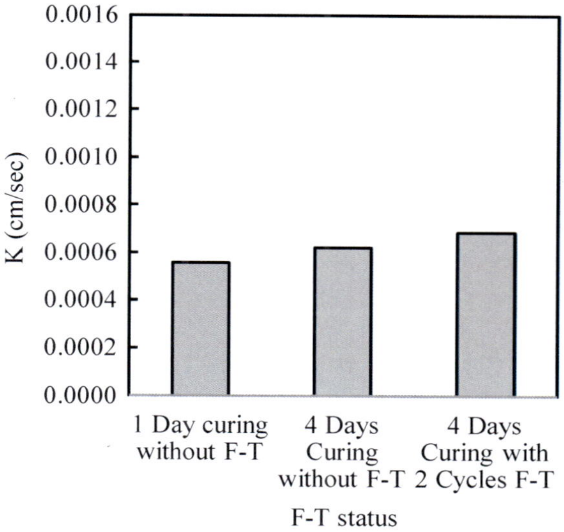

Fig. 11 Variation of parent soil

The improvement in the permeability could be due to the proper densification and pozzalanics reaction between the cement and wood ash in the presence of water.

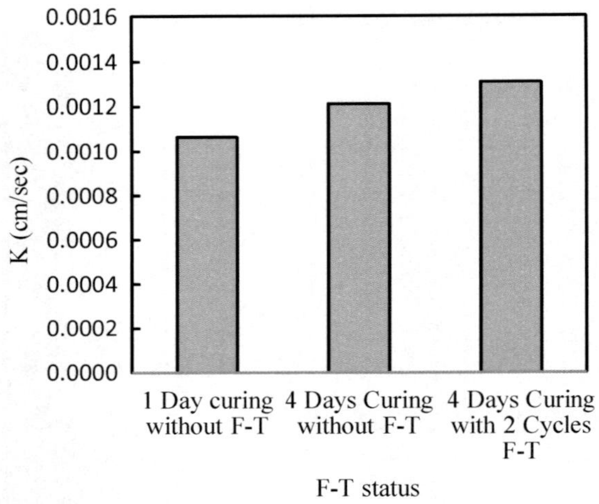

Fig. 12 Variation of 2% cement

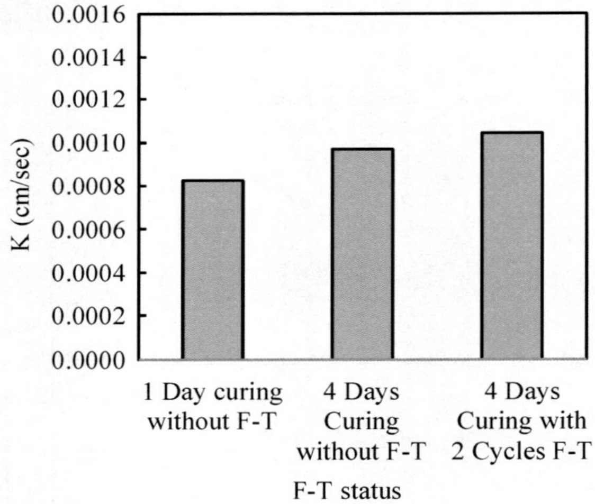

Fig. 13 Variation of 4% cement

3.5 Effect of Freezing–thawing and Unconfined Compressive Strength (UCS)

UCS tests were performed to investigate the strength characteristics of different mix samples constituting various percentages of soil, wood ash and cement subjected to defined curing periods and freeze–thaw cycles. The mix samples were first tested without curing and then tested after 3, 7, 14 and 28 days curing period using desiccators at room temperature. Further, to study the effect of freezing thawing, the samples were subjected to several freeze–thaw cycles by keeping the mix samples in deep

Fig. 14 Variation of 6% cement

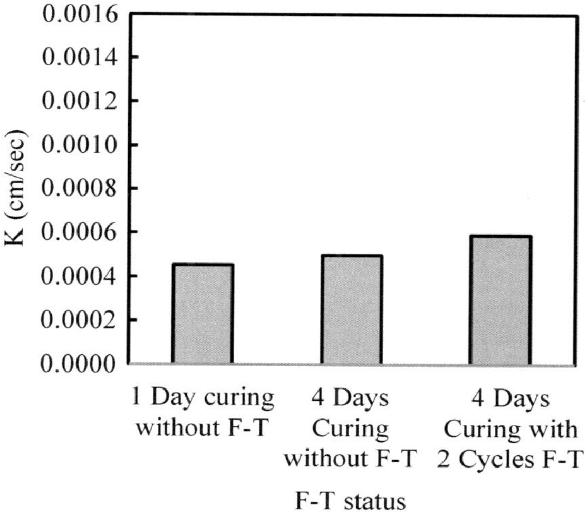

Fig. 15 Variation of 5% wood ash

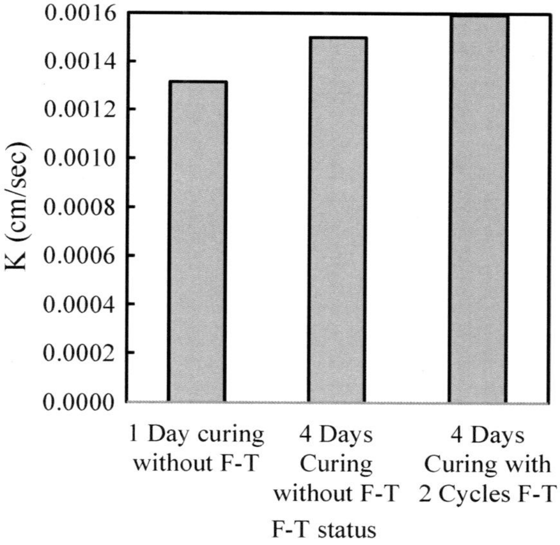

freezer at −18 °C for 12 h and then allowing it to thaw at room temperature of about + 20 °C for 12 h for one freeze–thaw cycle. As such, the samples were tested after 3, 7, 14 and 28 numbers freeze–thaw cycles with curing.

It was observed that UCS value of the mix samples increases with the increase in percentage of cement and wood ash as well as curing period. The UCS of parent soil with 28 days of curing in normal condition and after passing through 28 numbers

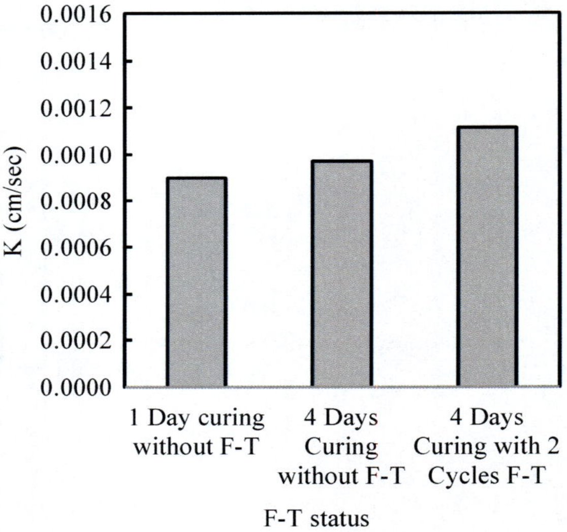

Fig. 16 Variation of 10% wood ash

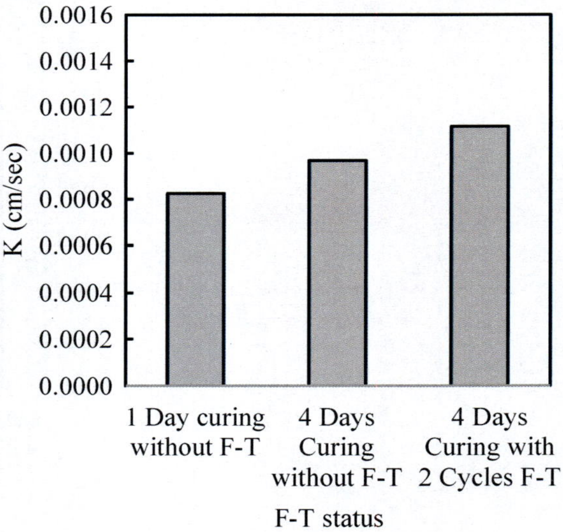

Fig. 17 Variation of 15% wood ash

freeze–thaw cycles was found to be 258 kPa and 237 kPa, respectively. Further, from Figs. 19 and 20,, it is observed that for S + %C and S + %WA mixes under the same conditions, the UCS values decrease when subjected to 28 numbers freeze–thaw cycles as compared to 28 days of curing of the test samples in normal condition. However, for S + %C + %WA mix samples, the UCS value is found to be increasing

Fig. 18 Variation of 20% wood ash

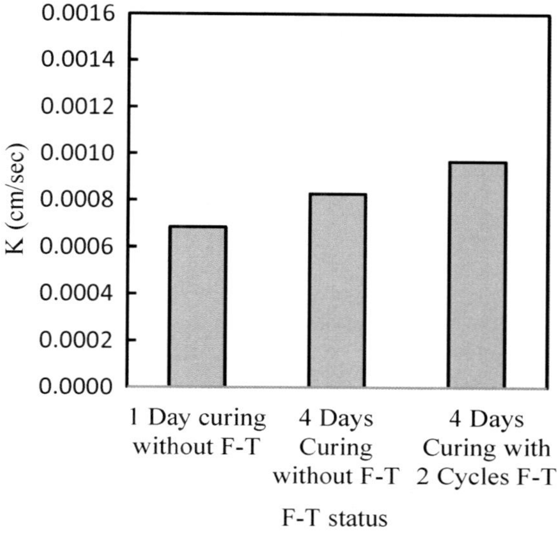

Fig. 19 Variation of UCS for S + different %WA

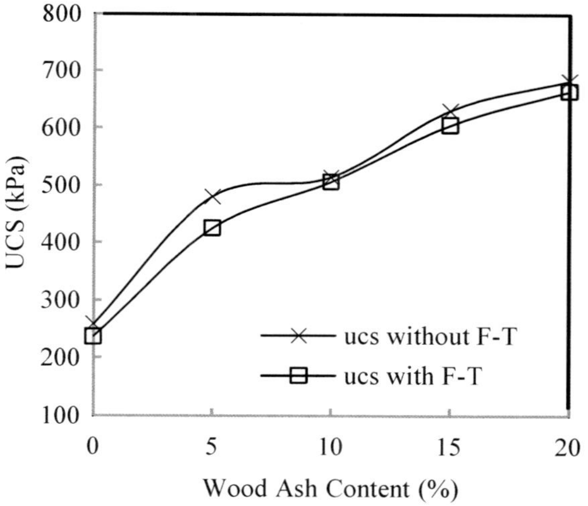

when subjected to 28 numbers freeze–thaw cycles as compared to the normal condition after 28 days of curing which can be attributed to the formation of cementitious products due to hydration of cement and wood ash with the presence of water.

It was found from Fig. 21 that mix proportion of S + 4%C + 20%WA at the same curing period and freeze–thaw cycles were found to be 2921 kPa and 3444 kPa, respectively, which shows substantial improvement in strength of mix samples subjected to freezing and thawing.

Fig. 20 Variation of cement content S + different %WA

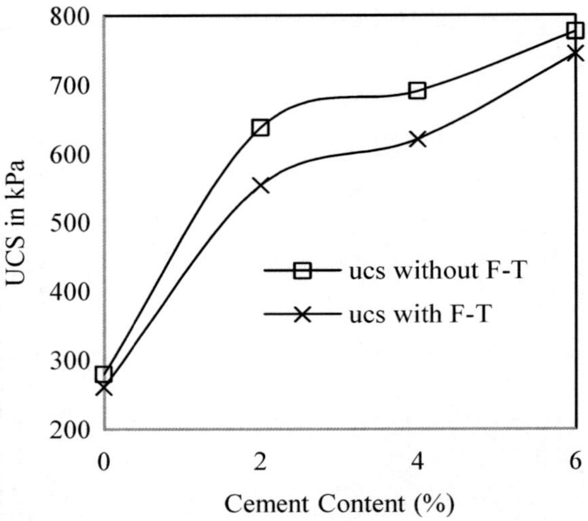

Fig. 21 Variation of UCS for S + 4%C + different %WA

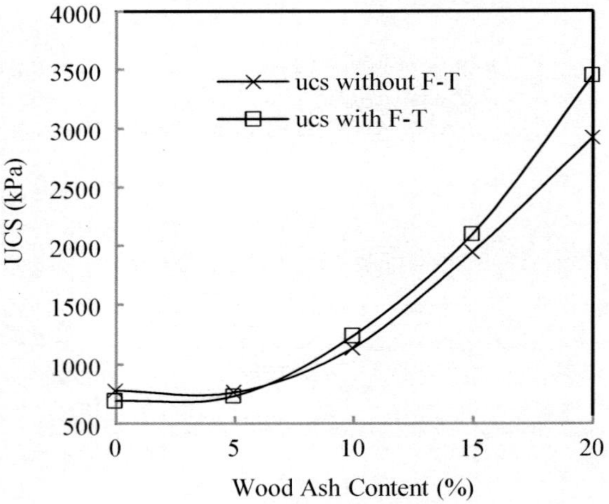

4 Conclusions

After conducting extensive experimental study on the various proportions of the admixtures, the major conclusions drawn on this study are as follows.

- The OMC of the soil decreases, and MDD increases with increase in cement content.

- The OMC of the soil increases, and MDD decreases with addition of wood ash content. This increase was more prominent for 4% cement with 20% wood ash-treated soil mixes.
- Freezing–thawing tests showed that when the soil was exposed to freeze–thaw cycles, permeability increases. Addition of cement and wood ash alone did not improve much on the permeability. However, when the soil was treated with cement and wood ash, noticeable improvement in the permeability was observed. It was found that mix proportion of S + 4%C + 20%WA and S + 6%C + 15% WA resulted into zero permeability or zero void ratio.
- The mix proportion of S + 4%C + 20%WA and S + 6%C + 15% WA shows highly impervious when exposed to freeze–thaw cycles. Therefore, the same mix proportions may be used for design of pavement where similar weather condition prevails.
- The UCS of parent soil at 28 days curing period and 28 numbers of freeze–thaw cycles were found to be 258 kPa and 237 kPa, respectively. The S + C and S + WA mixes had also shown less UCS value after 28 numbers of freeze–thaw cycles as compared to 28 days curing under normal condition. But when the soil is mixed in the proportion of S + 4%C + 20%WA, the UCS was found to be 2921 kPa and 3444 kPa for same curing period and freeze–thaw cycles, respectively, which shows substantial increase in strength due to addition of cement and wood ash.

References

1. Lambe TW, Kapler CW (1971) Additives for modifying the frost susceptibility of soils part-1. CORPS of Engineers, U.S. Army Cold Regions Research and Engineering Laboratory Hanover, New Hampshire, Technical Report 123, Pt.1
2. Edwin JC, Anthony JG (1979) Effect of freezing and thawing n the permeability and structure of soils. Engineering Geology, 13:73–92, Elsevier Scientific Publishing Company, Amsterdam, Printed in the Netherlands. https://doi.org/10.1016/0013-7952(79)90022-X.
3. Etitgni L, Campbell AG (1991) Physical and Chemical characteristics of wood ash. Bioresour Technol 37:173–178. https://doi.org/10.1016/0960-8524(91)90207-Z
4. Hermansson A, Guthrie WS (2005) Frost heave and water uptake rates in silty soil subject to variable water table height during freezing. Cold Regions Sci Technol 43:128–139. https://doi.org/10.1016/j.coldregions.2005.03.003
5. Yarbesi N, Kalkan E, Akbulut S (2007) Modification of the geotechnical properties as influenced by freeze-thaw, of granular soils with waste additives. Cold Regions Sci Technol 48:44–54. https://doi.org/10.1016/j.coldregions.2006.09.009
6. Li G, Ma W, Zhao S, Mao Y, Mu Y (2012) Effect of freeze-thaw cycles on mechanical behavior of compacted fine-grained soil. ASCE Cold Region Eng 72–81. doi: https://doi.org/10.1061/9780784412473.008.
7. Lai U, Zhang S, Yu W (2012) A new structure to control frost boiling and frost heave of embankments in cold regions. State Key Laboratory of Frozen Soil Engineering, Cold and Arid Regions Environmental Research Institute, Chinese Academy of Sciences, Lanzhou 730000, China. https://doi.org/10.1016/j.coldregions. 2012.04.002
8. Rahman MT, Tarefder RA (2015) Freeze-thaw durability of lime stabilized clayey subgrade soils. Computer Methods and Recent Advances in Geomechanics-Oka, Murakami, Uzuoka&

Kimoto (Eds.)© Taylor & Francis Group, London, ISBN 978–1–138–00148–0. https://doi.org/10.1201/b17435-205
9. **Peteris S Wood fly ash stabilization of unbound pavement layers**
10. Conference: 19th International Conference on Soil Mechanics and Geotechnical Engineering, ICSMGE 2017, At Coex Convention CenterSeoul; South Korea (2018)

Improvement of Strength Behaviour of Fly Ash and Cement Stabilized Soil with Glass Fibre Reinforcement

Phurba Dorjee Philley, Ajanta Kalita, and Yachang Omo

Abstract During road construction, on many occasions, weak soil is encountered which is not suitable as subgrade material. In such cases, it becomes imperative to improve the strength properties of such soil for undertaking any type of construction activity. In this research investigation, local weak subgrade soil is first stabilized with class F fly ash and cement and then the mixer is reinforced with glass fibre to study the improvement in strength characteristics. Varying percentages of fly ash (10, 30, 50%), cement (0.5, 1, 1.5, 2%) and glass fibre (0.25, 0.5 and 1%) had been used to prepare different mix proportions which were then tested to evaluate their index properties, compaction and shear strength parameter using triaxial shear test. The result showed substantial improvement in compaction and shear strength parameters with increase in the percentage of all the three materials up to an optimum limit. The compaction characteristics, stress–strain relationship and shear strength parameters were calculated to study the improvement in the mix. The result of this study showed that the improvement in strength of the parent subgrade soil will have vast applications in the field of geotechnical engineering like construction of subgrade, embankments, low-cost rural roads, etc.

Keywords MDD · OMC · Deviatoric stress · Shear strength parameters · UU test

1 Introduction

On many occasions, strength of parent subgrade soil is found to be not conforming to the required design parameters. Such cases have compelled engineers to explore

P. D. Philley (✉) · A. Kalita
NERIST, Nirjuli, India

A. Kalita
e-mail: ajanta@nerist.ac.in

Y. Omo
CIT, Kokrajhar, India
e-mail: y.omo@cit.ac.in

A. K. Choudhary et al. (eds.), *Advances in Geo-Science and Geo-Structures*, Lecture Notes in Civil Engineering 154,
https://doi.org/10.1007/978-981-16-1993-9_14

possibilities of use of industrial waste products such as fly ash, glass fibres, waste tyres, blast furnace slag, China clay, cement kiln dust as admixtures for an economic as well as environment-friendly solution to this problem.

It has also been observed that the engineering properties of soils could further be improved by reinforcing it with suitable materials at optimum ratio. The primary objective of soil reinforcement is to increase its stability, improve bearing capacity and reduce lateral deformation and settlements. Reinforcement of soils is done by either inclusion of reinforcing material (e.g. strip, sheet or bar) in a defined pattern or mixing discrete fibres randomly. The most commonly used fibres are coir, jute as natural fibres and polyester, polypropylene, polyethylene and glass fibres as synthetic fibres. The fibres are added in dry soil and mixed randomly as cement, lime or other additives are mixed. Potential weak planes which are normally developed in case of aligned reinforcement can be limited by using randomly distributed fibres.

Many studies have revealed that the use of fly ash and cement helps in stabilizing the soil by improving the strength behaviour. Studies have also proved that the use of industrial waste like glass fibre helps in reinforcing the soil and reduces its brittleness and helps the soil in carrying tensile load. Shooshpasha and Alijani Shirvani [1] tried stabilizing soil by adding 2.5, 5 and 7.5% of cement and observed changes in compaction parameters which are affected by higher specific gravity of cement, alteration in the grain size distribution of the mixture and reduction in moisture content. Kalita and Singh [2] conducted an experimental study to investigate the use of sand and red soil amended with varying percentages of fly ash (20, 30 and 50%) and cement (1–5%) as subgrade and found that there is improvement in strength behaviour of soil when optimum quantity of fly ash and cement is used. Mali and Singh [3] conducted triaxial tests to study the influence of fibre content on the strength behaviour of poorly graded fine sand with glass fibre reinforcement. It was found that with increase in fibre content the deviatoric stress at failure increased at same confining pressure, whereas the stiffness increased only at high fibre contents.

The main objective of this study was to use the industrial wastes like fly ash and glass fibre with nominal quantity of cement to improve the properties of locally available weak soil in terms of their strength and other parameters which make the parent soil suitable for use as subgrade in a more economical way as well as solve the disposal problem of fly ash.

2 Experimental Programme

2.1 Materials

The soil used in these experiments was collected from the shoulder of NH 415 passing through the Karsingsa area of Arunachal Pradesh. The physical properties of the soil are shown in Table 1. From the grain size analysis (Fig. 1), the soil had been classified as poorly graded sand (SP).

Table. 1. Physical properties of soil

Property	Value
Specific gravity	2.6
Plastic limit	Nil
Liquid limit	22.89
Plasticity index	Nil
Gravel (10–4.75 mm)	14.63%
Coarse sand (4.75–2 mm)	9.20%
Medium sand (2 mm–425 μm)	32.07%
Fine sand (425–75 μm)	32.06%
Uniformity coefficient, C_u	4.61
Coefficient of curvature, C_C	1.15
Soil classification	SP

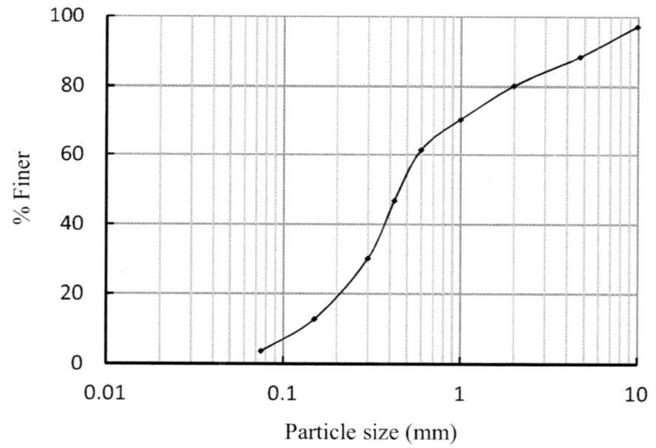

Fig. 1 Particle size distribution of soil

The ordinary Portland cement (OPC) used was of 53 grades with specific gravity 3.15. The class F fly ash for this study was collected from the Satyam Ispat factory, Banderdewa, Arunachal Pradesh. The fly ash was found to be non-plastic, pozzolanic in nature, and the specific gravity was 2.6. Glass fibre was collected from the local market available in Itanagar, Arunachal Pradesh. It had a whitish colour obtained in bundles known as strands.

While preparing the mixed specimen, the parent soil was first dried and then the required percentage of fly ash was added and mixed thoroughly to obtain a uniform soil-fly ash mix. Thereafter, cement in required percentage was added to the soil-fly ash mix in dry state. The mix was then further reinforced with glass fibre to make a homogeneous mix. After attaining a uniform mix, water was added to the mix to obtain the MDD corresponding to the OMC.

The designations S, C, FA and GF indicate soil, cement, fly ash and glass fibre, respectively. 20% FA means 20% by dry weight of fly ash.

3 Results and Discussions

3.1 Compaction Behaviour of Soil, Cement and Fly Ash Mixes

The standard proctor test was carried out on parent soil as well as soil mixed with varying cement percentages of 0.5, 1, 1.5 and 2%, and it was found that, as the cement content increases, the MDD increases and OMC decreases. Generally, it is difficult to perform compaction tests on pure sand, but the instant parent soil sample constituted of ample amount of fine sand particles (approx. 32%) mixed with coarse sand which made the compaction test feasible as fine sands get readjusted and re-orient themselves in between the large particles to make the sample more compact and denser. Soil + 2% cement gave OMC of 12.98% and MDD of 19.15 KN/m^3, respectively. The behavioural changes in compaction parameters are due to the higher specific gravity of cement than that of soil, modification in the grain size distribution of the mixture and reduction in moisture content (Al-Aghbari et al. [4], Zabielska [5]). The amount of modification in dry density is more prominent at the lower percentage of stabilizer content.

The MDD was found to increase from 18.63 to 19.15 KN/m^3 on varying percentages of cement from 0.5 to 2% as shown in Fig. 2 and Table 2 which can be ascribed to the higher specific gravity of cement. The OMC of the mixes decreases with the increase in cement content as more water is utilized during the initial stage of hydration process. The variation of OMC with the increase in cement is shown in Fig. 2.

The MDD and OMC of the soil used in this study were found to be 18.33 KN/m^3 and 13.69%, respectively, and for fly ash, the same had been found to be 14.11 KN/m^3 and 18.34%, respectively. The variation of MDD and OMC of soil with

Fig. 2 MDD and OMC of soil + cement mixes

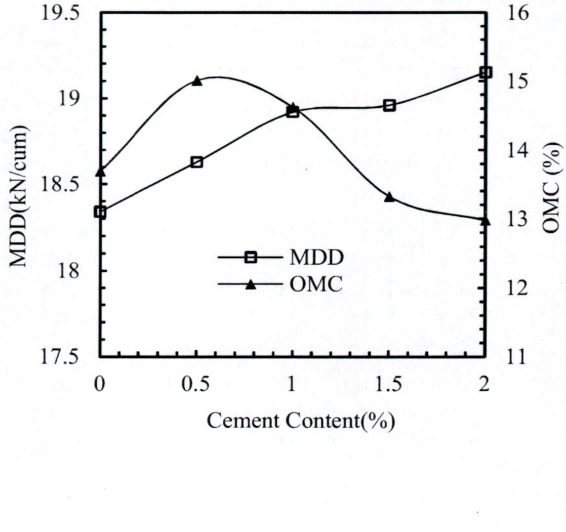

Table. 2. MDD and OMC of soil–cement-fly ash mixes

Mix proportions	MDD (KN/m^3)	OMC (%)
Soil	18.339	13.69
FA	14.11	18.34
S + 0.5%C	18.63	15.01
S + 1%C	18.92	14.623
S + 1.5%C	18.96	13.32
S + 2%C	19.15	12.98
S + 10% FA	17.6	13.5
S + 30%FA	17.18	14.2
S + 50%FA	15.29	15.48
S + 10%FA + 0.5%C	18.27	15.09
S + 10%FA + 1%C	18.69	13.74
S + 10%FA + 1.5%C	19.04	10.99
S + 10%FA + 2%C	19.44	8.92
S + 30%FA + 0.5%C	17.04	16.22
S + 30%FA + 1%C	17.16	13.92
S + 30%FA + 1.5C	17.79	11.49
S + 30%FA + 2%C	18.29	9.96
S + 50%FA + 0.5%C	16.01	17.12
S + 50%FA + 1%C	16.43	15.29
S + 50%FA + 1.5%C	17.08	13.96
S + 50%FA + 2%C	16.9	11.33

varying percentages of fly ash is shown in Fig. 3. The MDD was found to decrease from 17.6 to 15.29 KN/m^3 with varying percentages of fly ash from 10 to 50% as shown in Fig. 3. This can be attributed due to the lower specific gravity of fly ash than that of the soil. The fly ash particles having hollow and thin-walled cenosphere structure tend to hold more moisture, thus increasing the OMC of the mix [6]. The variation of MDD and OMC of the mixed specimen with 30% fly ash and varying percentage of cement is shown in Fig. 4.

3.2 *Deviatoric Stress Behaviour of Soil-Fly Ash–Cement Mixed Specimens*

Several unconsolidated-undrained (UU) triaxial tests were conducted to determine the shear strength parameter, i.e. cohesion (C_u), and angle of internal friction (ϕ_u) of the soil stabilized with cement, fly ash and glass fibre at different proportions. UU triaxial tests on pure sand are difficult, but the instant parent soil sample is comprised of fine sand particles to the tune of 32% which has made the UU tests

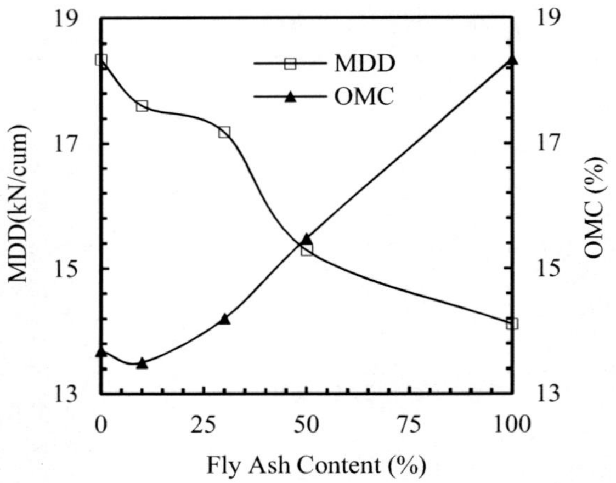

Fig. 3 MDD and OMC of soil + FA mixes

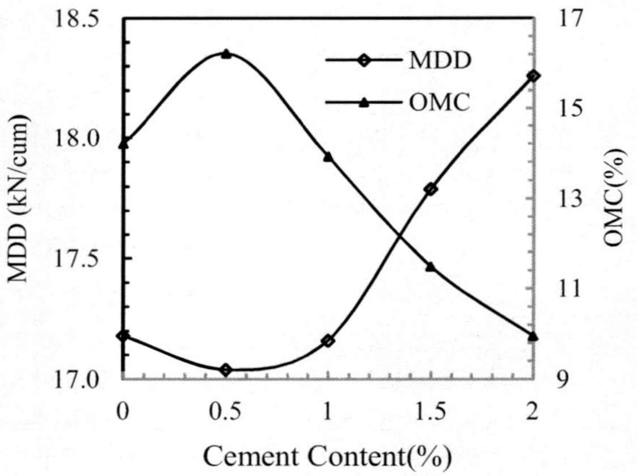

Fig. 4 MDD and OMC of soil + 30% FA + varying cement content

practicable. Further addition of admixtures has made the mix specimens more stable for conducting UU tests. Various tests were conducted at different confining pressures, viz. 0.4, 0.6 and 0.8 kg/cm^2. Each sample has been prepared taking into account the MDD and OMC of the mix samples. All samples were kept in desiccators for 7, 14 and 28 days curing in order to evaluate the effect of curing on shear parameter of soil.

3.2.1 Effect of Cement Content on Soil

In the UU tests, the increase in cement content from 0 to 1.5% had slight increase in deviatoric stress values; however, when the cement was increased to 2% it had

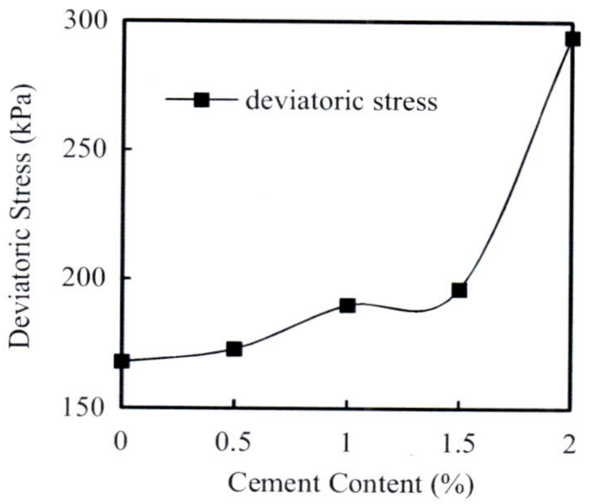

Fig. 5 Deviatoric stress of soil cement mixes at 0.8 kg/cm^2 confining pressure and 28 days of curing period

a significant increase in the deviatoric stress of the mixes. It was observed that the highest 28 days deviatoric strength was obtained at mix proportion of S+2%C at 0.8 kg/cm^2 cell pressure (Fig. 5). The addition of cement up to 2% has modified the properties of soil significantly, thereby increasing the deviatoric strength.

3.2.2 Effect of Glass Fibre Reinforcement on Soil-Fly Ash–Cement Mixes

The results of the UU tests of S+C+FA+GF mix samples have shown that the highest deviatoric stress at failure was observed at mix proportion of S + 30%FA + 0.25%GF + 1.5%C with 0.8 kg/cm^2 confining pressure after 28 days of curing (Fig. 6). The addition of glass fibre content has improved the strength behaviour, brittle failure nature and other shear strength parameters of the mix specimens. The improvement in strength can be attributed to the interlocking of sand and fibre which contributes in the improvement of resisting force due to mobilization of tensile strength of fibres in the mix specimens. It was also observed that 0.8 kg/cm^2 confining pressure gave higher deviatoric strength value compared to 0.6 and 0.4 kg/cm^2 confining pressure. This is due to increase in lateral support as well as apparent cohesion which is evident from the test results of different mix specimens.

3.3 Shear Strength Parameters of the Soil

All the specimens had been compacted considering their respective MDD and OMC. For all test specimens, the UU test was conducted and Mohr's circle was drawn for determination of shear strength parameters, namely cohesion intercepts (C_u) and

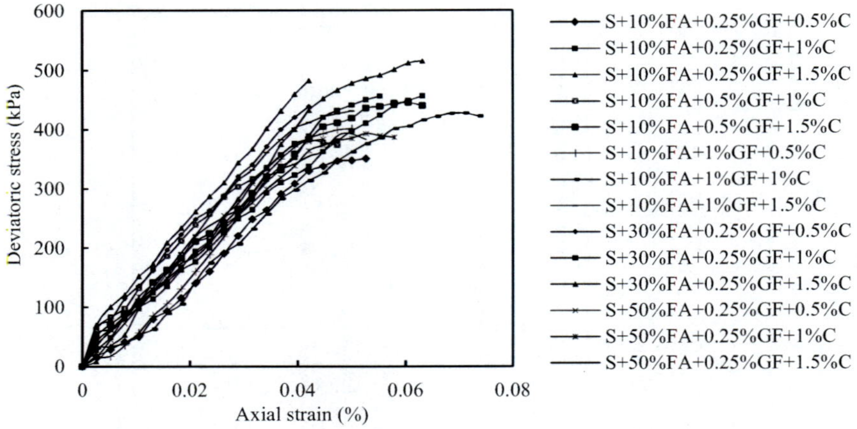

Fig. 6 Deviatoric stress versus strain at 0.8 kg/cm^2 confining pressure for S + FA + GF + C mixes

angle of shearing resistance ϕ_u. The cohesion (C_u) and friction angle (ϕ_u) values of soil-cement and soil-cement-fly ash-glass fibre mixes for various curing periods were evaluated. It is found that addition of cement to the soil increases the cohesion intercept (C_u) substantially (Fig. 7), but the friction angle is reduced at higher cement contents. Addition of fly ash to the soil had increased the C_u (Fig. 8) and ϕ_u parameters of the samples. The increase in cohesion reflects the increase in cementation caused by fly ash and cement. Although dry fly ash shows no cohesion, the surface tension of the pore water present in the moist fly ash causes the sample to exhibit some amount of apparent cohesion. It was further observed that the test specimen containing S + 30%, FA + 0.25% and GF + 0.5%C shows significant improvement in shear strength parameter in both cohesion (C_u) and angle of internal friction ϕ_u (Fig. 9).

Fig. 7 Cohesion (C_u) versus curing periods for soil + varying % of C

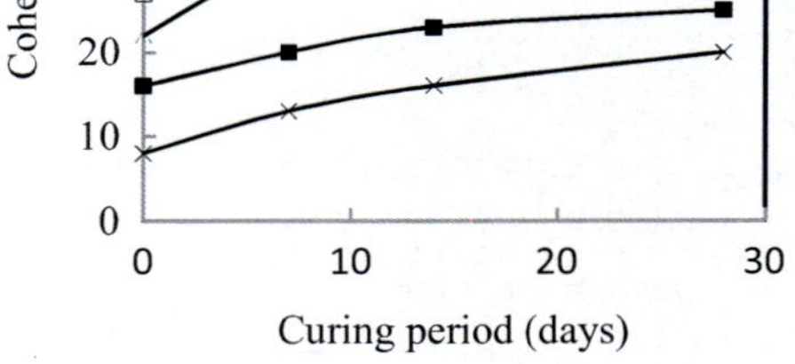

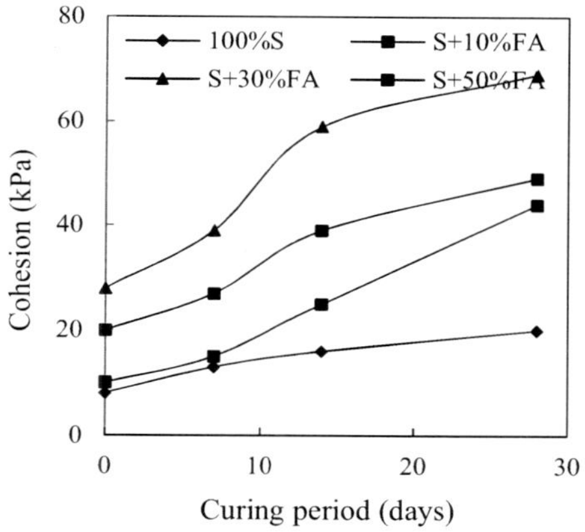

Fig. 8 Cohesion (C_u) versus curing periods for soil + varying% of FA

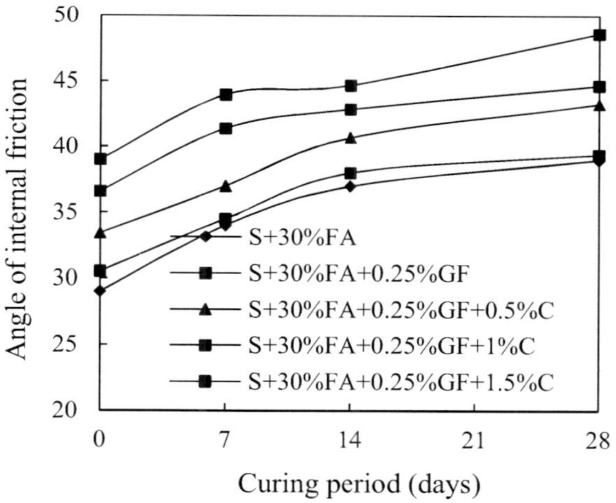

Fig. 9 Angle of internal friction versus curing periods for S + 30%FA + 0.25%GF + varying % of C

The improvement in friction component is attributed to the addition of glass fibre which is good in providing tensile strength and supports the soil in bearing resistance.

4 Conclusions

Based on the test results and analysis performed in this study, it is concluded that on increasing the cement content and curing period the shear strength and its associated parameters improved. On increasing the fly ash content, the OMC of S + FA mix increases and MDD decreases. The fly ash shows pozzolanic property when mixed with water which is also a time- dependant property and increases with curing period.

Glass fibre being strong in tension provides stronger bond between its surface and soil particles when load is acted upon and thus delaying the shear failure. The shear strength parameter improved with addition of cement, fly ash and glass fibre, and it was observed that at a mix proportion of S + 30%FA + 0.25%GF + 0.5%C the cohesion (C_u) and friction angle value had increased substantially.

References

1. Shooshpasha I, Alijani Shirvani R (2015) Effect of cement stabilization on geotechnical properties of sandy soils, Department of Civil Engineering, Faculty of Civil Engineering, Babol university of Technology, Iran
2. Kalita A, Singh B (2009) Experimental study on strength behavior of cement blended soil-fly ash mixtures. In: Students symposium on research in civil engineering, Issue 2009, pp 1–6
3. Mali S, Singh B (2013) A study on shear strength of sand reinforced with glass fibres. Int J Sci Eng Res 4(5):285–235
4. Al-Aghbari M.Y., Mohamedzein, Y.E.A. and Taha, R. (2009) "Stabilization of desert sands using cement and cement dust". Proceedings of the Institution of Civil Engineering, Ground Improvement, 162(3), 145–151
5. Zabielska K (2008) "Laboratory compaction of fly ash and fly ash with cement additions". J. Hazard. Mater. 151(2–3), 481–489
6. Deb T, Pal SK (2014) Effect of fly ash on geotechnical properties of local soil-fly ash mixed samples. Int J Res Eng Technol 3(5), 507–516

Indigenous Circular Shape Tamper for Dynamic Compaction

Vijay Kumar Singh and Syed M. Ali Jawaid

Abstract Many innovative soil improvement techniques have been developed in the last few decades, and dynamic compaction using heavyweight tampers is one of those. It is performed by repeatedly dropping a heavyweight on a grid pattern prepared on the ground surface. In this study, a circular shaped tamper is designed indigenously for the dynamic compaction of alluvial deposits. The improvement in alluvial deposits in terms of in situ density and depth of improvement has been monitored using static cone penetration test (SCPT). This study is an attempt to analyze the effectiveness of tamper in compaction of loose alluvial deposits. Based on this study, it is found that the dynamic compaction is a useful method of ground improvement and can be used as an effective technique for mitigation of earthquake in such deposits.

Keywords Dynamic compaction · Lowlands · Tamper · CPT · Alluvial

1 Introduction

The soil transported and deposited by water is called alluvial soil. As flowing water (stream or river) looses velocity, it tends to deposit some of particles that it was carrying in suspension or by rolling, sliding, or skipping along the river bed. Coarser or heavier particles are dropped first. Hence on the higher reaches of a river, gravel and sand are found. When this loose alluvial material is deposited or cemented into a lithological unit, it is called an alluvial deposit [1]. Most of the regions with alluvium deposits lie in Zone IV of the seismic zone in India. The alluvial lowlands consist of soft fine-grained materials, mostly SM/ML in the upper horizon followed by sand (SP/SM). Strata with adequate or required bearing capacity are often available only below 1.00–4.0 m from the natural ground surface. Due to non-availability of

V. K. Singh (✉) · S. M. Ali Jawaid
Madan Mohan Malaviya University of Technology, Gorakhpur, U.P., India

S. M. Ali Jawaid
e-mail: smajce@mmmut.ac.in

A. K. Choudhary et al. (eds.), *Advances in Geo-Science and Geo-Structures*,
Lecture Notes in Civil Engineering 154,
https://doi.org/10.1007/978-981-16-1993-9_15

adequate bearing strata, conventional shallow foundations are difficult to construct. There is need of time to improve the top strata so that conventional shallow foundation may be utilized. Dynamic compaction equipment with heavy tampers is not available in this region. To overcome this problem, a new circular tamper is designed and tested.

2 Literature Review

An approach for deep dynamic compaction was given by Menard and Broise in 1975. In this method, a tamper of weight 5–30 tones falls from a height of 10 to 30 m improves the strength of soil deposits up to the depth of 3–9 m [2]. Soil improvement can be assessed using tests like standard penetration test (SPT), cone penetration test (CPT), and pressuremeter test (PMT) before and after the compaction process [3–8].

Empirical evidence suggests that the effective depth of influence increases with the impact energy, and the greatest degree of improvement is usually observed at about half the effective depth of influence [3]. Dynamic compaction has been studied experimentally by several researchers, notably Menard and Broise [9], Mayne et al. [3], etc. These studies have given the soils ranges for which the dynamic compaction is most effective. An empirical expression for the effective depth of ground improvement is suggested as

$$D = 0 : 5\surd(\mathrm{WH}) \text{ to } 1.0\surd(\mathrm{WH}) \tag{1}$$

3 Experimental Program

3.1 Tamper Specification

The tamper used in the dynamic compaction consists of two separate parts joint together using the nut-bolt system to obtain a flat circular base for impact (Fig. 1). The weight of tamper is 85 kg, and its base diameter is 19 cm.

3.2 Height of Drop

The height of the drop of tamper that had been achieved using tripod assembly for dynamic compaction is 2 m. Though during the densification process because of the formation of the crater, the height of drop was increased gradually to keep a constant height of the drop.

Fig. 1 Indigenous circular shape tamper

3.3 Grid Spacing

The grid spacing or drop spacing is usually varied from 1.5 to 2.5 times the diameter or width of tamper. The center to center distance between each impact point was taken as 28.5 cm, calculated using 1.5 *d* formula where *d* is the diameter of tamper used.

3.4 Site Characterization and Subsoil Condition

A pit of size 3.0 m × 3.0 m × 3.0 m is prepared in front of Civil Engineering Block II, M M M University of Technology, Gorakhpur. Before the dynamic compaction of the in situ, soil properties were determined with an aim to check the suitability of dynamic compaction process. The groundwater level was found below the depth of 15.0 m, making it a suitable site for DC. The soil properties are documented in Table 1.

Table 1 In situ soil properties

Soil type	Inorganic silt (ML)
Soil fraction retained on 75 μm sieve	79.55%
D_{10}	6.20 μm
D_{30}	11.2 μm
D_{60}	24.36 μm
Liquid limit	26.90%
Plastic limit	23.80%
Plasticity index	3.10
Natural moisture content	$w = 17.26\%$
Consistency index	$I_C = 3.11$
Bulk density	
(a) Loose soil	1.76 g/cc
(b) Compacted soil	3.35 g/cc
Dry density	
(a) Loose soil	1.50 g/cc
(b) Compacted soil	2.85 g/cc
Avg. Cone resistance q_c (up to 6 0.0 m depth, before DC)	6.41 kgf/cm^2

4 Methodology

The conventional approach for carrying out the dynamic compaction has two steps:

Step 1. Area compaction
Step 2. Ironing compaction.

The area compaction was further divided into two passes (Pass I and II), which involved dropping of tamper from a height of 2.0 m on each point of the grid. The crater used to fill with the same soil after each passes and leveled. During ironing compaction, the tamper was dropped from the same height of 2.0 m. The adjacent tamper imprints was overlapped.

5 Result and Discussion

Static cone penetration tests (SCPT) were performed to check the effectiveness of dynamic compaction using the indigenous circular shape tamper.

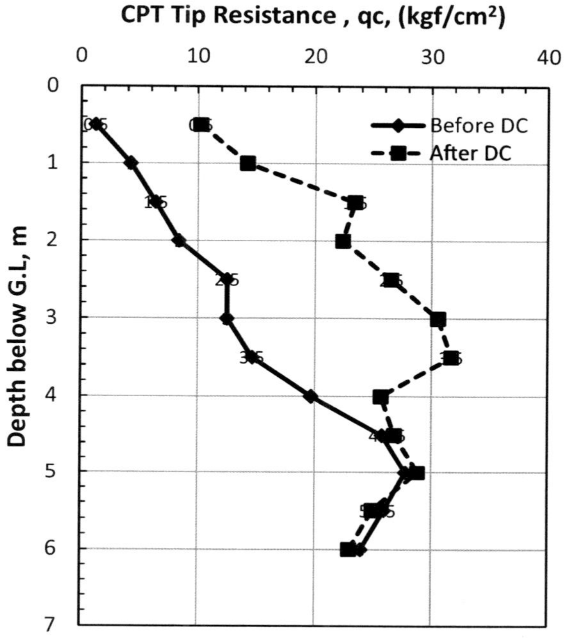

Fig. 2 Variation of cone tip resistance with depth

5.1 *Effect of Dynamic Compaction on Cone Tip Resistance*

The cone tip resistance of untreated and treated soil (after DC) is shown in Fig. 2. Due to dynamic compaction, the value of cone resistance of the treated soil is greater than that of loose soil. But it is only up to a depth of 4.0 m below ground level after which the cone resistance value of compacted and loose soil coincides with each other indicating the zone up to which the dynamic compaction is effective in context with the tamper used.

5.2 *Effect of Dynamic Compaction on Frictional Resistance of Cone*

The frictional resistance obtained by the static cone penetration test in case of untreated and treated soil (after DC) is shown in Fig. 3. Due to dynamic compaction, the value of frictional resistance of the treated soil is greater than that of untreated soil. But it is only up to a depth of 4.0 m below ground level after which the frictional resistance value of compacted and loose soil coincides with each other indicating the zone up to which the dynamic compaction has its influence.

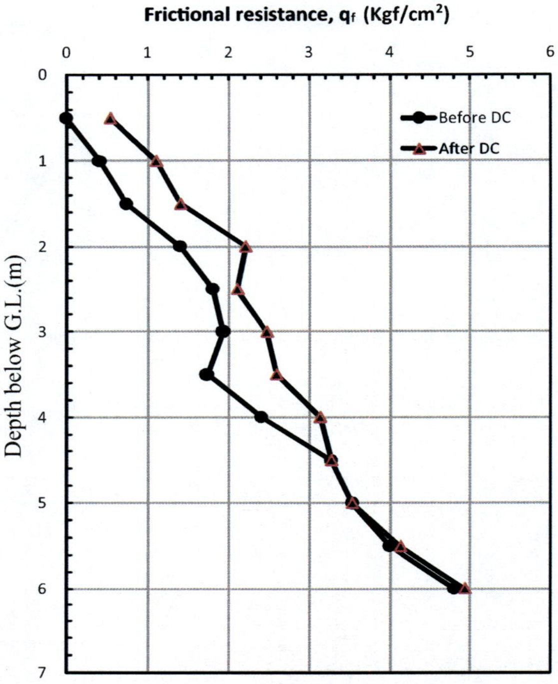

Fig. 3 Variation of frictional resistance with depth

5.3 Effect of Dynamic Compaction on the Frictional Ratio

The variation of frictional ratio with depth for untreated and treated soil (after DC) is shown in Fig. 4. Friction ratio is the ratio of frictional resistance to that of cone resistance. Due to dynamic compaction, the value of cone resistance increases more in comparison of the frictional resistance (though the frictional resistance of compacted soil is greater than that of loose), due to which the friction ratio value of compacted soil is less in comparison to that of loose soil. Here, its value is less up to 4.0 m depth, and after that it nearly coincides with that of loose friction ratio value indicating the influence zone of compaction.

6 Conclusion

Based on this study, the following conclusions are drawn:

1. Dynamic compaction (DC) can be used as a ground improvement technique for alluvium lowlands.
2. The strength of the soil is improved up to a depth of 4.0 m below ground level.

Fig. 4 Variation of frictional ratio with depth

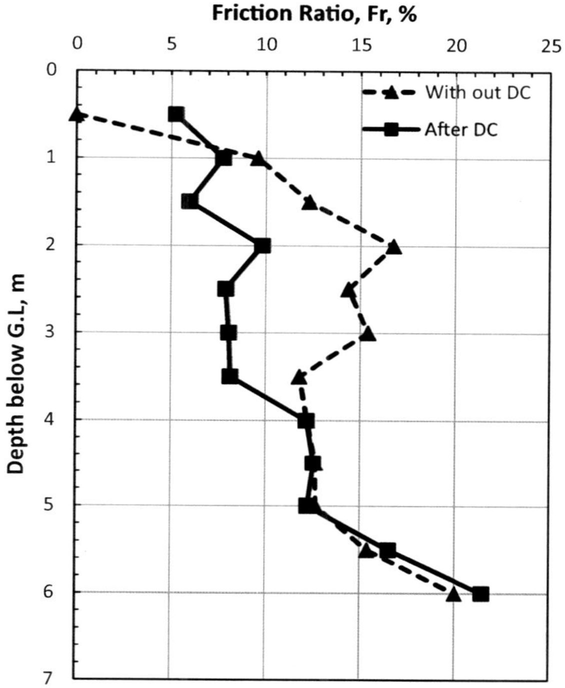

3. The value of the friction ratio of compacted soil is less in comparison of loose soil indicating the increase in frictional resistance of soil due to DC.
4. After 4.0 m depth, there is no significant change in the strength of soil after the DC process. It indicates that dynamic compaction is ineffective after the depth of 4.0 m for this particular tamper weight and fall.
5. This influence range of DC depends on factors like height of drop of tamper, the weight of tamper, grid pattern and spacing, etc. The tamper used with the selected pattern can compact the soil up to the depth of 4.0 m.

References

1. Ames D (1998) Formation of the soils. Soil Survey of Jerome County and Part of Twin Falls County, Idaho, Natural Resources Conservation Service, USDA, p 238
2. Lukas RG (1995) Dynamic compaction. Geotechnical Engineering Circular No. 1, Federal Highway Report FHWA-SA-95-037, USA
3. Mayne PW, Jones JS, Dumas JC (1984) Ground response to dynamic compaction. J Geotech Eng 110(6):757–774
4. Rollins KM, Kim J (2010) Dynamic compaction of collapsible soil based on U.S. case studies J Geotech Geo-environ Eng 136(9):1178–1186

5. Pan JL, Selby AR (2002) Simulation of dynamic compaction of loose granular soils. Adv Eng Softw 33(7):431–440
6. Gu Q, Lee FH (2002) Ground response to dynamic compaction. Geotechnique 52(7):481–493
7. Ghanbari E, Hamidi A (2015) Improved parameters in dynamic compaction adjacent to slope. J Rock Mech Geotech Eng 2(2):233–236
8. Mostafa KF, Liang RY (2011) Numerical modelling of dynamic compaction in cohesive soil. Geo-Front Congr: Adv Geot Eng I:738–747
9. Menard L, Broise Y (1975) Theoretical and practical aspects of dynamic consolidation. Geotechnique 15(1):3–18

Numerical Modeling of Centrifuge Experiment on Vacuum Consolidation of Soft Clay

Sridhar Gangaputhiran

Abstract This paper presents the finite element (FE) modeling of centrifuge experiments on vacuum consolidation of soft clay deposits. FE analysis of centrifuge tests as a model scale was performed, and the comparison is made with the centrifuge tests modeled as a prototype. The effect of different material parameters namely unit weight of water and density, initial void ratio and permeability of clay at model and prototype were studied. FE analysis confirmed that the use of unit weight of pore water at 1-g as an input parameter results in an incorrect distribution of vertical effective stress. To correctly compute vertical effective stress, unit weight of water at *N*-g should be used. In order to model the centrifuge test exactly it would be appropriate to give the values of initial dry density, initial void ratio and the initial vertical stress corresponding to those at 1-g so that the final conditions after self-weight consolidation results in proper distribution of void ratio and other field variables. To correctly model the increase in seepage velocity with gravity in centrifuge tests, permeability must be increased by the same factor as the gravity. Four pore water pressure boundary conditions to model the vacuum pressure were analyzed through numerical runs. The numerical model at the model scale considering all factors analyzed earlier were used in modeling vacuum consolidation of soft clay. It was found that to accurately model the vacuum pressure, the vacuum pressure should be applied only at the top of the soil layer for accurate prediction of the field variables.

Keywords Finite element modeling · Centrifuge test · Vacuum consolidation · Soft clay · Ground improvement

1 Introduction

As the demand for land increases for urban development, the problematic soils-like soft clays having low-bearing capacity and high compressibility characteristics have to be used to support foundations of large-scale infrastructures. Vacuum preloading

S. Gangaputhiran (✉)
National Institute of Technology Karnataka, Surathkal, Mangalore, Karnataka 575025, India
e-mail: sridhar@nitk.edu.in

A. K. Choudhary et al. (eds.), *Advances in Geo-Science and Geo-Structures*,
Lecture Notes in Civil Engineering 154,
https://doi.org/10.1007/978-981-16-1993-9_16

along with PVDs is being used as a viable option nowadays to improve soft clay deposits and has several advantages over the conventional surcharge preloading [1]. This method was first introduced by Kjellman [2]. In this form of preloading, the vacuum pressure is applied to the pore phase through a sealed membrane system. The vacuum creates negative pore water pressure in the soil and causes water to drain out from the soil [3]. There is an increase in effective stress equal to the magnitude of the induced negative pore water pressure, without the increase in total stress.

Plane strain, axisymmetric and 3-D FE analysis of vacuum preloading problem have been carried out in the past by various researchers [4–6]. Literature review shows that vacuum pressure has been modeled numerically by application of equivalent surcharge pressure [4] or by applying negative pore water pressure at the top of the model only or by applying both at the top and along the drain [5]. The vacuum pressure is either considered constant along depth if there is no vacuum loss or considered linear with maximum value at the top of the drain and zero at the bottom of drain. Hence different ways of modeling the vacuum pressure have been proposed in literature and the comparison of different pore water pressure boundary conditions is scarce in literature. Therefore, modeling of the vacuum pressure requires further attention and is presented in this paper.

FE modeling of centrifuge tests involving variety of geotechnical problems including vacuum consolidation of soft clay have been carried out in the past [7, 8]. The FE models of centrifuge tests are either at model scale [8] or at prototype scale [7].The FE results obtained using prototype scale can be interpreted as such, while the FE results from model scale have to be extrapolated to prototype scale. In order to overcome this limitation, a FE procedure using ABAQUS Version 6.12 [9] to model and interpret results at model scale is presented in this paper.

2 Constitutive Model and Material Parameters

The stress–strain behavior of kaolin clay used in the present study is assumed to follow Modified Cam-Clay Model [10]. All the critical state parameters except M, for kaolin clay are obtained through laboratory testing. The parameters λ, κ, e_0 and N_0 were obtained from 1-D consolidation tests (Table 1). The initial values of void ratio are obtained from e versus $\ln p'$ for the given initial stress condition. The critical state parameter M is obtained as 0.9 [11]. The value of angle of internal friction is obtained as 23.5° from M through the relationship

$$M = \frac{6 \sin \phi'}{3 - \sin \phi'} \tag{1}$$

The empirical relationship given by Jaky [12] has been used to find K_{0NC}. The empirical relationship is

$$K_{0NC} = 1 - \sin \phi' \tag{2}$$

Table 1 Summary of material parameters used in FE analysis

Symbol	Parameter	Value
λ	Slope of 1-D CL on e-ln p'	0.197
κ	Slope of RCL on e-ln p'	0.062
N_0	Void ratio at $p' = 1$ kPa on 1-D line	2.825
M	Ratio q/p' at critical state	0.9
e_0	Initial void ratio	2.2719
K_{ONC}	Coefficient of earth pressure	0.6
Φ'	Critical state friction angle	23.5
γ	Poisson's ratio	0.34
k (m/s)	Permeability of clay	Equation 4

Based on the above relationship the value of K_{ONC} is obtained as 0.6. Poisson's ratio is obtained as 0.34 through the relationship

$$v = \frac{K_{ONC}}{(1 + K_{ONC})} \tag{3}$$

By falling head permeability test, the permeability of kaolin clay was determined at the end of each effective confining pressure of 1-D consolidation test. The permeability is assumed to be isotropic and is allowed to vary with void ratio based on the equation

$$k = 3.615 \times 10^{-8} * \ln(e) - 1.419 \times 10^{-8} \tag{4}$$

3 FE Modeling of Centrifuge Test

3.1 Finite Element Mesh

A two-dimensional plane strain finite element model is used in these studies. Five numerical runs (NR) were carried out. For NR 1 (represent prototype) the width of the mesh is 1 m and height of the mesh is 10 m. The FE mesh used in the NR 1 is shown in Fig. 1. The mesh consists of a 20 by 4 matrix of 8-node quadrilateral pore pressure elements with bi-quadratic displacement and bilinear pore-water pressure with reduced integration (CPE8RP). Dimensions of each element are identical (0.25 m × 0.5 m). The FE mesh used in the NR 2, 3, 4 and 5 (represent model) is shown in Fig. 1. The width of the mesh is 0.1 m and height of the mesh is 0.2 m. The mesh consists of a 10 by 5 matrix of 8-node quadrilateral pore pressure elements with bi-quadratic displacement and bilinear pore-water pressure with reduced integration (CPE8RP). Dimensions of each element are identical (0.01 m × 0.01 m).

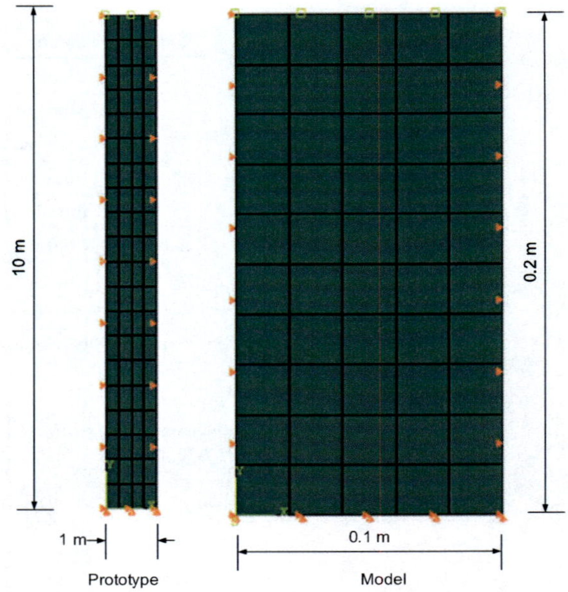

Fig. 1 Finite element mesh and boundary conditions for prototype and model

3.2 Boundary Conditions

The boundaries are laterally restrained by using a roller support. The base of the soil model is restrained in both vertical and horizontal direction used fixed support condition. The pore-water pressure boundary conditions on top of the FE mesh is always drained. In each numerical run, two vertical boundaries are either drained or undrained depending on the parameters to be studied. All the initial conditions namely effective stresses, pore water pressure and initial void ratios are set at 1-g condition.

3.3 Effect of Unit Weight of Pore Water

As mentioned earlier, NR 1 represent prototype scale with zero pore-water pressure maintained on the top of the mesh. Bottom of soil is undrained. At steady-state, it is expected that pore-water pressure follows a hydrostatic pore-water pressure distribution with zero pore-water pressure on the top of the clay. For a centrifuge test at 50 g, a layer of soil with a thickness of 0.2 m should model a prototype soil thickness of 10 m. NR 2 and NR 3 are two possible ways to model the centrifuge test at 50 g. Unit weight of pore water γ_w is 10 kPa in NR 2 while that in NR 3 is 500 kPa. If unit weight of pore water is automatically adjusted in ABAQUS in accordance to gravity load applied to soil elements, NR 2 should result in the same vertical effective stress distribution and pore-water pressure distribution as in NR 1. If unit weight of pore water is not automatically adjusted, NR 2 should result in a distribution of vertical effective stress and pore-water pressure different from as in NR 1.

The variations of pore water pressure along depth in NR 1 to NR 3 in prototype scale are shown in Fig. 2. The pore water pressure increased with depth in numerical NR 1 to NR 3, but the slope in NR 2 is only 1/50 of that in NR 1 and 3. As the pore-water pressure distribution is hydrostatic, the theoretical pore water pressure along the depth may be evaluated using the equation

$$P = \gamma_w * z \tag{5}$$

where z is the depth below soil surface and γ_w is the unit weight of pore water. For NR 1 (Prototype scale), the unit weight of pore water is 10 kN/m^3. Pore water pressure at the bottom of soil is 100 kPa as expected. If ABAQUS adjusts unit weight of water in accordance to variations in gravity automatically, NR 2 should adjust unit weight of pore-water at 50 g to 500 kPa automatically. From Fig. 2, it can be observed that the pore water pressure distribution in NR 2 is different from NR 1. The pore water pressure at the bottom of soil is only 2 kPa. Hence the unit weight of pore water is only 10 kPa, and is the value given in the ABAQUS input file. In contrast, in NR 3, unit weight of pore water is 500 kPa and hence the pore water pressure at the bottom of soil is 100 kPa. This implies that the unit weight of pore water is not adjusted automatically in accordance to change in the gravity load applied to soil elements. As a consequence of incorrect input of unit weight of pore water in NR 2, the vertical effective stress distribution is also getting affected as shown in Fig. 3. NR 1 and 3 results in identical distribution of vertical effective stress and is the same as the theoretical value. However, NR 2 results in a larger vertical effective stress than the theoretical value. The above analysis confirmed that the use of unit weight of

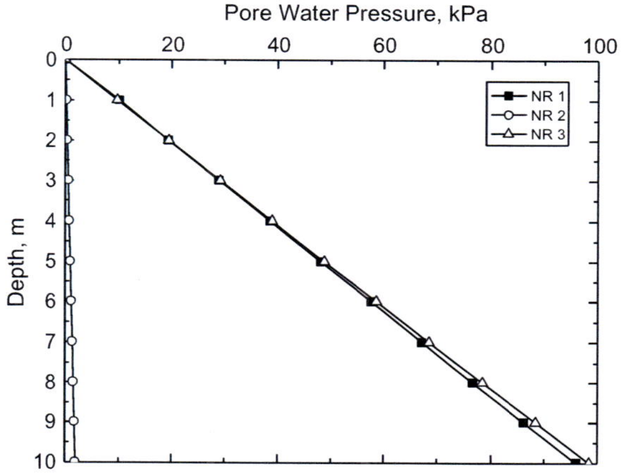

Fig. 2 Variation of pore water pressure along depth

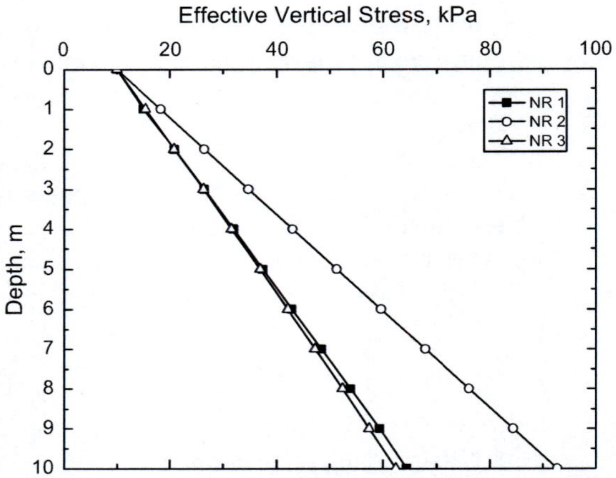

Fig. 3 Variation of effective vertical stress along depth

pore water at 1-g as an input parameter results in an incorrect distribution of vertical effective stress. To correctly compute vertical effective stress, unit weight of water at Ng should be used.

3.4 Effect of Density and Initial Void Ratio of Clay

NR 4 was carried out to study the effect of initial dry density, initial void ratio and initial stress condition at model scale. The initial dry density for the NR 3 is 0.8307 at the top and 0.9315 at the bottom of the clay and the corresponding initial stress is −10 kPa at the top of clay and −65 kPa at the bottom and the corresponding void ratio is 2.2519 and 1.8985. These initial conditions are similar to that of prototype and it results in zero settlement during gravity or self-weight consolidation. In NR 4 the initial dry density is 0.8307 at the top and 0.8307 at the bottom of the clay and the corresponding initial stress is −10 kPa at the top of clay and −11 kPa at the bottom of clay and the corresponding void ratio are 2.2519 and 2.2519, respectively. These initial conditions are that of model scale and it results in 13 mm settlement during gravity or self-weight consolidation, i.e., in prototype scale the settlement corresponds to 65 cm. Therefore, NR 4 results in settlement of 65 cm since the initial conditions are at 1-g, while the NR 3 results in almost zero settlement since the initial conditions of clay are those corresponding to 50-g. The variation of void ratio with depth for NRs 1, 3 and 4 are shown in Fig. 4. From Fig. 4, it can be observed that the void ratio distribution under NRs1 and 3 are almost similar since the initial conditions are those at the prototype scale, while for NR 4 the initial conditions are those corresponding to model scale. Therefore, in order to model the centrifuge test exactly it would be appropriate to give the values of initial dry density, initial void ratio and the initial vertical stress corresponding to those at 1-g so that the final

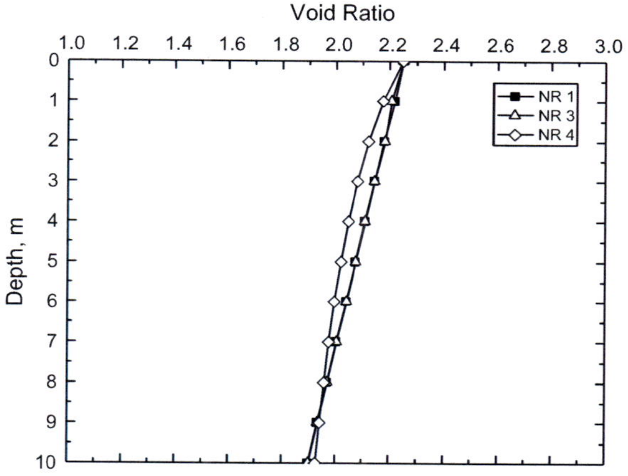

Fig. 4 Variation of void ratio along depth

conditions after self-weight consolidation results in proper distribution of void ratio and other field variables.

3.5 *Effect of Permeability of Clay*

In centrifuge model testing, the velocity of flow is N times greater in the model than the prototype [13], i.e., when gravity is increased by N times in a centrifuge model test, seepage velocity is increased by N times when compared with a model test carried out at 1-g. From ABAQUS, the seepage velocity is given by

$$v = -\frac{k}{\Delta L}\Delta\left(z + \frac{P_w}{\rho_w g}\right) \tag{6}$$

where, v is the seepage velocity, k is the permeability specified in ABAQUS input file, ΔL is the length of flow, z is the elevation, P_w is the pore water pressure, ρ_w is the unit weight of water, and g is the acceleration due to gravity. In Sect. 3.3, it was shown that the unit weight of pore water must be correctly specified in proportion to the acceleration field. To model a centrifuge test at N-g, unit weight of water must be specified as N times of unit weight of water at 1-g. Compared with a model at 1-g, pore water pressure P_w in a model at N-g is increased by N times. However, $\rho_w g$ is also increased by N times. The elevation z remains the same with gravity. Therefore, from Eq. 6, to correctly model increase in seepage velocity with gravity, permeability must be increased by the same factor as gravity. Numerical NR 5 was performed to check this aspect of permeability. The time settlement plot for NR 4 and NR 5 are shown in Fig. 5. From Fig. 5, it can be observed that the accelerated rate of consolidation in NR 5 compared with NR 4.

Fig. 5 Time settlement plot

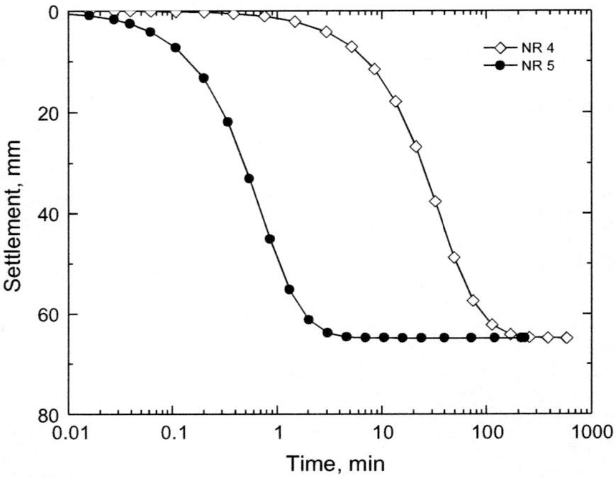

4 Effect of Boundary Conditions on Behavior of FE Model

As summarized in the literature review, the pore water pressure boundary conditions that should be given under vacuum preloading at the drainage boundaries is not comprehensively addressed in the literature. Hence an attempt has been made here to study the effect of giving different pore water pressure (vacuum pressure) at the drainage boundaries (top and/or along the drain). Four pore water pressure boundary conditions were analyzed through numerical runs. The refined model from the previous section, i.e., the numerical model at the model scale considering all the factors studied in the previous section is used for this study. The configuration for the numerical runs are detailed below.

(a) The BC-1 is a combination of 10 kPa seating pressure at the top drainage boundary of the soil during gravity and surcharge consolidation and 80 kPa surcharge pressure at the top drainage boundary of clay during surcharge preloading.
(b) The BC-2 is a combination of 10 kPa seating pressure at the top drainage boundary of the soil during gravity and vacuum consolidation and 80 kPa vacuum pressure at the top drainage boundary of clay during vacuum preloading.
(c) For BC-3 is a combination of BC-2 and constant 80 kPa vacuum pressure along the vertical drainage boundary (sides of the model) during vacuum preloading.
(d) For BC-4 is a combination of BC-2 and varying 80 kPa vacuum pressure (taking into account the hydrostatic pore water pressure distribution) along the vertical drainage boundary during vacuum preloading.

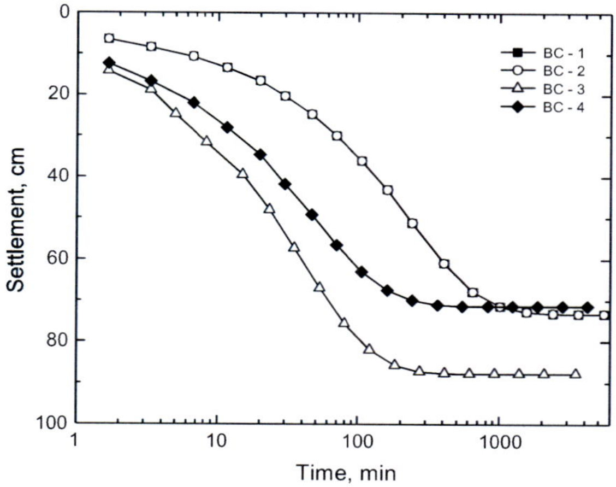

Fig. 6 Time- settlement response for different boundary conditions

4.1 *Results and Discussion*

The time settlement plot for different pore water pressure boundary conditions are shown in Fig. 6. The time settlement plot does not include the settlement under gravity and 10 kPa consolidation. From Fig. 6, it can be observed that the rate of consolidation and the total settlement for BC-1 and BC-2 are the same. This implies that the surcharge or vacuum pressure does not affect the rate of consolidation and the total settlement. It should be noted that here that the surcharge pressure application was instantaneous and not as step loading. The settlement under BC-1 and BC-2 is 73.3 cm. The BC-3 was studied to see the effect of the pore water pressure boundary conditions along the vertical drainage boundary without considering the hydrostatic pore water pressure. From Fig. 6, it can be observed that the rate of consolidation and the total settlement for BC-3 are higher compared to those from BC-1. This is due to the assumption that the vacuum pressure is constant along the vertical drainage boundary, which is incorrect since the hydrostatic pore water pressure should also be taken into account, while defining the pore water pressure boundary conditions along the vertical drainage boundaries. The settlement under BC-3 is 88 cm. The BC-4 was studied to see the effect of the pore water pressure boundary conditions along the vertical drainage boundary considering the hydrostatic pore water pressure. From Fig. 6, it can be observed that the total settlement for BC-3 is almost equal to those from BC-1. But the rate of consolidation is faster compared to those from BC-1. This is due to the assumption that the vacuum pressure is immediately developed at the drainage boundary. But in field, the soil takes times to stabilize to the final pore pressure values. The settlement under BC-4 is 72.4 cm.

The pore water distribution along the depth and at the end of consolidation for different pore water pressure boundary conditions are shown in Fig. 7. For BC-1, the thickness of clay is 8.62 m at the end of surcharge consolidation. Therefore, the hydrostatic pore water pressure at the end of consolidation and at the bottom

Fig. 7 Variation of pore water pressure along depth

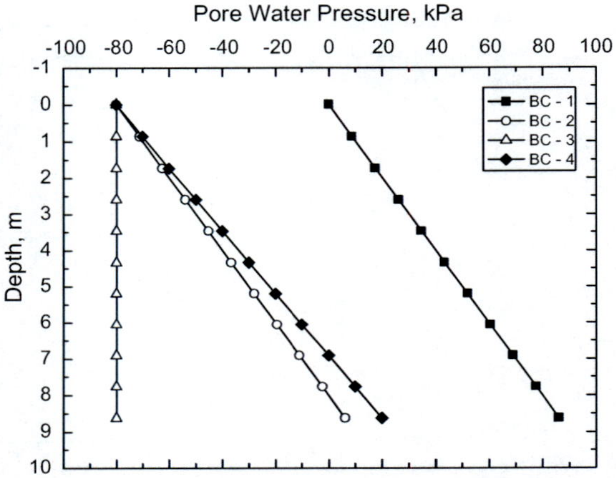

of the clay is 86.2 kPa, which is the value obtained for the BC-1 and it can be observed in Fig. 7. For BC-2 also, the thickness of clay is 8.62 m at the end of vacuum consolidation. Therefore, the hydrostatic pore water pressure at the end of consolidation and at the bottom of the clay is 6.2 kPa (taking in account −80 kPa due to vacuum pressure), which is the value obtained for the BC-2 and it can be observed in Fig. 7. For BC-3, the thickness of clay is 8.4 m at the end of vacuum consolidation. But, since we predefined a value of −80 kPa at the drainage boundary, the same can be observed from Fig. 7. For BC-4, the thickness of clay is 8.6 m at the end of vacuum consolidation. But, since we have predefined values pore pressures (taking into account, the hydrostatic pore water pressure at the end of gravity consolidation along with vacuum pressure) at the drainage boundary, the same can be observed from Fig. 7. Since the predefined values pore pressures are slightly different from the actual values, the difference can be observed between BC-4 and BC-2.

The effective vertical stress distribution along the depth and at the end of consolidation for different pore water pressure boundary conditions are shown in Fig. 8. For BC-1, the thickness of clay is 8.62 m at the end of surcharge consolidation. Since a total surcharge pressure of 90 kPa was applied at the top of the clay, the same can be observed from Fig. 8 at zero depth. The effective vertical stress at the bottom of the clay is 142.3 kPa which is reasonable if we assume an average saturated unit weight of 16 kPa over the depth of 8.62 m. For BC-2, the thickness of clay is 8.62 m at the end of surcharge consolidation. The effective vertical stress at the top and at the bottom of the clay are 90 kPa and 142.3 kPa, respectively. These are the values obtained for BC-1 as well. This implies that the constant surcharge or vacuum pressure at the top of the clay does not affect the effective vertical stress distribution in the soil. For BC-3, the effective vertical stress at the top and at the bottom of the clay are 90 kPa and 336 kPa, respectively. Also the entire effective vertical stress distribution is higher compared to BC-1 and BC-2. The higher effective vertical stress distribution in clay is due to the fact that we have predefined a value of −80 kPa at the drainage

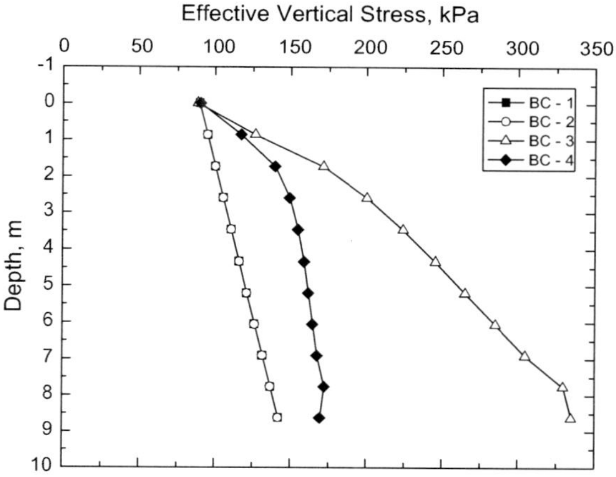

Fig. 8 Variation of effective vertical stress along depth

boundary without considering the hydrostatic pressure. The other reason may be due to non-uniform stress distribution under horizontal consolidation (vacuum pressure along the vertical drainage boundary, i.e., sides of the model). For BC-4, the effective vertical stress at the top and at the bottom of the clay are 90 kPa and 172 kPa, respectively. Also the entire effective vertical stress distribution is slightly higher compared to those from BC-1 and BC-2. The higher effective vertical stress distribution in clay is due to non-uniform stress distribution under horizontal consolidation (vacuum pressure along the vertical drainage boundary, i.e., along the side of the model). Hence, from above discussion we can conclude that too accurately model the vacuum pressure, the vacuum pressure should be applied only at the top of the soil layer for accurate prediction of the field variables.

5 Conclusion

A 2-D finite element analysis was performed to simulate the centrifuge test on vacuum consolidation of soft clay deposit. FE analysis of centrifuge tests as a model scale was performed and a comparison was made with the centrifuge tests modeled as a prototype. To correctly simulate centrifuge test at model scale, it was found that the N-g values should be used for unit weight of water and the permeability of pore water. Also, the values of initial dry density, initial void ratio and the initial vertical stress corresponding to those at 1-g should be used. An attempt was made to study the effect of giving different pore water pressure (vacuum pressure) at the drainage boundaries (top and/or along the drain). The numerical analysis at the model scale considering all factors analyzed earlier were used in modeling vacuum consolidation of soft clay. It was found that to accurately model the vacuum pressure, the vacuum pressure should be applied only at the top of the soil layer for accurate prediction of the field variables.

References

1. Gangaputhiran S, Robinson RG, Karpurapu R (2016) Properties of soil after surcharge or vacuum preloading. Proc Inst Civ Eng—Gr Improv 169(3):217–230. https://doi.org/10.1680/jgrim.15.00028
2. Kjellman W (1952) Consolidation of clay soil by means of atmospheric pressure. In: Proceedings, conference on soil stabilization MIT, Cambridge, pp 258–263
3. Mitchell JM, Jardine FM (2002) A guide to ground improvement, Report RP 373. Construction Industry Research and Information Association, London
4. Park CL, Jeong HJ, Park JB, Lee SW, Kim YS, Kim SJ (1997) A case study of vacuum preloading with vertical drains. In: DaS (ed) Proceedings of the 3rd international conference on ground improvement geosystems, London, pp 68–74
5. Indraratna B, Rujikiatkamjorn C, Sathananthan I (2005) Analytical and numerical solutions for a single vertical drain including the effects of vacuum preloading. Can Geotech J 42(4):994–1014
6. Rujikiatkamjorn C, Indraratna B, Chu J (2007) Numerical modelling of soft soil stabilized by vertical drains, combining surcharge and vacuum preloading for a storage yard. Can Geotech J 44(3):326–342. https://doi.org/10.1139/T06-124
7. Sharma JS, Bolton MD (2001) Centrifugal and numerical modelling of reinforced embankments on soft clay installed with wick drains. Geotext Geomembr 19(1):23–44. https://doi.org/10.1016/S0266-1144(00)00009-1
8. Hu L (2010) Consolidation behaviour of soft soil Subjected to on-land and underwater Vacuum preloading. Ph.D. The Hong Kong University of Science and Technology
9. Abaqus (2012) ABAQUS user's and theory manuals, Version 6.12. Hibbit, Karlsson and Sorensen, Inc., Rhode Island
10. Roscoe KH, Burland JB (1968) On the generalized stress-strain behaviour of wet clay. In: Engineering plasticity, Cambridge, pp 535–609
11. Goh TL (2003) Stabilization of an excavation by an embedded improved soil layer. Ph.D. thesis, National University of Singapore, Singapore
12. Jaky J (1944) The coefficient of earth pressure at rest. J Soc Hung Archit Eng, pp 355–358
13. Taylor RN (1995) Geotechnical centrifuge technology. Chapman and Hall, London

Deterministic and Probabilistic Analysis of Effects of Fibre Reinforcement on Strength and Deformation of Soil—A State-of-the-Art Review

Anasuya Goswami, Shailen Deka, and Arunav Chakraborty

Abstract Erection of a structure on poor quality soil is very risky because such soil is prone to low bearing capacity, high settlement and low shear strength. Soil reinforcement using fibres is a technique to achieve the desirable properties of soil. In this paper, an attempt has been made to review both deterministic and probabilistic approaches of analysis of soil reinforcement using different types of fibres. The existing literature review shows good agreement between deterministic and probabilistic approaches. The probabilistic approach, if properly modelled, can give easy methods of strength and deformation predictions of fibre reinforced soils.

Keywords Bearing capacity · Deterministic · Probabilistic · Deformation

1 Introduction

Erection of a structure on low strength soil is very risky because such soil is prone to low bearing capacity, high settlement and low shear strength. Soil reinforcement can be defined as a process to enhance some engineering properties of soil thereby increasing its strength, bearing capacity and stability. Fibres are considered to perform as effective soil reinforcing materials due to its ease of availability, better economy and easy adaptability.

Vidal introduced the use of fibres as soil reinforcing materials in modern times [1]. Fibres can be either randomly distributed in the soil as discrete reinforcing elements or systematically oriented in a particular direction, in layers or in the form of geosynthetics (geotextiles, geomembranes, geocells, geogrids, geonets, etc.). Although experimental analyses of soil reinforcement have been significantly reviewed in the past, limited review has been found in the existing literature on numerical and probabilistic approaches of soil reinforcement using fibres. Some natural and

A. Goswami (✉)
Department of Civil Engineering, Sonitpur Polytechnic, Sonitpur, Assam, India

S. Deka · A. Chakraborty
Department of Civil Engineering, Tezpur University, Tezpur, Assam, India

A. K. Choudhary et al. (eds.), *Advances in Geo-Science and Geo-Structures*, Lecture Notes in Civil Engineering 154,
https://doi.org/10.1007/978-981-16-1993-9_17

synthetic fibres and their effects on different types of soils are briefly described in the subsequent sections considering both deterministic and probabilistic approaches.

2 Literature Review

The deterministic approach helps to comprehend the computed experimental data while the probabilistic approach takes into account the uncertainties in the measured data. Both the approaches are discussed in this paper.

2.1 *Deterministic Approach*

Experimental Analysis of Randomly Distributed Fibre Reinforced Soil. The effect of reinforcement of a CL soil with three different randomly distributed synthetic fibres of 20 mm length was studied at fibre content of 1% by volume of soil—a spun nylon string unwound into a fine lint, polypropylene rope fibre and a polypropylene olefin concrete reinforcement fibre trade-named Fibre mesh on UCS of the soil, and increase in strength and toughness of the soil was observed [2]. Triaxial compression tests were performed on an SP-SM soil specimen reinforced with discrete, randomly distributed plastic fibres with fibre content varying from 1 to 4% by weight of soil at different aspect ratios (60–120), and it was inferred that the shear strength increased with an increase in aspect ratio and fibre content, the increase not being appreciable beyond 2% [3]. The effect of reinforcement of an SP soil with randomly distributed strips of reclaimed HDPE was studied at aspect ratios of 4–12 with strip contents from 0 to 4% by weight of soil and observed an increase in shear strength, CBR, secant modulus and resilient modulus in the reinforced soil [4]. The viability of using randomly distributed sisal and coconut fibres to reinforce a CL soil was studied by conducting compression test at an optimum water/soil ratio of 28%, and an increase in the compression strength and ductility of the reinforced soil was observed, the optimum fibre content being 4% in case of both the fibres [5]. UCS tests were conducted on combination of fly ash with sand and silt stabilised with cement alone and combination of cement and randomly distributed polyester fibres at aspect ratio 267, their contents being 3% and 1%, respectively, by combined dry weight of fly ash and soil, after different periods of curing, and an increase in the shear strength and UCS and decrease in brittleness of the fly ash soil specimen caused by the inclusion of the fibres was observed [6]. Compaction and triaxial compression tests were conducted on a CL soil randomly reinforced with sisal fibre at four different fibre contents ranging from 0.25 to 1% by weight of raw soil and four different lengths from 10 to 25 mm and a remarkable enhancement in the and shear strength of the reinforced soil was observed [7]. Reinforcement of an SP soil with randomly distributed reinforcement mesh increased the ultimate bearing capacity by almost 1.8 times [8]. The influence of inclusion 24 mm long polypropylene fibres on the behaviour of two SM and one SP soils at of 0.5% fibre by weight of soil by conducting bender element, ring and standard triaxial tests and enhancement in the ultimate strength was observed [9]. Silty sand soils were randomly reinforced with sisal fibres stabilised with cement or cactus pulp, and it was found that the reinforcement enhanced the tensile behaviour the unbaked soil. The durability of the structure is increased when the fibres are included in combination with the stabilising agents [10]. The effect of inclusion of a mixture of polypropylene fibre at 0.05–0.25% and lime at 2, 5, 8% by weight of soil on the properties of a CL soil was studied by conducting UCS, direct shear, swelling and shrinkage tests, and it was concluded that fibre-lime reinforced soil gains more strength as compared to only lime-stabilised soil [11]. Reinforcement of uncemented and cemented CL soil with randomly distributed short polypropylene fibres of average length 12 mm and 0.05–0.25% of fibre content and 5 and 8% of cement content by weight of soil showed an increase in the UCS, shear strength and axial strain at failure [12]. The influence of randomly distributed polyester fibres on fly ash–black cotton soil–lime mixtures was investigated at the optimum percentage contents obtained for lime and fly ash (8% and 15%, respectively) with 0 to 2% plain and crimped polyester fibres by total dry weight of the mixture by conducting UCS and split tensile strength tests, and it was inferred that the inclusion of the fibres caused an increase in the UCS and split tensile strength [13]. UCS tests were conducted on an SM soil randomly reinforced with 80 mm long coir fibres and 20 mm long polypropylene fibres at 0.5–2.0% fibre content by weight of dry soil at an aspect ratio of 400, and it was concluded that inclusion of both the fibres increased the UCS value of the soil, the optimum fibre content of coir and polypropylene being 0.75% and 1%, respectively. Also, the performance of coir fibre was found to better than synthetic fibre [14]. The influence of inclusion of randomly distributed polypropylene fibres of length 24 mm to an artificially cemented SM soil was studied at a fibre content of 0.5% by weight of the sum of dry soil and cement, and an increase in the UCS of the reinforced cemented soil was observed for all considered cement contents [15]. Triaxial compression tests were conducted on an SM soil randomly reinforced with oil palm empty fruit bunch (OPEFB) fibres of lengths ranging from 15 to 45 mm at 0.25 and 0.5% fibre content by weight of dry soil, and an increase was observed in shear strength of the reinforced soil. Also, there was an increase in shear strength when the soil is reinforced with coated OPEFB fibres than with uncoated fibres [16]. A CL soil was reinforced with of 4.0 mm long randomly distributed nylon fibres at an aspect ratio of 2 with 10–30% of fibre, and it was observed that the addition of the fibres cause remarkable increase in shear strength of the soil [17]. Direct shear, UCS and CBR tests were conducted on a CL soil randomly reinforced with polypropylene fibres with aspect ratios from 75 to 125 and fibre content varying from 0 to 1% by the weight of dry soil, and increase in peak and residual shear strength, UCS and CBR value of the reinforced soil was observed, with the optimum fibre content being 0.4 to 0.8% at an aspect ratio of 100 [18]. The advantages of using randomly distributed wheat straw treated with modified polyvinyl alcohol to reinforce a saline clayey soil were presented by conducting UCS and triaxial tests on the soil samples [19]. The effect of fibre content, fibre aspect ratio and confining stress on shear strength two CL and one OH soils reinforced with randomly distributed coir fibres was studied at aspect ratios of 50–150 and fibre contents of 0.5–2% by weight of soil, and it was inferred that shear strength of the reinforced clays was sufficiently increased [20]. The effect of including randomly distributed palm fibres of lengths 20 and 40 mm at fibre contents of 0.5–2.0% by weight of soil on the CBR of an SP soil was studied, and a significant increase was found in the CBR strength of the reinforced sand specimens, with an optimum amount being 1% of 40 mm long fibres [21]. The influence of reinforcing of kaolin clay composite with paper was studied by considering paper length from 15 to 40 mm and paper content as 5 and 10% by weight of soil, and it was inferred that an increase in paper content and length caused increase in OMC and small decrease in MDD [22]. The mechanical properties of a silty clay randomly reinforced with sisal fibres of lengths—5 to 15 mm at 0.5–1.5% fibre content by weight of dry soil were studied by conducting triaxial shear tests, and an increase of 20% in the strength of the reinforced soil at a fibre length and content of 10 mm and 1.0%, respectively, was observed as compared to that of the unreinforced soil [23]. Two types of synthetic fibres, Fibre Mesh® and Fibre Cast® of lengths 6 and 12 mm and fibre content varying from 0.2 to 0.6%, were evaluated for randomly distributed reinforcement of a CH soil stabilised with lime, and it was observed that the enhancement in CBR usually increased with increase in both fibre contents and lengths, and the effect was significant on addition of lime [24]. Extensive experimental works have been carried out on soil randomly reinforced with both natural and synthetic fibres, and it can be seen that the addition of fibres improve the strength and deformation characteristics of soil and that fibre content, length and aspect ratio and use of admixture are the governing factors in such improvement.

Numerical Analysis of Randomly Distributed Fibre Reinforced Soil. Numerical analysis involves various techniques like finite element method (F.E.M.) and finite difference method (F.D.M.). Numerical simulations for analysing the behaviour of randomly distributed coir fibre reinforced sand were performed using FLAC^{3D}, and results indicated that the reinforcement caused the stress concentration to be more diffuse and restricted the shear band formation [25]. A model using FLAC^{3D} and numerical simulations using results of triaxial compression tests on randomly distributed coir fibre reinforced clay specimens were developed with fibre content varying from 0 to 2% at confining pressures from 50 to 150 kPa, and it was concluded that the experimental results and numerical simulations were in close agreement and that reinforcing the soil with fibres improved the strength of the soils [26]. Modelling of footing resting on randomly distributed polyester fibre reinforced CH soil was done using PLAXIS 2D, and good correlation was observed between the analytical and experimental results that showed a remarkable improvement in the bearing capacity of the reinforced soil [27]. Comparison of unreinforced soil and soil reinforced with randomly distributed fibre using PLAXIS 2D showed that fibre reinforcement had considerable influence in enhancing the soil properties. The settlement and lateral displacement of laterally loaded pile was found to be reduced [28]. Chemically treated randomly distributed coir fibres and lime-treated OH soil was found to perform better as a pile-supported earth platform by using ABAQUS. Also, the numerical analysis and experimental results were in close agreement [29]. With the advancement of computational methods, methods like F.E.M. are used extensively to study the effect of soil reinforcement using randomly distributed fibres, and it can be seen that the results obtained by such methods are in good agreement with those obtained using experimental results.

Experimental Analysis of Systematically Oriented Fibre Reinforced Soil. Laboratory model tests were conducted on SP soil reinforced with systematically placed metal strips in layers and with an increase in the strip length, the bearing capacity ratio was found to increase [30]. Performances of various types of reinforcement in sand beds under strip loading were compared, and the geocell reinforcement was found to be the most superior method of soil reinforcement among those that were investigated [8]. The potential of using woven coir geotextiles for strengthening the highly compressible waterfront clay dykes was investigated, and they were found to act satisfactorily as reinforcing material [31]. Vegetable fibre geotextiles improved the properties soft compressible ground for the construction of embankments by increasing the tensile strength [32]. UCS tests were performed on fibre reinforced cemented sand by including PVA fibres of length 12 mm in different number of layers, and an enhancement in the strength of the reinforced soil was caused by increasing the number of fibre layers [33]. The utilisation of woven coir geotextiles to reinforce a double-layer pavement section was studied, and an increase in the bearing capacity of thin sections was observed [34]. A study was conducted on sand reinforced with woven coir geotextile, and it was observed that even one layer of the geotextile increased the strength by three times [35]. The use of different forms of geosynthetics as soil reinforcing material has studied in published works, and they are found to successfully improve the engineering properties of soil, the material of the geosynthetics used, number of layers and their placement being important factors governing the results.

Numerical Analysis of Systematically Oriented Fibre Reinforced Soil. The feasibility of using a layer of woven polypropylene geotextile as a reinforcement at base of an embankment on soft marine deposits was studied using F.E.M. and an increase in the factor of safety (F.O.S.) at the local position, under reinforcement, and not in the overall F.O.S., and a decrease the maximum settlement was caused due to reinforcement [36]. A geogrid reinforced soil slope was modelled using PLAXIS-2D, and the internal angle of friction of the backfill soil was found to be the significant variable that influenced the F.O.S. of the reinforced slope [37].

2.2 *Probabilistic Approach*

Randomly Distributed Fibre Reinforced Soil. The effect of fibre and soil characteristics, density and confining stress on the shear strength of the reinforced soils was studied by conducting triaxial compression tests on cohesionless soils randomly reinforced with discrete, randomly distributed plastic, coir and bhabar fibres of varying fibre content (0–4%) and aspect ratio (50–125) at confining stress of 50–400 kPa;

and a mathematical model was developed to derive the effect of these factors by conducting regression analysis of the test results. The shear strength was seen to remarkably increase, and the increase was found to be a function of aspect ratio, fibre weight fraction and grain size of soil [38]. A consistency between the shear parameters of a sandy soil randomly reinforced with coir fibres predicted by methods such as discrete frame-work, regression analysis, and energy-based homogenisation was observed [39]. An artificial neural network (ANN) model was developed based on results of UCS tests conducted on reinforced lightweight soil (RLS) comprising of dredged clayey soil, waste fishing net, air-foam and cement, air-foam at five different water contents that could fairly estimate the UCS of RLS at a given mixing ratio [40]. Polypropylene fibres were considered to study the improvement of the crack resistance of an initially saturated CL soil specimen subjected to desiccation, by random reinforcement of the fibres. Linear regression analysis showed a remarkable reduction in the quantity of desiccation cracks as a result of fibre reinforcement [41]. Regression models were developed for approximate prediction of shear strength of two CL and one OH soils reinforced with randomly distributed coir fibres at aspect ratios of 50–150 and fibre contents of 0.5–2% by weight of soil based on consolidated undrained triaxial compression tests by considering soil characteristics, aspect ratio, fibre content and confining stress as the input parameters [20]. Two different types of synthetic fibres, Fibre Mesh® and Fibre Cast® of lengths 6 and 12 mm and fibre content varying from 0.2 to 0.6%, were evaluated for randomly distributed reinforcement of a CH soil stabilised with lime, and it was observed that the probabilistic approach determined the effect of fibre content and length in CBR strength [24]. Analysis of variance (ANOVA) conducted on artificially cemented SC soil randomly reinforced with monofilament polypropylene fibres of content 0 and 0.50% by weight of the sum of dry soil and cement and lengths 6–24 mm showed that the porosity, fibre length and cement content influenced the computed changes in UCS and split tensile values remarkably [42]. Multi-gene genetic programming (MGGP) was used to develop relationships between UCS of a silty sand randomly reinforced with fibres extracted from water hyacinth and fibre content, soil density, soil moisture and unreinforced soil strength as the inputs, and it was found that the MGGP model could satisfactorily infer the UCS of the reinforced sand beyond the given input range [43]. With time, probabilistic approach is gaining popularity in analysis of soil reinforcement using different types of randomly distributed fibres, as unlike the deterministic approach, it takes into account the uncertainties in the measured data. The probabilistic approach, if properly modelled, can give easy methods of strength and deformation predictions of fibre reinforced soils.

Systematically Oriented Fibre Reinforced Soil. Reliability analysis procedure was found to be successfully applied for analysis of retaining walls reinforced with geogrids [44]. An equation to obtain the enhancement in strength of sand reinforced with woven coir geotextile was presented using multiple linear regression analyses [35]. Limited review has been found in existing literature on numerical analysis of soil reinforcement using geosynthetics.

3 Conclusion

There are a lot of experimental as well as numerical studies on fibre reinforced soils. But very little information exists on probabilistic studies on their strength and deformation behaviour. The existing literature review shows good agreement between deterministic and probabilistic approaches. There is ample scope of study in this area. The probabilistic approach, if properly modelled, can give easy methods of strength and deformation predictions of fibre reinforced soils.

References

1. Vidal H (1969) The principle of reinforced earth. Highw Res Rec 1–16
2. Freitag DR (1986) Soil randomly reinforced with fibers. J Geotech Eng 112:823–826
3. Ranjan G, Vasan RM, Charan HD (1994) Behaviour of plastic-fibre-reinforced sand. Geotext Geomembr 13:555–565
4. Benson CH, Khire MV (1994) Reinforcing sand with strips of reclaimed high-density polyethylene. J Geotech Eng 120:838–855
5. Ghavami K, Filho RDT, Barbosa NP (1999) Behaviour of composite soil reinforced with natural fibres. Cem Concr Compos 21:39–48
6. Kaniraj SR, Havanagi VG (2001) Behavior of cement-stabilized fiber-reinforced fly ash-soil mixtures. J Geotech Geoenviron Eng 127:574–584
7. Prabakar J, Sridhar RS (2002) Effect of random inclusion of sisal fibre on strength behaviour of soil. Constr Build Mater 16:123–131
8. Dash SK, Rajagopal K, Krishnaswamy NR (2004) Performance of different geosynthetic reinforcement materials in sand foundations. Geosynth Int 11:35–42
9. Heineck KS, Coop MR, Consoli NC (2005) Effect of microreinforcement of soils from very small to large shear strains. J Geotech Geoenviron Eng 131:1024–1033
10. Mattone R (2005) Sisal fibre reinforced soil with cement or cactus pulp in bahareque technique. Cem Concr Compos 27:611–616
11. Cai Y, Shi B, Ng CWW, Tang C (2006) Effect of polypropylene fibre and lime admixture on engineering properties of clayey soil. Eng Geol 87:230–240
12. Tang C, Bin S, Gao W, Chen F, Cai Y (2007) Strength and mechanical behavior of short polypropylene fiber reinforced and cement stabilized clayey soil. Geotext Geomembr 25:194–202
13. Kumar A, Walia BS, Bajaj A (2007) Influence of fly ash, lime, and polyester fibers on. J Mater Civ Eng 19:242–248
14. Chauhan MS, Mittal S, Mohanty B (2008) Performance evaluation of silty sand subgrade reinforced with fly ash and fibre. Geotext Geomembr 26:429–435
15. Consoli NC, Bassani MAA, Festugato L (2010) Effect of fiber-reinforcement on the strength of cemented soils. Geotext Geomembr 28:344–351
16. Ahmad F, Bateni F, Azmi M (2010) Performance evaluation of silty sand reinforced with fibres. Geotext Geomembr 28:93–99
17. Estabragh AR, Bordbar AT, Javadi AA (2011) Mechanical behavior of a clay soil reinforced with nylon fibers. Geotech Geol Eng 29:899–908
18. Pradhan PK, Kar RK, Naik A (2012) Effect of random inclusion of polypropylene fibers on strength characteristics of cohesive soil. Geotech Geol Eng 30:15–25
19. Li M, Xi S, Yuan H, Pu H, Wei L (2012) Feasibility of saline soil reinforced with treated wheat straw and lime. Soils Found 52:228–238

20. Maliakal T, Thiyyakkandi S (2013) Influence of randomly distributed coir fibers on shear strength of clay. Geotech Geol Eng 31:425–433
21. Sarbaz H, Ghiassian H, Heshmati AA (2013) CBR strength of reinforced soil with natural fibres and considering environmental conditions. Int J Pavement Eng
22. Chegenizadeh A, Nikraz PH (2013) Paper reinforcement and soil. Adv Mater Res 608–609:1741–1745
23. Wu Y, Li Y, Niu B (2014) Assessment of the mechanical properties of Sisal fiber-reinforced silty clay using triaxial shear tests. Sci World J 2014:1–9
24. Moghal AAB, Chittoori BCS, Basha BM (2017) Effect of fibre reinforcement on CBR behaviour of lime-blended expansive soils: reliability approach. Road Mater Pavement Des 19:690–709
25. Babu GLS, Vasudevan AK, Haldar S (2008) Numerical simulation of fiber-reinforced sand behavior. Geotext Geomembr 26:181–188
26. Babu GLS, Chouksey SK (2010) An international model for analysis of fiber-reinforced clayey soil. Geomech Geoengin 5:277–285
27. Maheshwari K, Desai AK, Solanki CH (2011) Application and modeling of fiber reinforced soil. In: Proceedings of Indian geotechnical conference, pp 1–4
28. Swarnalatha V, Divya P (2013) Improving the engineering response of subsoil by fibre reinforcement technology. Int J Eng Res Technol 2:3044–3048
29. Anggraini V, Asadi A, Huat BBK, Nahazanan H (2015) Performance of chemically treated natural fibres and lime in soft soil for the utilisation as pile-supported earth platform. Int J Geosynth Gr Eng 1:1–14
30. Fragaszy RJ, Lawton E (1984) Bearing capacity of reinforced sand subgrades. J Geotech Eng 110:1500–1507
31. Lekha KRÃ, Kavitha V (2006) Coir geotextile reinforced clay dykes for drainage of low-lying areas. Geotext Geomembr 24:38–51
32. Mwasha A (2009) Using environmentally friendly geotextiles for soil reinforcement: a parametric study. Mater Des 30:1798–1803
33. Park S-S (2009) Effect of fiber reinforcement and distribution on unconfined compressive strength of fiber-reinforced cemented sand. Geotext Geomembr 27:162–166
34. Subaida EA, Chandrakaran S, Sankar N (2009) Laboratory performance of unpaved roads reinforced with woven coir geotextiles. Geotext Geomembr 27:204–210
35. Vinod P, Bhaskar AB (2012) Model studies on woven coir geotextile-reinforced sand bed. Gr Improv 165:53–58
36. Zhang N, Shen S-L, Wu H-N, Chai J-C, Xu Y-S, Yin Z-Y (2015) Evaluation of effect of basal geotextile reinforcement under embankment loading on soft marine deposits. Geotext Geomembr 43:506–514
37. Thongpetch I, Chmoye W (2019) Behavior of reinforced soil slope by finite element methods. Int J Eng Res Technol 8:159–163
38. Ranjan G, Vasan RM, Charan HD (1996) Probabilistic analysis of randomly distributed fiber-reinforced soil. J Geotech Eng 122:419–426
39. Babu GLS, Vasudevan AK (2007) Evaluation of strength and stiffness response of coir-fibre-reinforced soil. Gr Improv 11:111–116
40. Park HI, Kim YT (2012) Prediction of strength of reinforced lightweight soil using an artificial neural network. Eng Comput Int J Comput Eng Softw 28:600–615
41. Tang C, Shi B, Cui Y, Liu C, Gu K (2012) Desiccation cracking behavior of polypropylene fiber—reinforced clayey soil. Can Geotech J 49:1088–1101
42. Festugato L, Menger E, Benezra F, Kipper E, Consoli NC (2017) Fibre-reinforced cemented soils compressive and tensile strength assessment as a function of filament length. Geotext Geomembr 45:77–82
43. Vardhan H, Bordoloi S, Garg A, Garg A, Sreedeep S (2017) Compressive strength analysis of soil reinforced with fiber extracted from water hyacinth. Eng Comput 34:330–342
44. Sayed S, Dodagoudar GR, Rajagopal K (2010) Finite element reliability analysis of reinforced retaining walls. Geomech Geoengin An Int J 5:187–197

Stability Prediction of a Two-Layered Soil Slope

Moumita Goswami and Arunav Chakraborty

Abstract In geotechnical engineering, it is always a herculean task and also a topic of worry to carry out the stability analysis of slopes, since a number of slope failures and heavy settlements are occurring causing enormous destruction of mankind and property. In this paper, two-layered soil slopes with diverse soil combinations are used to predict the factor of safety (FOS) using statistical and machine learning techniques. The research is then taken to a different level to develop some design charts based on finite element method (FEM). Finally, the prediction models and design charts are validated by comparing the results with independent data sets which are not used during the construction of the model. The results obtained by ANN are in close agreement with those obtained by FEM having a higher R value of 0.92 and lesser MSE and MAE values of 0.08 and 0.04, respectively, compared to other methods.

Keywords Geotechnical engineering · Slope stability · Finite element method · Machine learning

1 Introduction

Stabilization of slopes is a very tough task and also a major concern, since a number of slope failures have occurred in the recent past causing destruction of life and property. The stabilization of slopes is an engineering dilemma which involves different type of parameters, and their impact can be fully tacit by the use of well-known machine learning techniques viz., regression analysis, neural networks, fuzzy logic, etc. These well-known methods can act as a strong tool for modeling. The practice of using finite element method (FEM) and limit equilibrium method (LEM) is tremendously growing in stability analysis of slopes. Due to rapid development in computer programs, FEM has gained very high demand over the conventional methods, particularly in problems of slope stability. The FOS values obtained by FEM give higher values compared to LEM [1–4]. Recently, the practice of using

---

M. Goswami · A. Chakraborty (✉)
Department of Civil Engineering, Tezpur University, Tezpur, Assam, India

A. K. Choudhary et al. (eds.), *Advances in Geo-Science and Geo-Structures*,
Lecture Notes in Civil Engineering 154,
https://doi.org/10.1007/978-981-16-1993-9_18

LEM and FEM to study the behavior of homogenous and layered soil slope is gaining high popularity. It has been observed that the failure surface formed will remain at the top layer if the foundation soil layer is stiffer than the upper layer [5–7]. Many researches were carried out to develop stability charts derived from FEM which allow the users to estimate the FOS to a higher degree of precision [8–11]. The conventional methods of stability analysis of slopes failed to describe the uncertainties present in the soil slope. Hence, the probability concept is getting more popular. Advanced slope stability methods have been established using various statistical and machine learning techniques like fuzzy logic, neural networks, optimization, etc. [12–18].

The key intention of this study has been organized as follows.

Firstly, a number of two-layered artificial slopes with different combination of soil properties and varying slope parameters viz., height of the slope (H) ranging from 10 to 35 m, shear parameter, cohesion (c) ranging from 5 to 60 kN/m^2, angle of internal friction (φ) ranging from 10° to 35°, slope inclinations (β) of 30°, 40°, and 50° are considered. Secondly, models are built up using MLR, MNLR, and ANN. Thirdly, separate design charts are prepared using the same values of data sets as used for model building process. Fourthly, by conducting the error analysis, i.e., mean square error (MSE) and mean absolute error (MAE), it has been computed for cases which are not used during model building process. Lastly, a comparison has been done between all the different methods used for the slope stability prediction.

2 Methodology

Using FEM, 620 artificial two-layered slopes having different geometrical and shear parameters are analyzed to determine the FOS. For developing the prediction models using MLR, MNLR, and ANN and to prepare the design charts, the analytical FOS values for 600 cases are used. The proposed models consist of seven input parameters. They are inclination of the slope (β), height of the slope (H), ratio of slope height to layer thickness (d/H), cohesion of layer 1 (c_1), cohesion of layer 2 (c_2), average angle of internal friction (φ_a), and average unit weight of the soil (γ_a). The FOS, on the other hand, is used as the output parameter.

Microsoft Excel 2010 has been used for developing multiple linear and nonlinear regression model.

The ANN model has been prepared in MATLAB 2011a, where 15 different network architectures are constructed considering 1 to 15 numbers of hidden layers and the error analysis for each network has been performed. Out of the 15 different network architecture, the one having architecture, $7-8-1$ (Fig. 1) is found to have lowest MSE and MAE values, and hence, this prediction model has been selected. Cross-validation technique is used by dividing the 600 data sets into training (80% data), testing (10% data), and validation set (10% data). The network is trained up using Levenberg–Marquardt method until the training error becomes negligible or the regression coefficient (R) of training, testing, and validation approaches to unity. If a

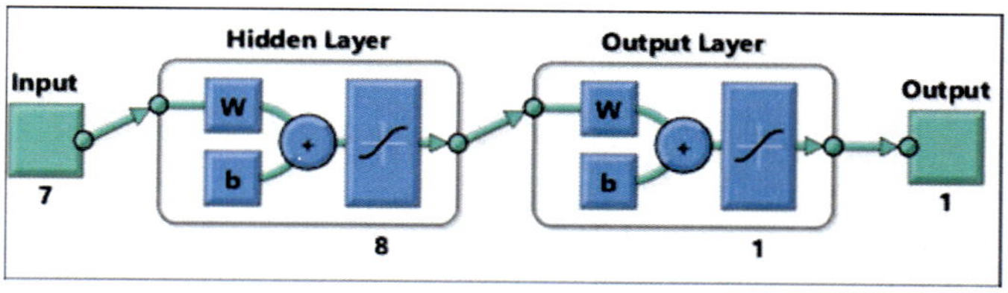

Fig. 1. 7 – 8 – 1 network architecture

proposed model accords $R > 0.8$, a strong relationship develops among the predicted values and the target values [19].

Further, design charts for stability analysis of soils are established by comparing various trendlines and data forecasting method. These design charts enable the user to arrive at a stable slope condition without going through long conventional iterative methods. Three different design charts for cohesive soils are prepared by considering two-layered artificial soil slopes having different values of slope angle. For preparing the design charts, the same 600 data sets are considered which are used for the preparation of MLR, MNLR, and ANN models.

For validation of the network models and design charts, 20 numbers of cases are considered which are not used during the development of the model, and finally, the results are compared.

3 Results and Discussions

The subsequent sections present the test results.

3.1 Multiple Linear Regression (MLR)

The test results of MLR for 600 artificial two-layered slope cases are given in Table 1. The value of R is found to be 0.76.

3.2 Multiple Nonlinear Regression (MNLR)

The test results of MNLR for 600 artificial two-layered slope cases are given in Table 2. The value of R has found to be 0.77.

Table 1 Test results of MLR for 600 artificial slope cases

Regression statistics				
R			0.758	
Observations			600	
Stability parameters	Coefficients	Standard error	t stat	p-value
Intercept	−0.154	0.243	−0.632	0.528
β	−0.003	0.001	−4.574	0.000
H	−0.002	0.001	−3.268	0.001
d/H	−0.266	0.025	−10.810	0.000
c_1	0.013	0.001	11.207	0.000
c_2	0.009	0.001	7.695	0.000
φ_a	0.042	0.002	18.251	0.000
γ_a	0.037	0.014	2.691	0.007

Table 2 Test results of MNLR for 600 artificial slope cases

Observations	600
R	0.774
MSE	0.015
Parameter	**Value**
pr1	−7.001
pr2	−0.013
pr3	−0.007
pr4	−0.823
pr5	0.016
pr6	−0.003
pr7	0.052
pr8	0.894
pr9	0.000
pr10	0.000
pr11	0.555
pr12	0.000
pr13	0.000
pr14	−0.001
pr15	−0.024

3.3 Artificial Neural Network (ANN)

The test results of ANN having R value for training, testing, and validation are shown in Fig. 2. From the regression data, it is seen that R value is 0.92 which is close to

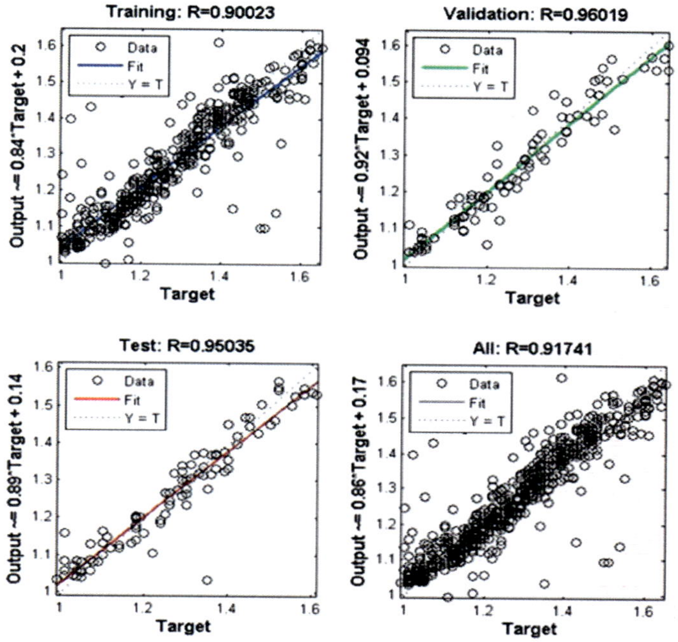

Fig. 2 Regression plot for the network model

unity, and hence, the prediction results provide a closer relationship with the input variables.

3.4 *Design Charts*

The design charts obtained for layered soil slope having slope angles 30°, 40°, and 50° are presented in Figs. 3, 4, and 5, respectively. Here, the results are represented as a function of $F/\tan(\varphi_a)$ versus $c_a/\gamma_a H\tan(\varphi_a)$ for d/H ratios ranging from 0.2 to 0.8.

A comparison has been made with the results of regression coefficient (R) with the prediction methods viz., ANN, MLR, MNLR, and design chart as shown in Fig. 6, and it is evident that the prediction model developed by ANN is found to have higher regression coefficient of 0.92 as compared to MLR, MNLR, and design chart having only 0.76, 0.77, and 0.89, respectively.

The network models are then validated by using 20 numbers of slope cases. The stability of the prediction models are thereafter checked for error analysis. From Fig. 7, it can be observed that the MSE values for ANN and design chart are low

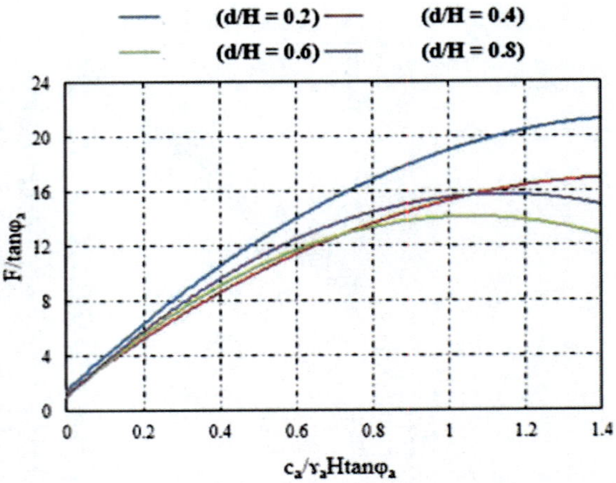

Fig. 3 Design chart for slope angle 30°

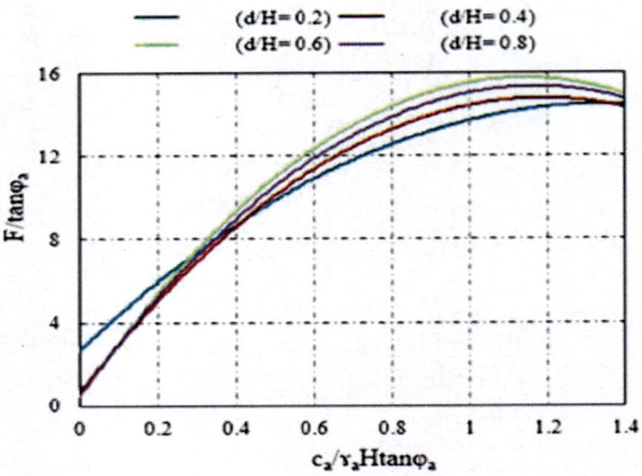

Fig. 4 Design chart for slope angle 40°

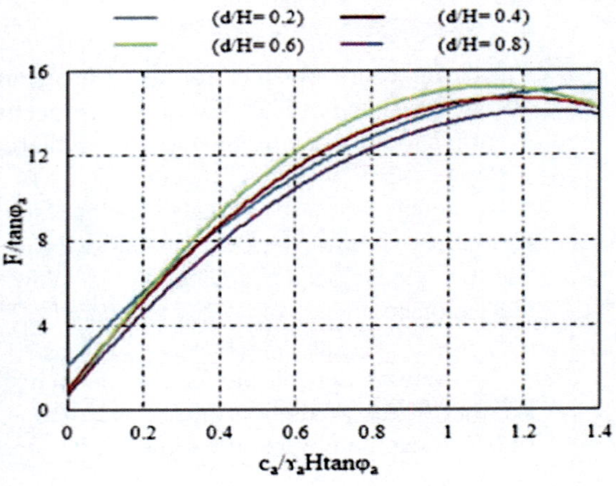

Fig. 5 Design chart for slope angle 50°

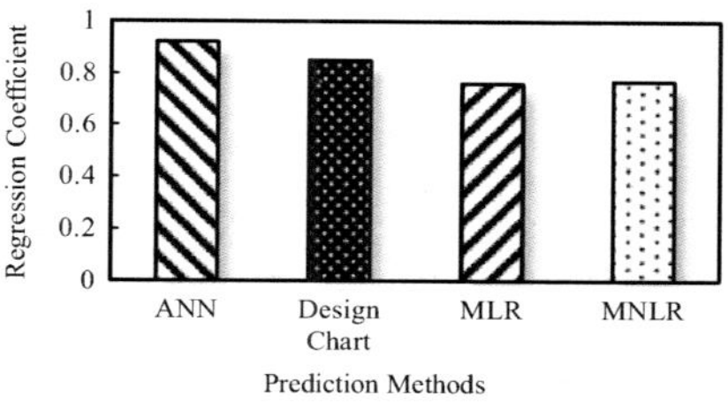

Fig. 6 Regression coefficient for different prediction methods

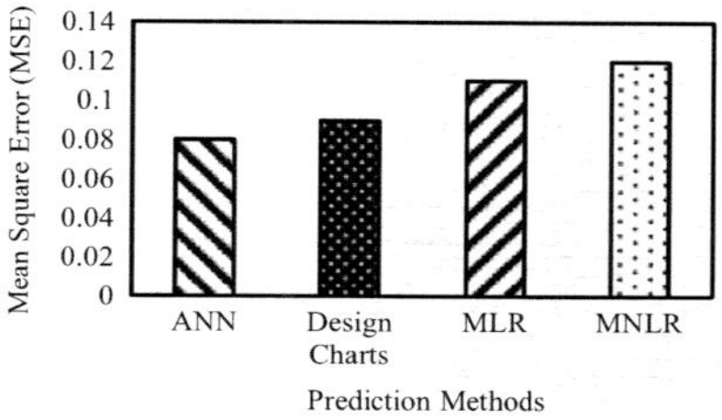

Fig. 7 MSE for different prediction methods

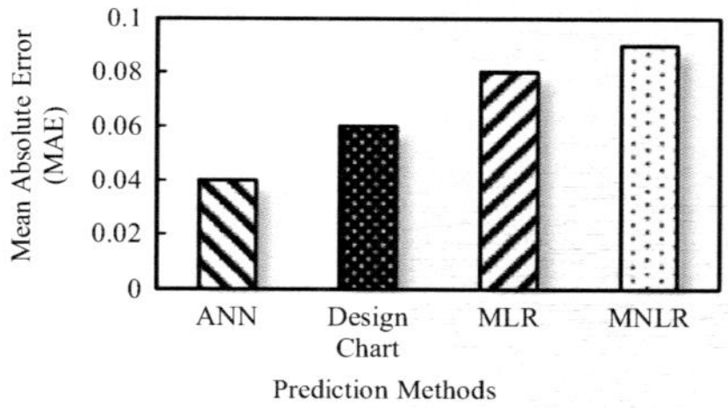

Fig. 8 MAE for different prediction methods

compared to MLR and MNLR, and similarly from Fig. 8, MAE value of ANN is found to be very low compared to design chart, MLR, and MNLR.

The ability of ANN to learn and model the nonlinear and complex relationships is very important because most of the relationships between inputs and outputs in real-life are nonlinear as well as complex. Unlike the other prediction techniques, ANN does not impose any restrictions on the input variables like how they should be distributed. Additionally, many studies showed that ANNs can better model heteroskedasticity. However, regression analysis is subjected to overfitting of the model, and the method is mainly limited to linear regressions only. Hence, it can be inferred that ANN is a better method, and the target values can be predicted with higher degree of accuracy.

4 Conclusions

In this paper, a study has been conducted for prediction of FOS of two-layered soil slopes having different combination of soil properties using MLR, MNLR, and ANN. The level of research is then boosted to develop some design charts based on FEM. The study involves the development of the prediction models by analyzing 600 artificial two-layered slopes using FEM. The models are validated by comparing the predicted results with the analytical results of 20 artificial case studies which are not used during the development of model. From the results discussed and presented, the following conclusions are drawn:

- MLR, MNLR, ANN, and design charts can be used to predict the stability performance of slopes.
- ANN is found to have higher R value of 0.92 as against design charts (0.85), MLR (0.76), and MNLR (0.77).
- ANN shows comparatively less MSE and MAE values of 0.08 and 0.04, respectively, compared to design charts (0.09, 0.06), MLR (0.11, 0.08), and MNLR (0.12, 0.09).
- Design charts may be utilized to arrive at a safe inclination of slope based on the desired magnitude of FOS. Moreover, the height of the slope can also be decided on the basis of the desired value of FOS.

References

1. Sazzad M, Moni M (2017) Stability analysis of slopes for homogeneous and layered soil by FEM. J Eng Sci 08(1):51–62
2. Baba K, Bahi L, Ouadif L, Akhssas A (2012) Slope stability evaluations by limit equilibrium and finite element methods applied to a railway in the Moroccan Rif. Open J Civ Eng 2:27–32
3. Griffiths DV, Lane PA (1999) Slope stability analysis by finite elements. Geotechnique 49(3):387–403
4. Matusai T, San K (1992) Finite element slope stability analysis by shear strength reduction technique. Soils Found 32(1):59–70
5. Chatterjee D, Murali Krishna A (2018) Stability analysis of non-homogeneous soil slopes using numerical techniques. Geotech Appl. Lect Notes Civ Eng, 13:219–227
6. Chatterjee D, Krishna AM (2018) Stability analysis of two-layered non-homogeneous slopes. Int J Geotech Eng
7. Qin C, Chian SC (2017) Kinematic stability of a two-stage slope in layered soils. Int J Geomech 17(9)
8. Michalowski F, Radoslaw L (2002) Stability charts for uniform slopes. J Geotech Geoenviron Eng 128(4):351–355
9. Michalowski F, Radoslaw L (2010) Limit analysis and stability charts for 3D slope failures. J Geotech Geoenviron Eng 136(4):583–593
10. Qian ZG et al (2015) Slope stability charts for two-layered purely cohesive soils based on finite element limit analysis method. Int J Geomech 15(3):1–14
11. Chakraborty A, Goswami D (2018) Three-dimensional (3D) slope stability analysis using stability charts. Int J Geotech Eng

12. Sakellariou MG, Ferentinou MD (2005) A study of slope stability prediction using neural networks. Geotech Geol Eng 23:419–445
13. Farrokhzad F, Jan Ali Zadeh A, Barari A (2008) Prediction of slope stability using artificial neural network (Case Study: Noabad, Mazandaran, Iran). In: International conference on case histories in geotechnical engineering
14. Choobhasti AJ, Zad F, Barari A (2009) Prediction of slope stability using artificial neural network (case study: Noabad, Mazandaran, Iran). Arab J Geosci 2(4):311–319
15. Mohamed T, Kasa A, Mukhlisin M (2012) Prediction of slope stability using statistical method and fuzzy logic. The Online J Sci Technol 2(4):68–73
16. Erzin Y, Cetin T (2013) The prediction of critical factor of safety of homogenous finite slopes using neural networks and multiple regressions. Comput Geosci 51:305–313
17. Chakraborty A, Goswami D (2017) Slope stability prediction using artificial neural network (ANN). Int J Eng Comput Sci 6(6):21845–21848
18. Chakraborty A, Goswami D (2017) Prediction of slope stability using multiple linear regression (MLR) and artificial neural network (ANN). Arab J Geosci
19. Smith GN (1986) Probability and statistics in civil engineering: an introduction. Collins, London. Nichols Pub. Co., New York

The Relevance of Expanded Polystyrene Beads for Ground Improvement: A Review

Rohan Deshmukh, Saivignesh Iyer, Prathamesh Bhangare, Muntazir Bhat, and Shantanu Upadhyay

Abstract An expanded polystyrene (EPS) beads are a lightweight synthetic material that is manufactured daily as a packing material. This study describes the engineering and mechanical behavior of the EPS beads under the actions of stress. The influence of size (diameter) and density of EPS beads on a stress–strain behavior have been well explained in the study. The application of EPS beads as geometrial in combination with fly ash, bottom ash, and cement has been discussed briefly in this study. A combination of 0.5% EPS beads with expansive soil reduces swelling pressure and increases the unconfined compressive strength. This study also discussed the relevance of EPS geofoam for ground improvement. EPS geofoam interface behavior with fly ash has been presented in this study. Results suggest that the influence of EPS geofoam density do not have a significant influence on its interface strength behavior. This study helps us understand EPS geofoam unit cell behavior with fly ash embedded in the core and used as a construction material. Field performance of EPS geofoam as infill and material for pavement has been discussed briefly in this study

Keywords EPS beads · Fly ash · Stress–strain behavior · Ground improvement

1 Introduction

The construction industry consumes lot of conventional materials at a larger scale which has a risk of its depletion. Hence, there is a need to explore alternative materials for construction purposes [6]. Expanded polystyrene (EPS) geofoam is utilized for construction in geotechnical engineering for decades. It has wider applications in the construction of retaining walls and road pavements as lightweight fill material. Many thermal power plants across the country produce an large quantity of fly ash which can be used as alternative material along with EPS beads for construction [25]. As there is a vast development of technology, there arises a need to reduce costs in

R. Deshmukh (✉) · S. Iyer · P. Bhangare · M. Bhat · S. Upadhyay
Department of Civil Engineering, Terna Engineering College, Nerul, Navi Mumbai 400706, India

A. K. Choudhary et al. (eds.), *Advances in Geo-Science and Geo-Structures*,
Lecture Notes in Civil Engineering 154,
https://doi.org/10.1007/978-981-16-1993-9_19

various civil engineering projects. This has led to the usage of lightweight materials viz. expanded polystyrene (EPS) that can be used for the enhancement of the soil properties [27]. The mechanical behavior and the utilization of EPS geofoam as a material for sub-base are influenced by its shrinkage behavior, chemical properties, thermal expansion, and water absorption [20]. Various research scholars carried out experimental studies and numerical studies on EPS Geofoam mix as a compressible inclusion material for checking its suitability for ground improvement [2, 8–10, 14, 16, 20, 29, 30]. The study focuses to determine the stress–strain stress and behavior of the EPS mix and to prepare its proportion dependent model [4]. Strength parameters of the EPS as packing material are stated by conducting laboratory tests that include unconfined compression test, direct shear test, and California bearing ratio test to analyze the effect of various compositions and mix ratios of EPS beads with cement and fly ash on its density, compressive strength, and initial tangent modulus of the geometrial [12, 17].

Unit cells of EPS geofoam were made for densities of 15, 20, 22, and 30 kg/m^3 with heights of 50, 75, and 100 mm. Fly ash was placed inside the unit cell with its optimum moisture content at 25 mm height in layers [21]. Numerical modeling and field tests were performed to check the suitability of EPS beads with fly ash and bottom ash [3, 5, 25].

2 Characterization of Material

Many researchers presented their studies by considering some characteristic EPS geofoam properties which are stated in Tables 1 and 2.

Table 1 Properties of EPS materials

Researcher	Properties		
	Density (kN/m^3)	Compressive strength (kPa)	Initial tangent modulus (MPa)
Horvath [9]	0.12–0.32	44–22	2.4–11.4
Qi et al. [23]	7–11	100–510	79–555
Padade and Mandal [18]	0.15 0.20 0.22 0.30	61.95 91.39 110.53 146.80	2.480 4 070 5.508 7.550
Padade and Mandal [17]	7.25–13.20	158–3290	51–500
Bejua and Mandal [4]	0.12 0.15 0.20	29 52 96	1.611 2.769 5.086

Table 2 Properties of EPS materials

Researcher	Properties		
	Density (kN/m^3)	Cohesion (kPa)	Modulus of elasticity (kPa)
Padade and Mandal [19]	0.15 0.20 0.22 0.30	33.75 38.75 41.88 62 00	2400 4000 5500 7800
Ram Rathan Lal et al. [25]	12.1 (dry unit weight)	1.5	4500

3 Experimental Studies

The influence of density, compressive strength, and initial tangent modulus of EPS beads on fly ash was investigated [17, 26]. The test setup consisted of a shear box having dimensions ranging from 60 × 60 × 25 to 305 × 305 × 175 mm. The direct shear test was conducted to obtain strength and characteristics of EPS beads.. It was concluded that addition of 1 mm (small beads) increased the value of cohesion due to which it showed significant intergranular friction of EPS beads with fly ash observed that CBR value of EPS reinforced fly ash changed from 4.9 to 25.64% when 1 mm EPS beads were added to fly ash, refer Fig. 1 [15].

Around 12 compressive strength tests were carried out on EPS geofoam mix having different heights [21]. The test was performed on the universal testing machine (UTM) which had a fixed deformation rate of 1 mm/min. It was concluded that high compressive strength and stiffness was seen for unit cells with lesser height. Unit cells of less height had undergone crushing failure while unit cells with tall height undergo buckling.

A test series were conducted on EPS geofoam having various densities (0.15, 0.20, 0.22, 0.33 kN/m^3) to find the effect of density and rate of stress–strain behavior and elastic modulus of EPS geofoam [13]. The result showed that an increase in strain rate, i.e., elastic modulus and compressive strength increases further elastic

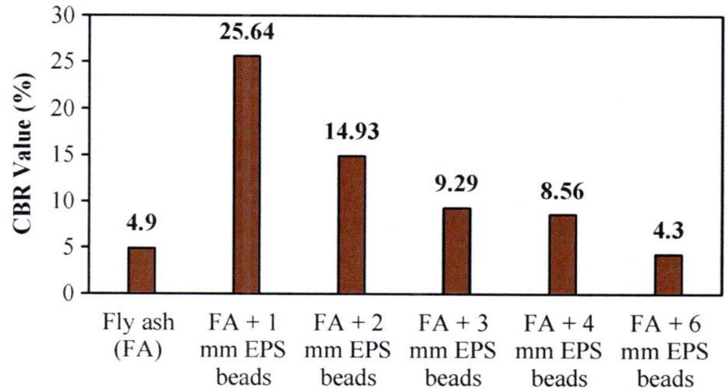

Fig. 1 CBR value variation with EPS beads addition in fly ash [15]

modulus of EPS lifts with increase in density of EPS Geofoam. Also, consolidated drained triaxial test was performed to determine the angle of internal friction and cohesion parameters [15]. It was concluded from the result that the addition of 1 mm EPS beads with fly ash increased the angle of internal friction. An unconsolidated undrained (UU) triaxial test was also performed with a specimen diameter 75 mm and height 150 mm.

It also concluded that cohesion value and peak deviator stress increased when EPS geofoam density increases. Consolidation parameters were determined by performing a consolidation test. The result showed permeability of fly ash increased sustainably with increase in beads size. All consolidation parameters increased with the increasing size of EPS beads size. The summary of all the experimental, numerical, and field study are given in Table 3.

Table 3 Summary of experimental, numerical, and field study

Researcher	Method of study	Outcome
Experimental analysis		
Ram Rathan Lal and Nawkhare [26]	Direct shear test	Addition of 1 mm EPS beads increased the value of cohesion
Padade et al. [21]	Compressive strength test	Higher compressive strength for less height of EPS unit sample
Nawghare and Mandal [15]	California bearing test	Addition of 1 mm EPS beads increased the CBR value
Nawghare and Mandal [15]	Triaxial test 1. Consolidated drained 2. Unconsolidated undrained	EPS beads addition to fly ash increased the value of the angle of internal friction Cohesion value and peak deviator stress increased when density of EPS Geofoam increases
Khalaj Omid et al. (2020) [13]	Compression test	Increasing the strain rate increases compressive strength and elastic modulus of EPS beads
Numerical analysis		
Padade and Mandal [18]	Numerical modeling in Plaxis 2D	Horizontal deformation and backfill settlement was decreased with a decrease in density
Field tests		
Amulya and Ravi Shankar [1]	Modified proctor test	EPS improve the mechanical characteristic of widening subgrade and decrease in differential settlement

4 Numerical Studies

4.1 Graphical Analysis

Density is a vital property to classify EPS geofoams characteristics. A compressive strength characteristic of EPS geofoams is majorly affected by its density and specimen dimensions. The compressive strength, elastic modulus, and initial tangent modulus of EPS samples increase as there is an increase in EPS geofoam density with increasing strain rate [4, 13, 26, 28]. The shear strength decreases with higher EPS inclusion [5]. Furthermore, peak deviator stress and cohesion value show an increment as EPS geofoam density increases [4, 18]. EPS also has good thermal insulation properties. To reduce sulfate has in subgrade sections thermal insulation can be readily used in EPS [7, 22]. The interface strength of EPS geofoam does not make any impact with the varied properties of EPS [19].

The EPS-soil composite properties are significantly influenced by its elements and mixt proportions [7, 26]. The increment in EPS beads content in the EPS-sand composite reduces the internal angle of friction and passive lateral Earth pressure (K_p). However, the lateral Earth pressure coefficient (K_0) increased [8, 23, 27, 28]. But if we add fly ash in the core of the geofoam unit cell, the compressive strength reduces with an increase in density and geofoam unit cell height [21]. The volumetric stress–strain relationship for EPS-sand composite with high EPS content showed substantial contraction during consolidation [5]. The compressive stress–strain relationship is most affected by the cement to fly ash ratio in EPS geomaterial with EPS beads, fly ash, and cement as constituents [17].

Reinforced expansive soil with 0.5% of EPS is considered to be the optimum percentage for field applications, as it showed high unconfined compressive strength and showed exponential decrease in the upward swelling pressure [5, 17]. The EPS beads reinforcement with fly ash of 1 mm size was found optimum and gave greatest values of internal angle of friction, shear strength, secant modulus, and CBR [15]. Better durability under dynamic loading and monotonic loading for higher duration makes EPS a suitable material for construction of pavements. It is especially very effective in cold regions and noise-friendly in the of pavement urban environment [7].

4.2 Software Analysis

A discrete element modeling (DEM) is necessary to simulate appropriately, the interaction mechanism governing the contact force chains formed during different loading conditions on EPS-soil composite [11]. The discrete element modeling (DEM) conducted on the EPS-sand mixture concluded that the soil particles with smaller size undergo more settlement as compared to larger size particles [28].

According to the finite element simulation perform on PLAXIS 2D, the horizontal deformation and backfill settlements on model walls decreased prominently with decreasing density and thickness of EPS geofoams [25]. Numerical modeling in PLAXIS 2D [18] concluded that horizontal deformation and backfill settlement was decreased with a decrease in density of EPS.

Numerical simulation of approach for EPS geofoam suggests that it can significantly reduce the lateral earth pressure in shored mechanically stabilized earth (MSE) walls. Hence, it can be beneficial in limited space construction and eventually lead to cost-saving constructions [8].

5 Field Tests

EPS geofoam is useful as an alternative material for ground improvement that can benefit many construction projects. Studies show that commercially available sizes of EPS beads have advantages as per engineering needs on-site conditions [15]. EPS were used in many geotechnical fields in areas of thermal insulation, compressible inclusion, damping vibration, and as a filler material behind RE walls and below slabs or beams at its foundation [24]. EPS beads reinforced lateritic soil of 0.5 and 1.0 Ms (silica modulus) were durable after 3, 7, and 28 days of curing. However, EPS reinforce black cotton soil was found durable with Ms 0.5 at modified Proctor density after 28 days of curing [1].

Three sections of pavement structure were studied by the Federal Highway Research Institute (BASt, BundesanstaltfürStrassenwesen) on the ground having soft silt. Two rigid EPS foams were used at the interface of the subgrade. The performances were calculated by Benkelman beam and falling weight deflector(FWD) layers. Results show that the efficiency of fine layers above EPS has a certain rigidity for load distribution, i.e., the pavement with long-term traffic load can resist the deformation.

6 Conclusions

The use of EPS beads has been fascinating to many modern infrastructure and construction developers due to its low-cost availability and surprising benefits as a geomaterial.

- This study reviews the research works related to EPS beads in ground improvement and further applications as a geo-composite with various other easily available by-products. The ultimate motive of the study is to present the summary of all the studies and results obtained to date on EPS as a geo-composite and as a whole.
- This study discussed various works related to the EPS in terms of experimental, numerical, analytical, and field performance. From the past findings, there are

many research gaps in the subject matter, especially in software analytics. A study with recent software utilization can considerably help us in filling these research gaps. When these gaps are studied and attempted to overcome, it will help to grasp the qualities of EPS beads to a great extent.

- Overall, EPS is found to be a good eco-friendly replacement for conventional construction materials. And in the future, the applications of EPS will prove more suitable in road pavements, railway subgrades, and foundations, retaining earth walls, MSE walls, and underground utilities.

References

1. Amulya S, Ravi Shankar AU (2020) Use of stabilized lateritic and black cotton soils as a base course replacing conventional granular layer in flexible pavement. Int J Geosynth Gr Eng 6. Article number: 5
2. Bathurst RJ, Zarnani S, Gaskin A (2007) Shaking table testing of geofoam seismic buffers. Soil Dyn Earthq Eng 27(4):324–332
3. Beinbrech G, Hillmann R (2018) EPS in road construction current situation in Germany, BASF AG
4. Bejua YZ, Mandal JN (2017) Expanded polystyrene (EPS) geofoam: preliminary characteristic evaluation. Procedia Eng 189:239–246
5. Deng AN, Yang X (2010) Measuring and modeling proportion-dependent stress-strain behavior of EPS-sand mixture. Int J Geomech 10(6):214–222
6. Ghafoori N, Buchole J (1998) Investigation of lignite-based bottom ash of structural concrete. J Mater Civ Eng, ASCE 8(3):128–137
7. Goa H, Liu J, Liu H (2011) Geotechnical properties of EPS composite soil. Int J Geotech Eng
8. Hatami K, Witthoeft AF (2008) A numerical study on the use of geofoam to increase the external stability of reinforced soil walls. Geosynth Int 15(6):452–470
9. Horvath JS (1997) The compressible-inclusion functions of EPS geofoam: analysis and design methodologies. Geotext Geomembr 15(1):77–120
10. Horvath JS (2004) Geofoam compressible inclusion: the new frontier in Earth retaining structures. In: Geotechnical engineering for transportation projects, GeoTrans ASCE, pp 1925–1934
11. Jamshidi Chenari R, Ebrahimi Khonachah R, Hosseinpour I, Khajeh A (2020) An experimental study for the cyclic interface properties of the EPS–sand mixtures reinforced with geogrid. Int J Civ Eng 18:151–159
12. Jamshidi Chenari R, Fatahi B, Ghorbani A (2020) A study on the effect of cement treatment on the behavior of EPS composite soils. Geomech Eng
13. Khalaj O, Kepka M, Kavalir T, Křížek M, Jeníček Š, Siabil SMAG, Tafreshi SNM (2020) The experimental investigation of behavior of expanded polystyrene (EPS). In: IOP conference, material science engineering 723.
14. Mandal JN, Nimbalkar SS (2000) Centrifuge tests for bearing capacity of EPS on Bearing capacity improvement on soft soil using expanded polystyrene (EPS) geofoam. Civ Eng J
15. Nawghare SM, Mandal JN (2020) Effectiveness of expanded polystyrene (EPS) beads size on fly ash properties. Int J Geosynth Gr Eng
16. Padade AH, Mandal JN (2012) Behavior of expanded polystyrene (EPS) geofoam under triaxial loading conditions. EJGE.
17. Padade AH, Mandal JN (2014) Expanded polystyrene-based geomaterial with fly ash. Int J Geomech 14(6):06014013

18. Padade AH, Mandal JN (2012) Expanded polystyrene-based geomaterial with fly ash. Int J Geomech, ASCE 6:1–7
19. Padade AH, Mandal JN (2012) Feasibility studies on expanded polystyrene (EPS) geofoam. In: International conference on ground improvement and ground technique. ICGI Wollongong, vol 2, pp 903–906
20. Padade AH, Mandal JN (2014) Interface strength behavior of expanded polystyrene EPS geofoam. Int J Geotech Eng 8(1):66–71
21. Padade AH, Dutta S, Nadaf MB, Ram Rathan Lal B (2016) Expanded polystyrene (EPS) geofoam unit cells with fly ash. Geo-Chicago, GSP 271
22. Shuquan P, Xibing Li, Fan W, Fan L (2020) Experimental research on employed expanded polystyrene (EPS) for lightened sulfate heave of sub-grade by thermal insulation properties. Geotext Geomembr, Elsevier 48(4):516–523
23. Qi S, Liu SH, Zhang Y, Xu GL (2012) Experimental study on capability of EPS beads-mixed lightweight soil. Appl Mech Mater 268–270:776–781
24. Ramli Sulong NH, Mustapa SAS, Abdul Rashid MK (2019) Application of expanded polystyrene (EPS) in buildings and constructions: a review. J Appl Polym Sci 47529
25. Ram Rathan Lal B, Padade AH, Mandal JN (2014) Numerical simulation of EPS geofoam as compressible inclusions in fly ash backfill retaining walls. In: Ground improvement and geosynthetics GSP 238 ASCE
26. Ram Rathan Lal B, Sonali Nawkhare S (2016) Experimental study on plastic strips and EPS beads reinforced bottom ash based material. Int J Geosynth Gr Eng 225
27. Rezaie B, Chenari RJ, Veiskarami M (2020) A study on the effect of cement treatment on the behavior of EPS composite soils. Switzerland AG
28. Tizpa P, Kazempour S, Jamshidi Chenari R, Farrokhi F, Ahmadi H (2019) Numerical and experimental investigations of the influence of grain size on the compressibility of sand–EPS mixtures. Int J Geosynth Gr Eng 5(4)
29. Zarnani S, Bathurst RJ (2007) Experimental investigation of EPS geofoam seismic buffers using shaking table tests. Geosynth Int 14(3):165–177
30. Zarmani S, Bathurst RJ (2008) Numerical modeling of EPS seismic buffer shaking table test. Geotext Geomembr 26(5):371–383

An Overview of Fiber-Reinforced Cemented Soil for Enhancing the Mechanical Properties of the Soil

Sagar Jaiswal, Ananya Srivastava, and Vinay Bhushan Chauhan

Abstract Over a previous couple of decades, there is a rise in research related to the use of synthetic fibers as soil reinforcement materials due to their availability, economy, and growth in recognition of sustainable development. As an outcome of rapid urbanization, an enormous amount of construction work is going on throughout the world, and due to the absence of hard soil strata, the geo-technicians are bound to use the available soft soil strata having a low bearing capacity which is responsible for excessive settlement. To overcome this situation, soil modification with cement along with the addition of fibers is an economical and well-established practice for the enhancement of mechanical properties of soil. The evolution of cement-based technology has led to the development of fiber-reinforced cemented soil (FRCS) to improve the compressive strength as well as the tensile strength of the soil with the addition of cementitious material and fibers. This paper is organized in various sections on properties, usages, etc., and provide a detailed review on up to date research going on to enhance the performance of synthetic fiber-reinforced cemented soil based on the addition of synthetic fibers such as polypropylene, polyethylene, and polyester as reinforcing materials and discussing the possibility of the further application of the fibers for the ground improvement.

Keywords FRCS · Fiber reinforcement · Synthetic fiber · Polymers · Mechanical properties

1 Introduction

Soil reinforcement is described as a technique to improve the physical properties and mechanical behavior of the soil. It has been found useful in developing the parameters that are similar to compressibility, durability, and shear strength. The trendy fiber-reinforced soil mass contains "indiscriminately distributed and discrete elements, i.e., fibers, which provide an improvement in the mechanical behavior of

S. Jaiswal · A. Srivastava · V. B. Chauhan (✉)
Department of Civil Engineering, Madan Mohan Malaviya University of Technology, Gorakhpur 273010, India

A. K. Choudhary et al. (eds.), *Advances in Geo-Science and Geo-Structures*, Lecture Notes in Civil Engineering 154,
https://doi.org/10.1007/978-981-16-1993-9_20

the soil composite" [1]. The soil characteristics are much influenced by environmental conditions and exhibit low tensile and shear strength. The idea of reinforcement is to include those properties into the soil which are otherwise lacking, thus, making the soil suitable for military and roadway applications and can withstand heavy transportation load. Hence, soil mass is meant to reduce lateral deformation and settlement and improve the stability and bearing capacity.

With the addition of cement to soil, a new cement-soil structure is created between soil particles that intensify the natural structure. This is accountable for the increase in strength of the original soil. With the formation of a reconstituted state (a new cement-soil structure) that takes place with the addition of cement, the behavior of soil is contrastingly different from the original soil and with an increase in the cement content in the soil matrix, compressive strength of the cement-treated soil enhances drastically. However, the inclusion of fiber elements into the cement-soil matrix improves the tensile strength, i.e., ductility is introduced which was absent in soil. Therefore, the usage of fibers together with cementitious material improves/alters the properties like compressibility, density, hydraulic conductivity, and shear strength of cohesionless soil with low bearing capacity and makes it a good choice for using the treated soil as a foundation bed for many geotechnical structures [2].

In civil engineering, the concept of incorporating fiber materials into excessive soil mass to improve mechanical behavior has always been notorious. For example, the Romans and Mesopotamians distinctly discovered that the strength of pathways to absorb traffic loads can be possibly increased by mixing the weak soils with a stabilizing agent such as pulverized sedimentary rock or metal. The walls of the temple Ziggurat Agar in the geographical region of Mesopotamia are an example of soil reinforcement that was engineered by exploiting layers of soil mixed with plant roots. Usage of straw and hay was prevalent in the prehistoric civilization times to create reinforced building blocks. The wool from a mammal called Llama was mixed by the Incas with soil as a reinforcement material for road construction. In the eighteenth century, wooden stick cushions were mixed with soil on the top of soft layers of soil strata that were used before embankments were built, and this was a standard resolution adopted by Brazilian [3].

Vidal [4] developed the principle of soil reinforcement in modern times and introduced the concept of reinforcing elements. More than 4000 structures were built in 37 countries with this technique of reinforcement.

The ultimate agenda of reinforcing the earth bed is to boost the engineering properties and stabilize the soil by using natural or artificial fibers. The concept of soil reinforcement was developed in antiquity, but due to its applicability in several areas of construction, it is gaining attention by geotechnical engineers.

2 Literature Review

In the recent past, synthetic fibers have become more fascinating as reinforcements for cementitious materials because they are proven to be inexpensive reinforcements for concrete as well as soil. Synthetic fiber can be described as a flexible, apparently homogenous body having a high aspect ratio (high ratio of length to thickness) and a small cross-sectional area. The properties of synthetic fibers vary widely, i.e., modulus of elasticity, which is a vital characteristic while its usages as reinforcement. The physical properties of various synthetic fibers that have been used as reinforcing material by various researchers in the past are shown in Table 1 [5]. Significant research attempts have been made to contribute to the theoretical as well as technological advancement in the knowledge about the performance of fiber-reinforced cemented soil (FRCS).

FRCS has numerous applications in the field of civil engineering. There are various types of fibers available commercially in different shapes and sizes, such as steel fibers and synthetic fibers (i.e., glass, carbon, polyvinyl and polyolefin, polypropylene, etc.). These can offer effective and relatively inexpensive soil reinforcement and on further optimization, and the same FRCS property improvement may well be achieved at a lower expense without increasing the reinforcement cost.

Due to an increase in demand for construction and less availability of land, the construction of soils with low bearing capacity is also necessary which may not be capable of supporting structures, for example—parkways, dams, extensions, residential buildings, etc. [6]. The improvement of the top layer of soil by adding cement-containing materials and also by including reinforcing materials such as fibers could be used in railways and roadways. In the above situation, the use of granular bases turns out to be impossible if the availability of the material is far away from the construction site.

The behavior of fiber-reinforced soil is somewhat similar to that shown by composite materials. To increase the tensile strength of the soil mass, the fibers of high tensile strength are mixed in a high amount (3–5% by weight) into the matrix of soil. The shear stresses of the soil mobilize the fibers for tensile reinforcements imparting higher strength to the soil. From the past studies, it is noted that the compressive strength of synthetically produced cemented soils by varying the water content, fiber content, dry density to investigate the cracking resistance, ductility, and reduction in brittleness of the soil mass. By performing plate load tests, it was also demonstrated that FRCS can increase and maintain the strength with continued

Table 1 Physical properties of synthetic fibers [5]

Fiber type	Elastic modulus (GPa)	Specific gravity, (G_s)	Tensile strength (MPa)	Ultimate elongation (%)
Polyester	18	1.35–1.40	897–1100	–
Polyethylene	5	0.95	210–300	3
Polypropylene	3.5–5.0	0.90	315–765	15
Acrylic	15–20	1.18	208–1000	8–50
Aramid I	61	1.45	3625	4.5
Aramid II	118	1.45	3625	2.6
Nylon	5.20	1.17	966	20

deformation, recommending a very ductile material. Also, "the failure mechanism of a fiber-reinforced soil depends upon the average effective stress acting on the soil mass" [7].

The quantity and behavior of cement, its structural properties, density, confining stress, and gradation are the parameters on which the performance of the FRCS is evaluated. Also, in previous literature, an extensive study was conducted to determine the factors that mostly alter the physical properties of cement-treated soils. The results of the study determined that the following points influence the mechanical behavior: grain-size distribution of soils, amount of cement, the grade of cement, compaction effort, and hardening time.

"A series of unconfined compression (UCS) and P-wave velocity tests on soil samples with different percentages of hay fiber and cement was studied" [8]. From the results obtained, it was determined that incorporation of hay fiber as reinforcement into non-cemented and cemented soil samples reported an increase in UCS and axial elongation before the incipient failure of the soil samples under testing. A decrease in *P*-velocity values and change in behavior of soil from brittle to ductile one was reported as well. It was recommended to stabilize the soil with both cement and fiber contents to gain a potential change in engineering properties of soil and also suggested to reinforce the soil mass with a low percentage of hay fiber (i.e., 0.5%) for more effective results.

3 Case Studies

3.1 Polypropylene (PP) Fibers

The polypropylene fiber, popularly named as PP fibers, is a synthetic fiber that has been transformed from 85% propylene by the process of polymerization. It is one of the most widely used synthetic reinforcing material that is used in the testing of reinforced soil for determining their feasibility along with suitability. Santoni and Webster [9] performed field experiments, in which cohesionless soil was reinforced with synthetically produced PP fibers. It is also advantageous for road applications and military airfields. Previous studies have shown that the sand fiber layers of 20.3 cm thickness were enough to support considerable amounts of moving load upon it. Field trials also indicated that "it is necessary to fix the surface using emulsion binder to prevent fiber pull-out under traffic below which the PP fiber is used as reinforcement" [9].

Tang et al. [10] investigated the mechanical interaction behavior between PP fibers reinforcement and soil particles by carrying a soil-fiber pull-out test and concluded that "the interfacial shear resistance of soil and fiber depends primarily on the rearrangement resistance of soil composition, soil particles, fiber surface roughness, and effective interface contact area." Also, a pull-out test apparatus was invented by the researchers which were useful to determine the rupture strength of fibers [10].

A series of drained triaxial tests on artificially cemented sand was performed by reinforcing the specimens with randomly oriented PP fibers. The test was brought about to comprehend the effect of cement content when mixed with soil. The results demonstrated that the inclusion of up to about 5% fiber reinforcement had supported in increasing the peak strength. An increase in ultimate strength and a change in the brittleness of the cemented sand to an extra ductile material have also been reported. "The peak strength of the soil mass under triaxial test increases due to fiber inclusion and found more effective for smaller amounts of cement, while the increase in ultimate strength is more significant when the fiber is added to sand improved with higher cement contents" [11].

It is worthy to note that the engineering properties of the swelling soil such as black cotton soil and silty sand can also be improved with reinforcement through the outcomes of a set of CBR tests, tensile strength tests, and triaxial tests [12]. The outcome from the experiment showed that both the black cotton soil and silty sand exhibited a slight fall in the angle of internal friction and a remarkable rise in the cohesion intercept and with the advancement in fiber content up to 3% by weight [12]. The basic properties of fibers that are used for reinforcing the soils in several field studies and laboratory tests on fiber-reinforced soils are shown in Table 2.

3.2 *Polyvinyl Alcohol (PVA) Fibers*

These are the high-performance artificial fiber that has been recently used for reinforcing the soil mass. PVA fiber has a wide variety of applications because of its crack fighting properties, chemical resistance, weather resistance, and tensile strength. These are monofilament fibers that are available in 6, 8, and 13 mm length. PVA fibers have considerably lower shrinkage on high temperatures in comparison with polyester and nylon. The addition of PVA fiber likely to result in an improvement in ductility and strength compared to different fibers beneath the same cementation of soil. It was also noted that adding 4% cement (by weight of matrix) and 1% of polyvinyl alcohol (PVA) fibers (by weight of matrix) to the sand resulted in a 100% increase in the unconfined compressive strength and the peak axial strain at failure, compared to the unreinforced soil sample [15]. Besides, the researchers reported that the deformability index values when adding only 1% fiber content, are greater than four, and it does not get affected by the cement ratios [22].

3.3 *Polyethylene (PE) Fibers*

Polyethylene (PE) fibers are synthetic fibers that are strong in tension. In the precedent days, the study on the viability of soil reinforcement with PE fibers was bounded to a constrained extent [23]. It has been reported that "the presence of a small amount

Table 2 Physical properties of natural and synthetic fibers reported by various researchers

References	[13]	[14]	[15]	[16]	[17]	[18]	[19]	[20]	[21]	[21]
Type	PET fibers	PP fibers	PVA fibers	PE fibers	PC fibers	Coir fibers	Glass fibers	Date palm fibers	Jute fibers	Nylon fibers
Average length, (mm)	20.0	24.0	12.0	12.0	3.0	15.0	10–15	295.0	7–9	6–18
Diameter, (mm)	0.075	0.023	0.10	0.035	0.015	0.25	0.02	0.42	0.005–0.025	0.003–0.01
Specific gravity, (G_S)	1.30	0.91	1.30	0.99	1.5	1.07	1.7	0.92	1.47	1.15
Strain at break/failure (%)	80.0	5.10	–	–	–	–	–	–	–	–
Tensile strength (MPa)	85–165	120	1078	600	500	102	300	123	331–414	300
Tensile modulus, (GPa)	1.45–2.50	3	2	50	–	–	–	–	–	2.4

of high-density polyethylene (HDPE) fibers can enhance the fracture energy of the soil" [24].

Its been informed that the increase in tensile strength of soil mass along with the addition of HDPE strips was not appreciable. Moreover, after the introduction of HDPE strips in the soil as reinforcement an increase in strain capacity was reported which resulted in a large increase in the toughness of the soil mass. Due to an increase in toughness, considerable performance benefits due to the incorporation of PE fibers were observed in the stress–strain behavior. Nonetheless, the researchers did not find any improvements in fatigue behavior [23].

Choudhary et al. [24] discovered that when recovered strips of high-density polyethylene (HDPE) were added to cohesionless sand, an increase in the California Bearing Ratio (CBR) and secant modulus were observed. It has also been investigated that the thickness of the base layer can be reduced when the fiber-reinforced sand is used as the basic construction material in the pavement construction. The researchers concluded that "maximum improvement in CBR and secant modulus were obtained when the strip content is 4% with a characteristic ratio of 3, which is about three times that of an unreinforced sand bed system" [24].

3.4 Polyester (PET) Fibers

The polyester (PET) fibers are the synthetic polymer composed of 85% ester by weight, which is responsible for the important desirable characteristics such as resistance to shrinking and abrasion. In India, a study was conducted on a soil-fly ash combination with a length of 20 mm mixed with 0.5% as well as 1% PET fiber content [13].

The effect of mixing the PET fiber in expansive clay with an optimized dose of fly ash and lime was studied [25]. A series of UCS test was performed on highly compressible clay by mixing the crimped polyester fiber ranging from 0 to 2% at an interval of 0.5% each. The researchers had chosen 3 mm long crimped fibers strips and for flat fibers, the segments of 3, 6, and 12 mm were preferred. The results of the aforementioned study have unveiled that the UCS value tends to improve with increasing fiber content and fiber length. The results obtained are well comparable to a very high extent with the finding of Tang et al. [26].

In a study, highly compressible clayey soil was reinforced with polyester fibers having a length of 12 mm, which varied between 0 and 1% [27]. The results showed a decrease in the settlement at the ultimate load and an increment in the ultimate load-bearing capacity, due to the random distribution of fibers. The results concluded that both the safe bearing pressure (SBP) and the soil bearing capacity increases with increasing the amount of fiber percentage up to 0.5% and decreases when the addition of fibers on a certain rate is continued [27].

4 Conclusion

In this paper, a review on the behavior of fiber-reinforced cemented soil (FRCS) has been presented showing the effect of the amount of cementitious material and fiber content on the engineering properties of the soil under varying conditions. The papers discussed above in the literature generally illustrate that the strength and stiffness of the treated soil are improved by the inclusion of fiber. Also, it is indicated that the strength of the soil increases with an increase in the aspect ratio of fiber, fiber content, and soil-fiber surface friction.

Moreover, with the help of unconfined compression tests (UCS), triaxial compression tests, and direct shear tests, researchers have demonstrated that the peak shear strength significantly increases with the inclusion of fiber percentage, also an increase in axial strain to failure was reported. The inclusion of fiber impedes the compaction process and the increase in fiber content results in reduction in the maximum dry density. The practice of fiber reinforcement is also useful for improving the bearing capacity of soft soil, for constructing low rise building the low-cost shallow foundation can be adapted instead of deep foundation. Also, the cohesionless soil can be treated with synthetic fiber to provide better results in the road applications and military airfields. The detailed literature review suggests that fibers effectively reduce the amounts of cracks when the soil is subjected to tension. The introduction of randomly distributed fibers for reinforcing a soil mass is an efficient method and can be applied to many earthen structures (e.g., in which stress cracks would occur), such as bottom liners of landfills, cover barriers, slopes, runway subgrades, earth-filled dams, and highway and railway embankments.

References

1. Li C (2005) Mechanical response of fiber-reinforced soil, Ph.D. thesis, Faculty of the Graduate School of the University of Texas at Austin
2. Kazemian S, Huat BB, Arun P, Barghchi M (2010) A review of stabilization of soft soils by injection of chemical grouting. Aust J Basic Applied Sci 4(12):5862–5868
3. Palmeira EM (1992) Geosynthetics: nature, and advances in the last years. Proc 1st Braz Conf Geosynth, pp 1–20.
4. Vidal H (1969) The principle of reinforced earth. Highw Res Rec (282):1–16
5. Balaguru PN, Shah SP (1992) Fiber-reinforced cement composites.McGraw-Hill: 116
6. Hassan WH (2010) Strength characteristic of soft soil reinforced with coir fibers (Doctoral dissertation, UMP)
7. Gray DH, Ohashi H (1983) Mechanics of fiber reinforcement in sand. J Geotech Eng 109(3):335–353. https://doi.org/10.1061/(ASCE)0733-9410(1983)109:3(335)
8. Aldaood A, Bouasker M, Khalil AA, Al-Mukhtar M (2014) Mechanical behavior of hay fiber-reinforced cemented soil, pp 9–15
9. Santoni RL, Tingle JS, Webster SL (2001) Engineering properties of sand-fiber mixtures for road construction. J Geotech Geoenviron Eng 127(3):258–268. https://doi.org/10.1061/(ASCE)1090-0241(2001)127:3(258)
10. Tang CS, Shi B, Zhao LZ (2010) Interfacial shear strength of fiber reinforced soil. Geotext Geomembr 28(1):54–62. https://doi.org/10.1016/j.geotexmem.2009.10.001

11. Consoli NC, Vendruscolo MA, Fonini A, Dalla Rosa F (2009) Fiber reinforcement effects on sand considering a wide cementation range. Geotext Geomembr 27(3):196–203. https://doi.org/10.1016/j.geotexmem.2008.11.005
12. Setty, KRNS, and Rao, SVG (1987) Characteristics of fiber reinforced lateritic soils. In: Proceedings Indian Geotech conference, Bangalore, India vol 1, pp 329–333
13. Kaniraj SR, Gayathri V (2003) Geotechnical behavior of fly ash mixed with randomly oriented fiber inclusions. Geotext Geomembr 21(3):123–149. https://doi.org/10.1016/S0266-1144(03)00005-0
14. Consoli NC, Festugato L, Heineck KS (2009) Strain-hardening behaviour of fiber-reinforced sand in view of filament geometry. Geosynth Int 16(2):109–115. https://doi.org/10.1680/gein.2009.16.2.109
15. Park SS (2009) Effect of fiber reinforcement and distribution on unconfined compressive strength of fiber-reinforced cemented sand. Geotext Geomembr 27(2):162–166. https://doi.org/10.1016/j.geotexmem.2008.09.001
16. Jha JN, Gill KS, Choudhary AK, Shukla SK (2015) Stress-strain characteristics of fiber-reinforced rice husk ash. Proceedings Geosynthetics, Portland, OR, USA, pp 134–141.
17. Khattak MJ, Alrashidi M (2006) Durability and mechanistic characteristics of fiber reinforced soil–cement mixtures. Int J Pavement Eng 7(1):53–62. https://doi.org/10.1080/10298430500489207
18. Babu GS, Vasudevan AK, Haldar S (2008) Numerical simulation of fiber-reinforced sand behavior. Geotext Geomembr 26(2):181–188. https://doi.org/10.1016/j.geotexmem.2007.06.004
19. Lovisa J, Shukla SK, Sivakugan N (2010) Shear strength of randomly distributed moist fiber-reinforced sand. Geosynth Int 17(2):100–106. https://doi.org/10.1680/gein.2010.17.2.100
20. Sarbaz H, Ghiassian H, Heshmati AA (2014) CBR strength of reinforced soil with natural fibers and considering environmental conditions. Int J Pavement Eng 15(7):577–583. https://doi.org/10.1080/10298436.2013.770511
21. Spritzer JM, Khachan MM, Bhatia SK (2015) Influence of synthetic and natural fibers on dewatering rate and shear strength of slurries in geotextile tube applications. Int J Geosynth Gr Eng 1(3):26.1–26.14
22. Park SS (2011) Unconfined compressive strength and ductility of fiber-reinforced cemented sand. Constr Build Mater 25(2):1134–1138. https://doi.org/10.1016/j.conbuildmat.2010.07.017
23. Sobhan K, Mashnad M (2002) Tensile strength and toughness of soil–cement–fly-ash composite reinforced with recycled high-density polyethylene strips. J Mater Civil Eng 14(2):177–184. https://doi.org/10.1061/(ASCE)0899-1561(2002)14:2(177)
24. Choudhary AK, Jha JN, Gill KS (2010) A study on CBR behavior of waste plastic strip reinforced soil. Emirates J Eng Res 15(1):51–57
25. Kumar A, Walia BS, Bajaj A (2007) Influence of fly ash, lime, and polyester fibers on compaction and strength properties of expansive soil. J Mater Civ Eng 19(3):242–248. https://doi.org/10.1061/(ASCE)0899-1561(2007)19:3(242)
26. Tang C, Shi B, Gao W, Chen F, Cai Y (2007) Strength and mechanical behavior of short polypropylene fiber reinforced and cement stabilized clayey soil. Geotext Geomembr 25(3):194–202. https://doi.org/10.1016/j.geotexmem.2006.11.002
27. Maheshwari KV, Desai AK, Solanki CH (2011) Performance of fiber reinforced clayey soil. Electron J Geotech Eng 16:1067–1082

Potential Use of Construction and Demolition Recycled Wastes in Geosynthetic-Reinforced Structures

Ananya Srivastava, Sagar Jaiswal, and Vinay Bhushan Chauhan

Abstract The advent of industrialization and unrestrained urbanization created a demand for the construction of heavy engineering infrastructures which caused extensive environmental deterioration. Also, the construction industry generates huge waste in the form of demolished material and recycled material. Therefore, the need of the hour is to explore all alternative approaches and materials which are durable, sustainable, and transcendent in comparison with the conventional building materials being used for construction. In view of the above, recycling and using the construction and demolition wastes have emerged as a suitable substitute for civil engineering applications as it utilizes the otherwise leftover waste product, saves natural resources, and reduces our environmental footprint. The geotechnical properties of the construction and demolition materials have shown low variability when compared with the guidelines of various agencies for their utilization as backfill materials in structures that are reinforced with geosynthetic materials. The adherence factor (ratio between the pull-out strength of the interface and the shear strength of the backfill) when calculated for geogrid construction and interfaces of the demolition materials were almost in a similar limit w.r.t the soil-geogrid interface values attained by few other researchers. Therefore, given the above, this study aims to present a detailed overview on the performance and evaluation of construction and demolition materials in mechanically stabilized earth retaining structures (using geosynthetic as a reinforcing layer), while highlighting its application as backfill material for retaining walls and building and construction material for the embankment creation.

Keywords Construction and demolition materials · GRS walls · Sustainable development · Backfill materials

---

A. Srivastava · S. Jaiswal · V. B. Chauhan (✉)
Department of Civil Engineering, Madan Mohan Malaviya University of Technology, Gorakhpur 273010, India

A. K. Choudhary et al. (eds.), *Advances in Geo-Science and Geo-Structures*,
Lecture Notes in Civil Engineering 154,
https://doi.org/10.1007/978-981-16-1993-9_21

1 Introduction

The increased global population has created more burden on nature and natural resources. Gradually land has been acquired for agricultural, residential, infrastructural purposes, and various commercial activities. This scenario has created a shortage of land space for the construction industry which is sprawling like wildfire to satisfy the requirements of an ever-increasing population. For the erection of heavy infrastructures, a huge amount of appropriate building materials like backfill material is needed which if not naturally available at the site of construction, is impossible to be transported every time. In such scenarios, not only the scarcity prevails, but also the pre-existing construction practices create havoc by producing tons and tons of waste. The waste so produced includes construction and demolition waste which if recycled suitably can develop as an alternative backfill material.

Among the mass variety of trashes produced by the building and construction industry, the wastes produced by the demolition of various obsolete civil engineering structures have the maximum potential for recycling and recovery. "They can be defined as the residues obtained by various activities including modernization, construction, alteration, repair and maintenance, extension, and demolition of building structures and other infrastructural facilities." Construction and demolition wastes are generally heterogeneous residues of the building or infrastructural elements with distinct materials which are used in construction processes [1].

The recycling of construction materials is a practice prevalent since the ancient era, where the Greeks, Egyptians, and Romans used it for their magnanimous structures. The modern era too utilized this sustainable practice thoroughly after World War II, beginning in Europe and slowly spreading throughout the world.

Constituents of construction and demolition materials and the general mechanical/engineering/physical characteristics of the construction and demolition materials are shown in Table 1.

As it is well established by many researchers by conducting numerous experimental investigations and numerical modeling, the geosynthetic-reinforced earth

Table 1 General composition of construction and demolition wastes materials

Elements	Composition range (%)
Concrete (part of slabs, beams, columns, and masonry units made up of concrete), mortar	40–42
Natural stone, aggregates (unbound and treated with hydraulic binders)	34–38
Masonry units containing calcium silicate and clay, aerated non-floating concrete	9–11
Bituminous constituents	0–0.5
Soils	10–11
Glass	1–1.5

structures are innovative design alternatives for the construction of heavy earth retaining structures such as retaining walls and steep slopes [2].

Various experimental studies have been conducted to gain a clear insight into these C and D materials. Many tests have been conducted worldwide to determine the strength and governing geotechnical properties of the C and D materials. A similar experimental investigation was conducted to investigate the competence of C&D materials when used as a substitute material for backfill in Mechanically Stabilized Earth (MSE) walls. The C&D wastes were obtained from demolished residential buildings in Yeddumaillaram, Telangana. Specific gravity and water absorption tests, X-ray fluorescence test, and modified proctor compaction test were performed to calculate the engineering and index properties to evaluate its potential. The maximum dry density exists nearly to 19.35 kN/m^3. The waste material is found to be alkaline and carries a Ph > 7. The hydraulic conductivity is around 2.1×10^{-5} which indicates that the material has poor drainage. The studies regarding the modeling of C&D waste materials include simple constitutive guidelines and laws. These laws either agree to the linear elastic isotropic nature or the perfect plastic behavior of the C&D materials. Macro and micro-modeling approaches are generally considered for this purpose. For the simulation of the behavior of interfaces between layers of geosynthetic and the fill material, the Mohr–Coulomb failure criterion is widely acknowledged. Unfortunately, only a handful of suitable studies have been performed w.r.t the numerical modeling of C&D wastes as backfill materials, and there lies a huge necessity to perform more numerical simulations to have a broad perspective on the performance of C&D waste materials.

2 Requirements of a Suitable Backfill Material Used in Geosynthetic-Reinforced Structures

There are primarily three bodies, namely National Concrete Masonry Association (NCMA), American Association of State Highway and Transportation Officials (AASHTO) and Federal Highway Administration (FHWA), whose recommendations and specifications are widely accepted for the harmless design and construction of structures reinforced with geosynthetic materials. Particularly, for reinforced soil slopes and GRS walls, the construction traditions of the North American region, focus on selecting soils which are cohesion less for erecting the backfill. AASHTO [3] recommends the backfill material to be used behind earth retaining structures, bridge abutment walls, and piers should be a granular (free-draining) material to dissipate any possible pore water pressure, with pre-defined gradation. Figure 1 shows the quantified gradation limits in materials to be used in the backfill of reinforced slopes [4, 5].

The backfill must not contain particles with poor durability like shale and must have an organic content lesser than 1%. The pH of the backfill material must vary from 4.5 to 9 for permanent applications. For the situation of temporary conditions, the pH

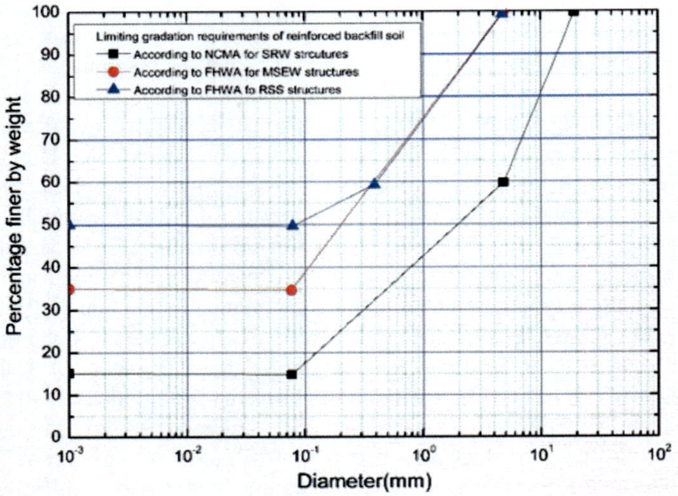

Fig. 1 Recommended particle size by FHWA [4] and NCMA [5]

may vary from 3 to 11. The NCMA lays down a rather extensive recommendation for the backfill materials. It classifies the soil by the Unified Soil Classification System. Here, the backfill soil is supposed to be free of any organic matter, debris, and must satisfy the specific limits of gradation (Fig. 1). NCMA prescribes the pH level to be inside the scope of 3–9. From a recent examination on the serviceability of geosynthetics used as a reinforcing element, it is noted that particle size distribution of the soil governs the effectiveness of the reinforcement. Furthermore, it is suggested that the greatest size of the particle for the backfill materials should not be more than 19 mm, except if appraisal tests were to be performed on the support mix with the particular or correspondingly evaluated huge size granular fill.

When compared to the recommendations laid down by FHWA [4], although the pH of construction and demolition materials had an alkali pH [6] as shown in Table 2, yet, construction and demolition materials satisfied the recommendations given by few

Table 2 Properties of construction and demolition materials [6]

Properties	Variability coefficient	Mean values
Dry unit weight	2.1	1.844 g/cm^3
California bearing ratio	–	60%
Cohesion	–	13 kPa
Specific gravity	3.1	2.819 g/cm^3
Optimum water content	13.3	14.9%
pH	4.3	9.1
Friction angle	–	41°

other researchers, in which geogrids made with polyester were used as reinforcements for the construction of an earth retaining structure.

3 Experimental Investigations on the Performance of C&D Wastes When Used as Backfill Materials

Arulrajah et al. [7] conducted many experimental investigations to evaluate the characteristics of geogrid reinforced construction and demolition materials. The outcomes so acquired by direct shear tests demonstrated a similar behavior as shown by granular materials. The unreinforced materials showed a better result than the geosynthetic-reinforced construction and demolition materials in terms of the interface shear properties. Be that as it may, this was credited to the absence of interlocking between the geogrids and the reused construction and demolition materials, and also to the fact that regular testing technique prompts a potential shear plane at zone between the upper and lower boxes where the geogrid is situated.

The aforementioned study was grounded on assessing the properties of shear strength of the geogrid-reinforced recycled construction and demolition materials, to evaluate the feasibility of their usage as an alternative backfill material. Recycled concrete aggregates (RCA), crushed bricks (CB), and broken asphalt pavement (RAP) were used for the experimental investigations performed in this study. Rahman et al. [8] conducted a similar study in which triaxial and biaxial geogrids were tested with analogous recycled construction and demolition materials. The geotechnical properties of various construction and demolition materials investigated by Arulrajah et al. [7] are discussed in Table 3.

Forwarding their previous study, Arulrajah et al. [9] conducted another experimental examination. This time they utilized an altered direct shear test apparatus to portray the interface shear quality properties of geogrid reinforced construction and demolition materials. This adjusted strategy utilizes a geosynthetic-cinching steel casing of 7 mm thickness joined to the head of the lower shear box (Fig. 2). Testing the interface with the changed shear box would initiate a shear plane 7 mm past the geogrid position level. The steel frame was 7 mm thick. "True field conditions" were created by avoiding a smooth interface and improved interlocking. This can also be explained as the modified method so proposed, induces better interface shear

Table 3 General properties of recycled concrete aggregates, crushed bricks, and broken asphalt pavement [7]

Various physical properties	CB	RCA	RAP
Max. dry density (g/m^3)	2.04	2.08	1.94
California bearing ratio (%)	135	172	39
Particle density-fine (g/cm^3)	2.48	2.60	2.33
Particle density-coarse (g/cm^3)	2.40	2.70	2.34
Optimum moisture content (%)	12.75	12.5	8.30

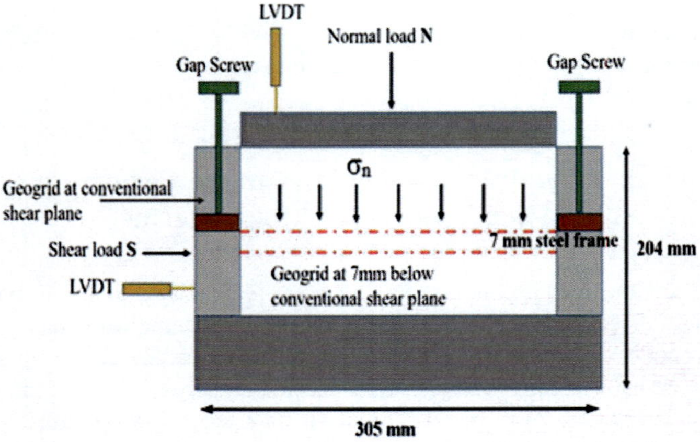

Fig. 2 Modified direct shear test box apparatus [3]

strength owing to the fact that the failure does not occur at the weaker planes. Then, a shearing plane which is 7 mm above the interface level is levied, as depicted in Arulrajah et al. [9], only the soil to soil friction is mobilized.

Vieira et al. [2] introduced the direct shear behavior of the geogrid strengthened reused construction and demolition material interfaces. The examination incorporated the fine-grained reused construction and demolition materials derived from the destruction of residential structures. The outcomes revealed that appropriately compacted reused construction and demolition materials showed better shear strength (in some cases more noteworthy) than the conventional backfill materials utilized in the development of geosynthetic-reinforced structures.

Jewell et al. [10] examined the impact of the particle size on geogrid–soil interaction in direct shear movement and stated that with an increase in the particle size of the soil, the coefficient of interaction increases. "It attains its maximum value when the grain size is similar to that of the apertures of the geogrid. When the grain size is equal to the size of the apertures of the geogrid, the bearing members and soil particles align against each other and the failure surface rises to the soil." The aperture sizes of the geogrids as studied by Arulrajah et al. [9] were 39 mm and 46 mm and 39 mm for the biaxial and triaxial geogrids, respectively.

The potential use of construction and demolition wastes as substitute materials was also conducted by Santos et al. through instrumentation, monitoring, and construction of three walls which were reinforced [11, 12]. One wall was constructed by using silty sand and the other walls were erected by the usage of reused construction and demolition materials. A collapsible foundation was taken for the walls. The walls which were constructed with construction and demolition materials were reinforced by a polyester geogrid and polypropylene nonwoven geotextile. The third wall incorporated a metallic grid for reinforcement which was built with sandy and silty backfill. The structures were monitored during wet and dry rainy seasons.

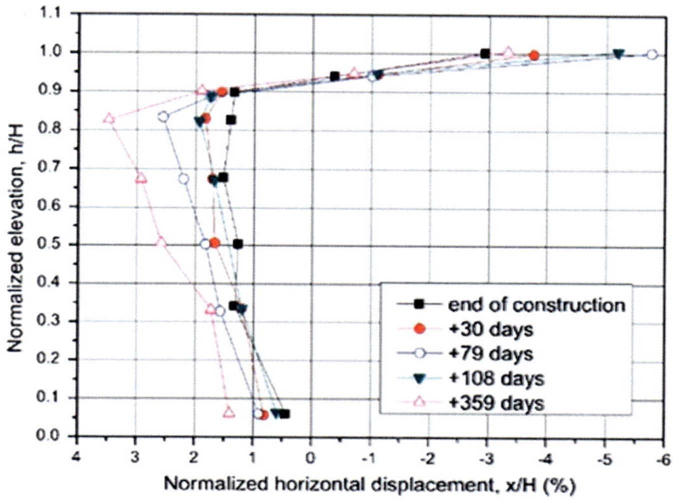

Fig. 3 Normalized lateral displacements of the wall face [13]

Figure 3 illustrates the lateral normalized displacements of the wrapped face high wall (3.6 m), constructed with reused construction and demolition backfill materials. This wall was reinforced with polyester geogrid [11]. As the construction ended, a maximum (normalized) lateral displacement of 1.4% was found at an elevation of 0.83H.

In continuation of the above experimental inquiries, pull-out capacity tests, and direct shear strength tests were directed by Santos et al. [13] to illustrate the performance of geogrid which were reinforced construction and demolition material interfaces. They employed the use of a biaxial geogrid, having a tensile strength of 30 and 61 kN/m on cross-section and in the direction of the machine. When compared with the findings of the investigations carried out by Lopes et al. [14], the results of the adherence factor derived for the structures which included geogrid and construction and demolition material interfaces were in the range similar to the range attained by Lopes et al. [14].

4 Conclusion

The concept of sustainable development has gained prime importance in present times and there is a lot that can be done by the construction industry to fulfill this purpose. The solution for attaining this goal is the use of reused construction and demolition materials in heavy earth retaining structures. "The partial or full replacement of soils and traditional aggregates by construction and demolition materials can provide much into the reduction of environmental impacts induced by the construction business, and thereby reducing our ecological footprint." Among the main conclusions in this review, the study reveals that the use of recycled construction and demolition wastes as back-filling material in geosynthetic-reinforced structures is a hopeful alternative from the conventional backfill materials. But a few more studies are still required to encourage this application.

References

1. EC-European Commission (2001) Competitiveness of the construction industry. Agenda for sustainable construction in Europe
2. Vieira CS, Lopes MD, Caldeira LM (2013) Sand-geotextile interface characterization through monotonic and cyclic direct shear tests. Geosynth Int 20(1):26–38. https://doi.org/10.1680/gein.12.00037
3. AASHTO LR (2012) Bridge design specifications, 3rd edn. AASHTO, Washington, DC
4. Berg RR, Christopher BR, Samtani NC (2009) Design and construction of mechanically stabilized earth walls and reinforced soil slopes. US Department of Transportation, Federal Highway Administration, National Highway Institute
5. NCMA (1997) Design manual for segmental retaining walls. In: Collin J (ed), 2nd edn, National Concrete Masonry Association, Herndon
6. Santos EC, Vilar OM (2008) Use of recycled construction and demolition wastes (RCDW) as backfill of reinforced soil structures. Proc 4th Euro Geosynth Con (EuroGeo 4), Edinburg, UK, pp 7–10
7. Arulrajah A, Piratheepan J, Aatheesan T, Bo MW (2011) Geotechnical properties of recycled crushed brick in pavement applications. J Mater Civil Eng 23(10):1444–1452. https://doi.org/10.1061/(ASCE)MT.1943-5533.0000319
8. Rahman MA, Imteaz M, Arulrajah A, Disfani MM (2014) Suitability of recycled construction and demolition aggregates as alternative pipe backfilling materials. J Clean Prod 66:75–84. https://doi.org/10.1016/j.jclepro.2013.11.005
9. Arulrajah A, Rahman MA, Piratheepan J, Bo MW, Imteaz MA (2013) Interface shear strength testing of geogrid-reinforced construction and demolition materials. Adv Civ Eng Mater 2(1):189–200. https://doi.org/10.1520/ACEM20120055
10. Jewell RA, Milligan GW, Dubois D (1984) Interaction between soil and geogrids, polymer grid reinforcement. Thomas Telford Publishing, pp 18–30
11. Santos EC, Palmeira EM, Bathurst RJ (2013) Behaviour of a geogrid reinforced wall built with recycled construction and demolition waste backfill on a collapsible foundation. Geotext Geomembr 39:9–19. https://doi.org/10.1016/j.geotexmem.2013.07.002
12. Santos EC, Palmeira EM, Bathurst RJ (2014) Performance of two geosynthetic reinforced walls with recycled construction waste backfill and constructed on collapsible ground. Geosynth Int 21(4):256–269. https://doi.org/10.1680/gein.14.00013
13. Santos EC, Vilar OM (2008) Use of recycled construction and demolition wastes (RCDW) as backfill of reinforced soil structures. In: Proceedings 4th Euro geosynthetics conference (EuroGeo 4), Edinburg, UK, pp 7–10
14. Lopes ML, Ladeira M (1996) Influence of the confinement, soil density and displacement rate on soil-geogrid interaction. Geotext Geomembr 10(14):543–554

A Study on Geotechnical Characteristics of Black Cotton Soil Treated with Red Mud and GGBS

Sweeti Singhadeo, Anil Kumar Choudhary, and Somenath Mondal

Abstract Disposal and storing of red mud, which is produced in a very large scale during bauxite refining, have become one of the most complex problem nowadays because of its high alkaline nature (pH > 11). To mitigate the negative impact of red mud on environment, it is essential to utilize red mud in different field of engineering. In this study, the efficacy of red mud and ground granulated blast furnace slag (GGBS) which is produced from steel manufacturing plant, in stabilizing black cotton soil. For finding out the change in engineering properties, three samples were prepared in which red mud was mixed as 10, 20, and 30% along with 20% GGBS. A remarkable decrease swelling property of black cotton soil after mixing red mud and GGBS was observed. Simultaneously, dry density and California bearing ratio was also increased when a certain percentage of red mud and GGBS was mixed. Improvement in engineering properties of black cotton soil after adding red mud and GGBS implies a successful application of industrial waste in geotechnical field after application of red mud and GGBS.

Keywords Black cotton soil · Red mud · GGBS · Free swell index · CBR

1 Introduction

Due to continuous population and rapid industrialization problems regarding land availability and degradation of soil are arising every day. To mitigate this problem, engineers and practitioners are continuously trying to use soft unused soil. In India, around 20% of the geographical area is covered by black cotton soil. Black cotton soil cannot be used to construct any structure because of its high swelling and shrinkage

S. Singhadeo (✉) · A. K. Choudhary · S. Mondal
Department of Civil Engineering, National Institute of Technology, Jamshedpur, India

A. K. Choudhary
e-mail: akchoudhary.ce@nitjsr.ac.in

S. Mondal
e-mail: smondal.ce@nitjsr.ac.in

A. K. Choudhary et al. (eds.), *Advances in Geo-Science and Geo-Structures*, Lecture Notes in Civil Engineering 154,
https://doi.org/10.1007/978-981-16-1993-9_22

properties. Improvement of engineering properties is a big challenge to geotechnical engineers. And in other side, due to industrialization, enormous quantity of waste is producing every year which needs proper disposal or storage; otherwise, it can be a great threat to environment. To reduce the negative impact of waste in environment, the waste products are further need to treat and use in different field. Nowadays, aluminum industries are one of the fastest growing industry and produced a huge amount of highly alkaline waste known as red mud. Red mud is obtained from bauxite refining during Bayer's process. During production of alumina, 35–40% of bauxite converted as red mud. In India, around 4.71 million tons per annum red mud is produced [1]. High alkalinity (pH 11–13) is one of the biggest problem of red mud [2, 3]. In addition, some toxic metals are also contained in red mud such as aluminum, silica, calcium, iron, and titanium. Some other constituents such as K, Na, Cr, Ba, Ni, Mn, Cu, Zn, and Pb are also contained in red mud [4]. The major drawbacks of red mud are high alkalinity, dispersiability, leaching, low compatibility, etc. [5] Red mud and cement mixed with clay can be used as clay liner because of its high compressive strength and low hydraulic conductivity [6]. Compression behavior of red mud was found to be similar as clayey soil and frictional behavior was same as sandy soil [7]. The specific gravity, angle of internal friction, and maximum dry density are generally maximum then local soil [8].

2 Experimental Study

2.1 Materials

Black Cotton Soil

In this research work, black cotton soil was taken from Nagpur, Maharashtra. The properties of this soil were evaluated by laboratory experiments corresponding to Indian standard code. From the experiment, it was found that specific gravity of this particular black cotton soil is found out as 2.61.

Red Mud

For this study, red mud was taken from Hindalco industries limited situated in Muri, Jharkhand. Utilization of red mud is the prime concern in this research work. Red mud used in this study has been shown in Fig. 1. The specific gravity of red mud is found out as 2.9.

Ground Granulated Blast Furnace Slag

GGBS is produced during production of iron and steel in steel plant. It is a glassy granular product, produced during quenching of iron slag which further ground into a fine powder. For this research work, GGBS was taken out from Tata steel plant in Jharkhand. In this study, 20% GGBS was mixed with each proportion of red mud

Fig. 1 Red mud

Table 1 Proportion of materials in sample to be tested

Sample designation	Proportion of material (%)		
	Black cotton soil (BCS)	Red mud (RM)	Ground granulated blast furnace slag (GGBS)
1	100	0	0
2	70	10	20
3	60	20	20
4	50	30	20

along with black cotton soil. The specific gravity of GGBS is found out as 2.88, and the compressive strength of GGBS is found out as 5 MPa.

2.2 Methodology

Preparation of Composite Sample

The collected red mud, GGBS, and black cotton soil were kept in oven for drying at a temperature of 105 °C for 24 h. After drying, the materials were crumbled and sieved through 4.75 mm sieve. Further, red mud and GGBS were mixed with black cotton soil in different proportion to achieve the expected result. Total four samples were prepared and tested including raw soil, which were shown in Table 1.

3 Results and Discussion

3.1 Particle Size Distribution

The particle size distribution or grain size distribution of black cotton soil and red mud was determined in laboratory by wet sieve analysis and hydrometer test as per IS: 2720 (Part-IV)-1985. The grain size distribution curve for black cotton soil and grain size distribution curve for red mud have been represented in Figs. 2 and 3.

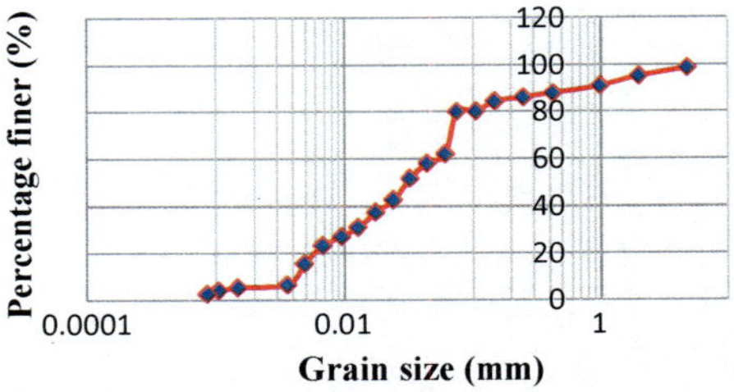

Fig. 2 Grain size distribution of black cotton soil

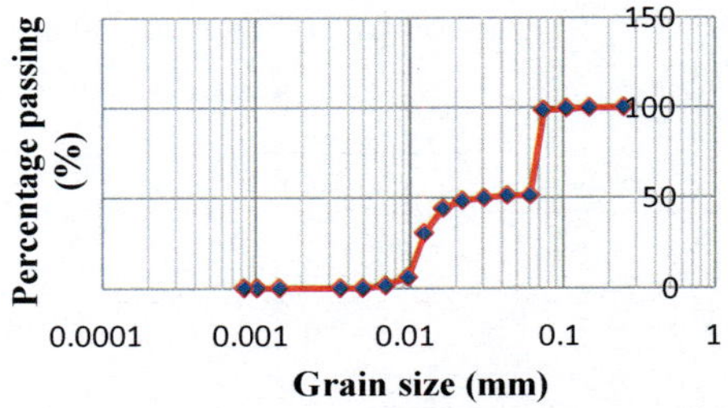

Fig. 3 Grain size distribution of red mud

Table 2 Properties of black cotton soil and red mud

Property	Black cotton soil	Red mud
IS Classification	MH	–
% of sand, silt and clay	19, 73.6, 6	1.6, 98.4, 0
D_{10} (mm)	0.0045	0.011
D_{30} (mm)	0.0126	0.012
D_{60} (mm)	0.0591	0.053
Coefficient of uniformity (C_u)	13.13	4.82
Coefficient of curvature (C_c)	0.596	0.246

The result obtained from the particle size distribution graph has been given in Table 2.

3.2 Variation in Free Swell Index due to Addition of Red Mud and GGBS

Black cotton soil possess high swelling characteristics due to high montmorillonite content. This expansive characteristic of black cotton soil is one of its biggest drawbacks. To eliminate this hindrance in construction work over black cotton soil, one needs to stabilize it first. In this study, industrial residue like RM and GGBS has been used to control swelling characteristics of black cotton soil. Hence, RM and GGBS

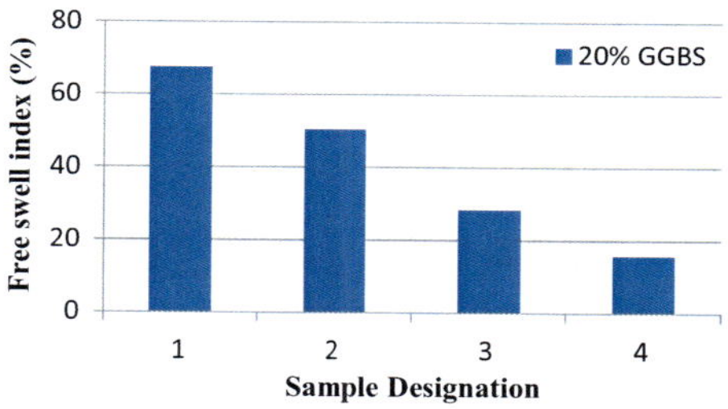

Fig. 4 Effect of RM and GGBS on free swell index of BCS

Table 3 Change in free swell index after adding RM and GGBS

S. No.	Composite samples	Free swell index (%)
1	100%BCS	67.5
2	70%BCS + 10%RM + 20%GGBS	50.3
3	60%BCS + 20%RM + 20%GGBS	28.5
4	50%BCS + 30%RM + 20%GGBS	15.8

were mixed in different proportion with black cotton soil. Figure 4 is showing the effect of RM and GGBS on expansiveness of black cotton soil.

From Fig 4, it can be concluded that when 30% RM and 20% GGBS were mixed with black cotton soil, a remarkable decrease in free swell index was obtained which dominate the use of RM and GGBS in black cotton soil. Due to decrease in expansiveness of black cotton soil by using RM and GGBS, it helps to take approach toward the use of RM GGBS treated black cotton soil in engineering field. The effect of different composition of RM and GGBS in black cotton soil is given in Table 3.

3.3 Variation in Compaction Test Result due to Addition of Red Mud and GGBS

Compaction test has been carried out for virgin black cotton soil as well as for composite samples to find out the variation in maximum dry density (MDD) and optimum moisture content (OMC). OMC obtained from compaction test further used to compact samples for CBR test. The test samples were compacted with 25 blows in five layers by a hammer of weight 4.9 kg. Dry density obtained for each water content was plotted in graph and formed a curve of bell-shaped. The peak point obtained from compaction curve is known as maximum dry density. The moisture content corresponding to MDD is known as optimum moisture content. Figures 5 and 6 show the variation in MDD and OMC for different samples.

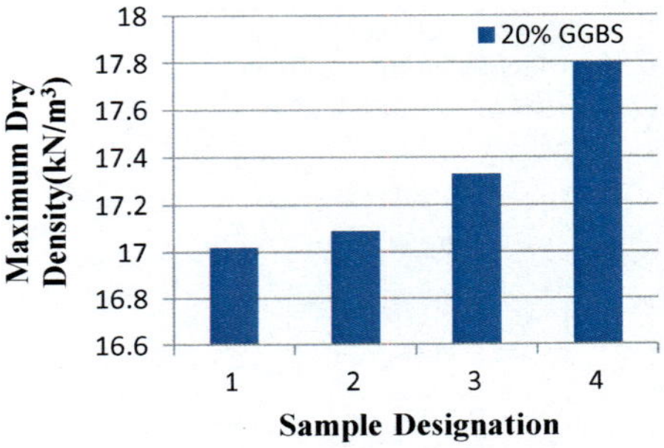

Fig. 5 Maximum dry density of different sample with 20% GGBS

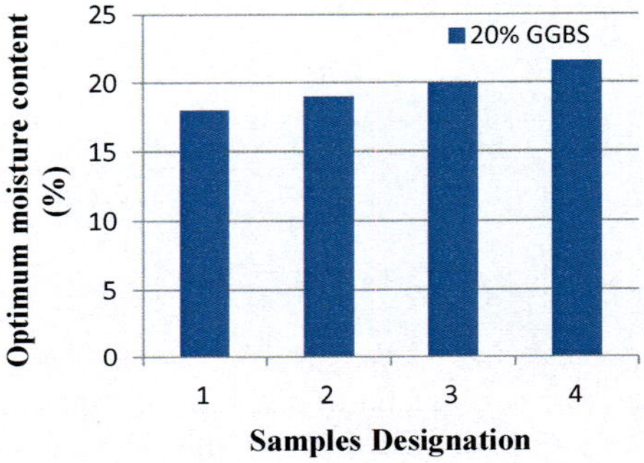

Fig. 6 Optimum moisture content of different sample with 20% GGBS

Figure 5 shows the variation in MDD for raw soil and for 20% GGBS mixed black cotton soil with varying percentage of red mud. MDD obtained from Fig. 5 indicates that maximum dry density is increasing with increase in red mud. By using 30% RM and 20% GGBS, MDD can be increased from 17.02 to 17.8 kN/m^3. In other side, it can be concluded that instead of decreasing, OMC was also increased on due to increase in red mud percentage. The OMC was increased from 18 to 21.6% when 30% red mud with 20% GGBS was mixed with black cotton soil. Table 4 shows the effect of RM and 20% GGBS on maximum dry density and optimum moisture content in tabulated form.

Table 4 Variation in MDD and OMC due to RM and GGBS

S. No.	Composite samples	MDD (kN/m^3)	OMC (%)
1	100%BCS	17.02	18
2	70%BCS + 10%RM + 20%GGBS	17.09	19
3	60%BCS + 20%RM + 20%GGBS	17.33	20
4	50%BCS + 30%RM + 20%GGBS	17.8	21.6

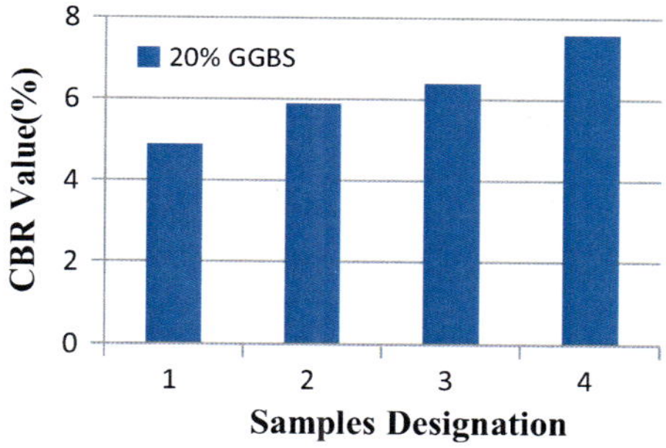

Fig. 7 Effect of different proportion of RM and 20% GGBS on CBR test result of BCS

Table 5 CBR value of BCS for different RM value and 20% GGBS

S. No.	Composite samples	CBR value (%)
1	100%BCS	4.9
2	70%BCS + 10%RM + 20%GGBS	5.9
3	60%BCS + 20%RM + 20%GGBS	6.4
4	50%BCS + 30%RM + 20%GGBS	7.6

3.4 Variation in CBR Test Result due to Addition of Red Mud and GGBS

CBR test is a very important parameter to find out the resistance against penetration or to find out strength characteristics. In this context, CBR test was conducted to determine the load carrying characteristics of raw black cotton soil and of different composite sample. The test was done as per the provision of IS: 2720 (Part-16)-1961.

For conducting CBR in unsoaked condition, the raw black cotton soil and composite samples were compacted at OMC, obtained by standard proctor test. The fresh compacted samples were than immediately tested. The result of the test has been shown in Fig. 7 for different RM and GGBS content. Table 4 is representing the test results for different samples.

From Fig. 7, it was concluded that use of 30% RM and 20% GGBS gives the best suited result for CBR. This composition increased CBR value from 4.9 to 7.6. Moreover, the CBR value increased up to 55% by using 30% RM and 20% GGBS. The other variations in CBR value for different composition of RM and GGBS were illustrate in Table 5.

4 Conclusions

1. The most important parameter that is expansiveness of soil get decreased after adding RM and GGBS. The effective decreasing of free swelling index, i.e.,

from 67.5 to 15.8% was observed for sample using 30% red mud and 20% GGBS.
2. The MDD and OMC of the composite sample get increased in comparison to virgin soil after mixing these additives. The MDD was increased up to a RM content of 30%. The increase in MDD is due to change in particle size distribution and specific surface area.
3. The optimum moisture content was also increased up to 21.6% due to addition of red mud up to 30% and due to addition of 20% GGBS due to change in surface area.
4. The value of CBR was also increased upto 7.6% with increase in red mud due to densely packing of composite samples by finer particles of additives.
5. Moreover, 30% RM and 20% GGBS is indicating a satisfactorily use, in stabilizing black cotton soil.

References

1. Parlikar UV, Saka PK, Khadilkar SA (2011) Technological options for effective utilization of bauxite residue (Red mud)—a review. In: International seminar on bauxite residue (RED MUD). Goa, India
2. Li LY (1998) Properties of red mud tailings produced under varying process conditions. J Environ Eng 124(3):254–264
3. Sundaram R, Gupta S (2010) Constructing foundations on red mud. In: 6th international congress on environmental geotechnics. New Delhi, India, pp 1172–1175
4. Summers RN, Guise NR, Smirk DD, Summers KJ (1996) Bauxite residue (red mud) improves pasture growth on sandy soils in Western Australia. Aust J Soil Res 34:569–581
5. Alam S, Das BK, Das SK (2018) Dispersion and sedimentation characteristics of red mud. J Hazard Toxic Radioact Waste 22(4):04018025. https://doi.org/10.1061/(ASCE)HZ2153-5515.0000420
6. Kalkan E (2006) Utilization of red mud as a stabilization material for the preparation of clay liners. Eng Geol 87(3):220–229
7. Newson T, Dyer T, Adam C, Sharp S (2006) Effect of structure on the geotechnical properties of bauxite residue. J Geotech Geoenviron Eng 132(2):143–151
8. Rout SK, Sahoo T, Das SK (2012) Utility of red mud as an embankment material. Int J Earth Sci Eng 5(6):1645–1651

Finite Element Analysis of Buried Pipe in Soil Slope

Rishi Ranjan, Awdhesh Kumar Choudhary, and Anil Kumar Choudhary

Abstract Buried pipelines are used from earlier stage of civilization for transporting drinking water, waste water, oil, natural gas, chemicals, etc. Due to large distribution system, buried pipes are needed to be installed in different geographical area. In this study, numerical analysis of buried pipes in soil slope was carried out in PLAXIS 3D. In addition, a strip footing was also installed to investigate the optimum vertical position of buried pipe which will have minimum effect on ultimate bearing capacity of footing. To find out the optimum vertical position of buried pipe, different parameters were considered such as slope angle, setback distance, and depth of pipe below lower surface of footing. From the result, it is observed that the footing's bearing capacity is maximum at slope angle 20° and at setback distance to footing's width 2. When the pipe is shifted away from the stress bulb of footing in a downward direction, the bearing capacity gets increased. Meanwhile, the safe vertical depth of buried pipe is adopted as 3.5 times the width of footing as the effect of buried pipes on bearing capacity of footing becomes minimum at this position.

Keywords Buried pipe · Slope · Strip footing · Finite element method · PLAXIS 3D

1 Introduction

Pipelines are extensively used to transport fluids in all over the world. Nowadays, most of the pipelines are distributed below ground surface due to unavailability of land. Pipes, which are laid down below the surface of earth, are termed as buried pipes. Different design of buried pipes infrastructure are required to satisfy the demand of

R. Ranjan (✉) · A. K. Choudhary · A. K. Choudhary
Department of Civil Engineering, National Institute of Technology, Jamshedpur, India

A. K. Choudhary
e-mail: awdhesh.ce@nitjsr.ac.in

A. K. Choudhary
e-mail: akchoudhary.ce@nitjsr.ac.in

A. K. Choudhary et al. (eds.), *Advances in Geo-Science and Geo-Structures*, Lecture Notes in Civil Engineering 154,
https://doi.org/10.1007/978-981-16-1993-9_23

communities to supply resources like oil, gas, etc., or to supply drinking water, sewage water, etc. Because of the large distribution system, buried pipes are needed to be passed through different geological and topographical area. Subsequently various difficulties occur during installation of pipelines as well as during its service life as the stability of buried pipelines get affected by soil motion or large ground movements or by slope failure.

First time, the analysis of the buried pipe was done on the basis of Terzaghi's theory to investigate the loads acting on the pipe [1]. Generally, failures in pipe may occur due to external forces, accidental pipe defects, or corrosion, so a pipe should have enough strength, so that it can resist soil load, foundation load, internal pressure, differential settlement, bending in longitudinal direction, etc. [2]. The stress developed in the pipeline caused by the failure of the slope due to landslides is not uniformly distributed, so it is very important to investigate the soil–pipeline interaction. The interaction between soil and pipe is normally studied in terms of maximum acting stress or force and lateral displacement at the upper limit of induced stress or force. To determine the displacement of pipe in lateral direction under loading condition, one of the most important parameter is stiffness of pipes [3]. Further, to investigate the deflection of a pipe in an infinite elastic medium, a new elastic solution was introduced [4]. From the field test conducted on polyurethane pipes, it was observed that displacement of pipe generally gets influenced by a factor known as pipe flexibility factor and is not proportional to the diameter of the pipe [5]. The horizontal strain of pipe is generally smaller than vertical strain for short term loading [6]. Throughout this study, the behavior of pipes embedded in sand slope and subjected to loads governed by strip footing was investigated by numerical analysis, i.e., finite element method in PLAXIS-3D.

2 Numerical Analysis

There are mainly two methods to analyze the complex problem by numerical modeling named as finite difference method and finite element method. The finite element method is done by splitting the complex structure into finite elements which can be interrelated by nodes. The numerical analysis analyzes the solution for the complex problem based on partial differential equation and integral equation. One of the commercial available program based on the finite element method is PLAXIS 3D which is used in the study for evaluation of deformation, stress, strain and failure aspect of the given problems.

This software is based on finite element method and is mainly developed for analyzing three-dimensional geotechnical activities. Geotechnical activity such as underground movement of water, deformation, and stability can form complex differential equation which can be solved by finite element method. In PLAXIS 3D, problems can be solved by forming mesh of different element. In PLAXIS 2D, the mesh is formed by triangular element, and in PLAXIS 3D, the mesh is formed by tetrahedron element.

Table 1 Properties of soil used in PLAXIS-3D

Parameters	Value	Parameters	Value
Unsaturated unit weight (γ_{unsat})	18.20 kN/m^3	Dilatation angle (ψ)	10°
Saturated unit weight (γ_{sat})	21.02 kN/m^3	Secant stiffness in standard drained triaxial test ($E_{50}{}^{ref}$)	60.00E3 kN/m^2
Initial void ratio (e_{init})	0.5000	Tangent stiffness for primary oedometer loading ($E_{oed}{}^{ref}$)	60.00E3 kN/m^2
Angle of friction of soil (φ)	38°	Young's modulus for unloading and reloading ($E_{ur}{}^{ref}$)	180.0E3 kN/m^2

Table 2 Properties of steel plate as footing in model tank

Parameters	Value	Parameters	Value
Width of footing (B)	0.2000 m	Young's Modulus (E)	210.0E6 kN/m^2
Unit weight (Y)	78.50 kN/m^3	Specific gravity of soil (G)	80.77E6 kN/m^2

Table 3 Properties of pipe

Parameters	Value	Parameters	Value
Diameter of pipe (D)	0.5000E−3 m	Young's Modulus (E)	933.0E3 kN/m^2
Unit weight (Y)	13.83 kN/m^3	Specific gravity of soil (G)	356.1E3 kN/m^2

2.1 Materials to be Modeled

For numerical modeling, the soil layer was simulated by hardening soil model (HS) in PLAXIS 3D. The properties of the soil layer are given in Table 1. After simulating soil layer, steel plate was simulated as strip footing and buried pipe. Properties of pipe as a strip footing and as a pipe have been given in Tables 2 and 3, respectively.

2.2 Details of Numerical Model

The test tanks used by Lee and Manjunath [7] during experimental study have been modeled in PLAXIS-3D. The dimension of the test tank is kept as 1.8 m × 0.9 m × 1 m. HS model was adopted for soil model to get more accurate result in PLAXIS 3D by assigning three stiffness parameters such as triaxial loading stiffness (E_{50}), the triaxial unloading–reloading stiffness (E_{ur}), and the oedometer loading stiffness (E_{oed}). Pipe having diameter of 75 mm is simulated in this model by pipe and installed

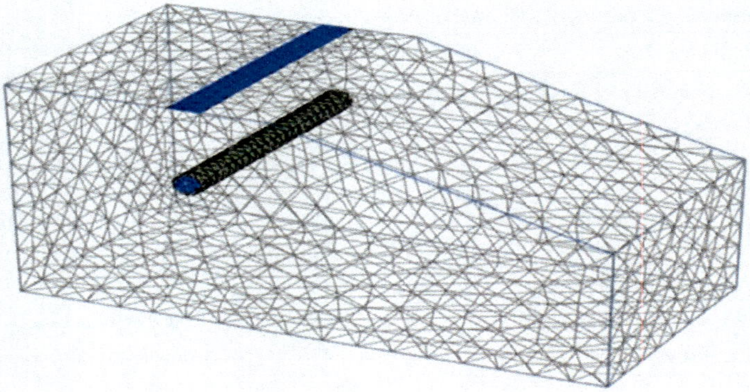

Fig. 1 Finite element mesh for model

at certain depth from footing. Pipe depth is varied to locate the safest position of pipe. The present model consists of 7944 triangular soil elements and 13,287 nodes. The generated mesh of this model is presented in Fig. 1.

3 Result and Discussion

In order to analyze the effect of buried pipe on the ultimate bearing capacity of footing in sand slope, a numerical analysis was done in PLAXIS 3D. The investigation is so performed to analyze the safest vertical position of buried pipe in soil slope and under the structural load. The variable parameters taken into consideration are sand slope angle (β), distance of footing from the crest (setback distance), relative density of sand, and vertical position of buried pipe. Numerical model is validated with an experimental results carried out by Lee and Manjunath [7].

Before installation of pipe in the slope, a non-dimensional factor has been introduced in this study which is known as bearing capacity reduction factor of footing. Bearing capacity reduction factor ($\boldsymbol{i_\beta}$) of footing is defined as the ratio of the ultimate bearing capacity of footing resting on soil slope (q_{slope}) to the ultimate bearing capacity of footing resting on the flat ground surface (q_{u}) without any pipe.

$$i_\beta = \frac{q_{\text{slope}}}{q_{\text{u}}}$$

The footing's bearing capacity on soil slope with embedded pipe has been also expressed in the term of non-dimensional variable is called bearing capacity ratio ($\text{BCR}_{\text{slope}}$). $\text{BCR}_{\text{slope}}$ is defined as the ratio of the footing's bearing capacity with pipe (q_{pipe}) to the footing's bearing capacity without pipe (q_{slope}) on slope soil.

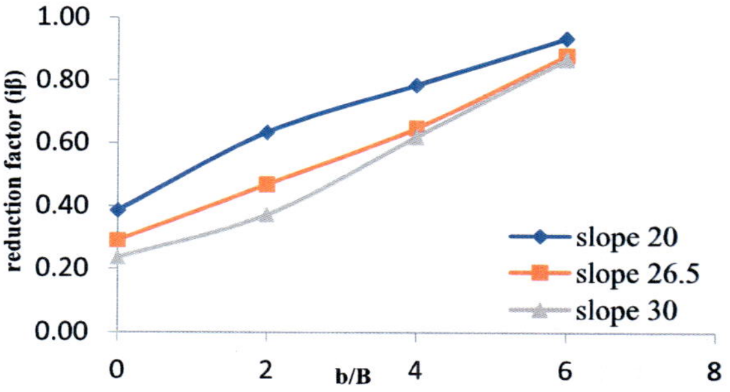

Fig. 2 Bearing capacity reduction factor for different slope angle

$$\mathrm{BCR}_{\mathrm{slope}} = \frac{q_{\mathrm{pipe}}}{q_{\mathrm{slope}}}$$

The ultimate bearing capacity of the footing was calculated by the curve of load–displacement. Generally, curves did not show a clear failure point on the load–displacement curves, so "double tangent method" was adopted to squeeze out the ultimate bearing capacity.

3.1 Variation in Bearing Capacity of Footing due to Change in Slope Angle

Effect of slope angle on footing behavior is analyzed in this study using different slope angles (β) (20°, 26.5°, and 30°). Further, the footing position was installed at four different setback distances for every slope angle, while the relative density (R.D) and footing's width were kept constant. The setback distance to the footing's width ratio b/B provided in this context were 0, 2, 4, and 6, and in other side, R.D = 85% and $B = 100$ mm were provided which were constant. The results are shown in Fig 2.

Figure 2 have shown that bearing capacity (q_{slope}) of footing decreases with increase in slope angle values. In other side, the bearing capacity of footing is increased when the footing position is shifted away from the slope crest (setback distance).

3.2 Influence of Footing Distance from the Slope Crest on Bearing Capacity of Footing

Several tests have been performed to find out the change in ultimate bearing capacity due to change in setback to footing's width ratio (b/B) of 0, 2, 4, and 6. In these analysis, the angle of slope (β) has been taken as 20°, the R.D of sand was taken as

Table 4 Bearing capacity and bearing capacity reduction factor due to change in setback distances

b/*B*	Bearing capacity of footing on slope (q_{slope})	Non-dimensional reduction factor (i_β)
0	560	0.39
2	900	0.62
4	1140	0.79
6	1280	0.88
Level ground	1450	1

85%, and the footing width was taken as 100 mm to find out the influence of footing position in respect to the slope crest. In addition, test on level surface ($\beta = 0°$) was also done to compare results.

The results obtained from the curves of load settlement for four different b/B ratios are given in Table 4. From the result, it can be concluded that when the footing position is shifted away from the slope, the bearing capacity get increased. The ultimate bearing capacity of the footing on soil slope beyond the *b*/*B* value of 6 gives the same result as on the level surface.

3.3 Influence on Bearing Capacity After Installation of Pipe in Slope

Effect of Vertical Position of Pipe on Ultimate Bearing Capacity Footing

The model test analysis is done for different ratio of vertical depth from lower surface of footing to the footings width, i.e., for embedment ratio (*H*/*B*) which are taken as 0.5, 1.5, 2.5, and 3.5. Throughout the analysis, the diameter of pipe (i.e., $D = 75$ mm) and footing's width ($B = 100$ mm). In addition, slope angle (β), relative density (R.D), and setback distance to the width of footing ratio (*b*/*B*) were taken as 20°, 85%, and 2.0.

The results of the different values of bearing capacity ratio (BCR_{slope}) for different embedment (*H*/*B*) ratios are squeezed out from load–settlement curves and its values are given in Table 5.

Table 5 Value of (BCR_{slope}) for different (*H*/*B*) ratios

H/*B*	BCR_{slope}
0.5	0.36
1.5	0.57
2.5	0.85
3.5	0.97
Without pipe in slope	1

It is found that when the vertical position of pipe from lower surface of footing (H) is increased, the value of bearing capacity ratio (BCR_{slope}) is also increased, because of that bearing capacity is directly get affected by a pipe installation within the stress bulb of footing.

A remarkable increase in bearing capacity was found when depth of embedment of pipe changes from $H/B = 0.5$ to $H/B = 2.5$. The increase rate in bearing capacity deceases after H/B value 2.5 and ultimately reaches near about 97% of case of without pipe. The footing's bearing capacity is directly getting affected when the pipe is installed within stress bulb. When the pipe position is shifted away from stress bulb of footing in vertical direction then the footing's bearing capacity gets increased. Contour for displacement of model are shown in Figs. 3 and 4. From the contours, it can be said that $H/B = 3.5$ is the best possible position of pipe in vertical direction.

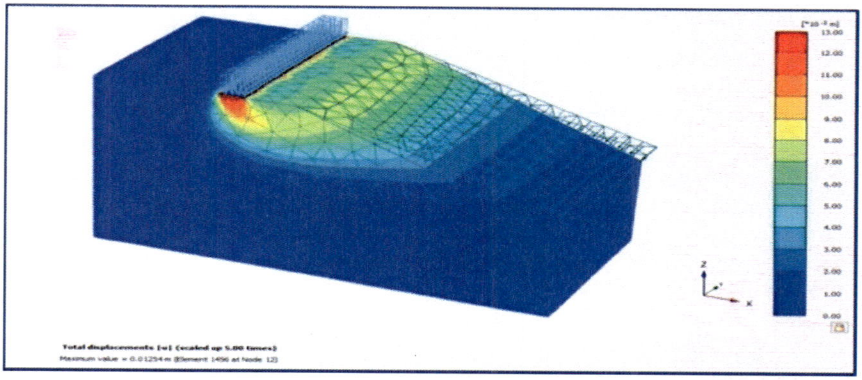

Fig. 3 Displacement contour without pipe

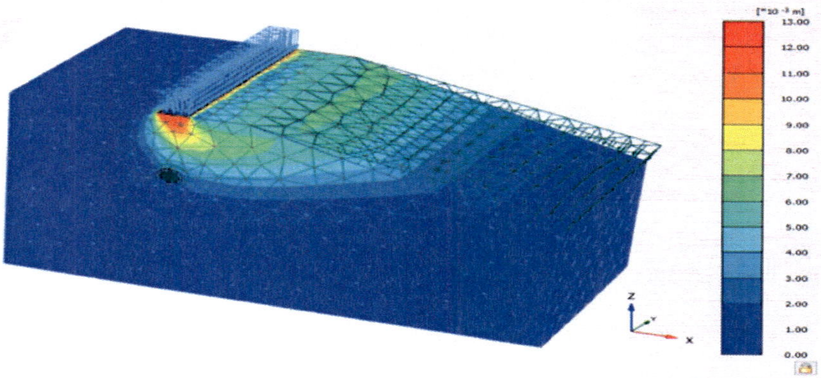

Fig. 4 Displacement contour with pipe: $H/B = 3.5$

At embedment ratio (H/B) 3.5 gives less displacement value and maximum bearing capacity ratio (BCR_{slope}), and this is the condition of more stable condition of buried pipe. Thus, based on numerical analysis observation, optimum position of pipe is fixed at H/B ratio 3.5 from the point of safety and serviceability of pipe.

4 Conclusions

1. From the test, it can be concluded that the load-carrying capacity of footing in soil slope was decreased with an increase in slope angle.
2. The result also indicated that, with increasing the setback distance (b), the bearing capacity of soil slope also increased up to $b/B = 6$, after that slope behaves like a level ground and corresponding bearing capacity is nearly equal to the bearing capacity on level ground.
3. It was also concluded that when the pipe depth was increased with respect to the lower surface of the strip footing, then the bearing capacity of the footing also get increased which signifies lesser effect on pipe.
4. Increasing the H/B ratio of pipe leads the way to an increase in the percentage rate of increment in bearing capacity. After a certain H/B ratio, the change in the bearing capacity of soil on the slope is negligible and BCR_{slope} tends to reach unity. Moreover, if the pipe is placed at $H/B = 3.5$, the bearing capacity tends to nearly equal to the same bearing capacity that in the case of without embedded pipe in soil slope. On the basis of present observation, optimum depth of pipe is fixed at H/B ratio 3.5.
5. Based on observation, the optimum position of pipe with respect to footing is fixed at H/B from the point of view of safety and serviceability.

References

1. Marston A, Anderson AO (1913) The theory of external loads on closed conduits in the iight of the latest experiments, Iowa State College Bulletin, No. 96, vol XXVIII. Iowa Engineering Experimental Station, Iowa State College
2. Moser AP, Folkman S (2008) Buried pipe design. McGraw-Hill, New York, NY, USA
3. Spangler MG (1941) Structural design of flexible pipe culverts, Bulletin No. 153. Iowa Engineering Experiment Station, IA, USA
4. Hoeg K (1966) Pressure distribution on underground structural cylinders. Technical Report No. AFWL TR 65–98, Kirtland Air Force Base, NM, USA
5. Hurd JO (1986) Field performance of corrugated polyethylene pipe culverts in Ohio. J Transp Res Board 1087:1–6
6. Adams DN, Muindi T, Selig ET (1989) Polyethylene pipe under high fill. Transp Res Record 1231:88–95
7. Lee KM, Manjunath VR (2000) Experimental and numerical studies of geosynthetic-reinforced sand slopes loaded with a footing. Can Geotech J 37(2000):828–842

8. Bildik S, Laman M (2015) Experimental investigation of the effects of pipe location on the bearing capacity. Geomech Eng 8(2):221–235. https://doi.org/10.12989/gae.2015.8.2.221
9. Bildik S, Laman M (2019) Experimental investigation of soil—structure—pipe interaction. Geotech Eng. KSCE J Civ Eng. https://doi.org/10.1007/s12205-019-0134-y
10. Acharyya R, Dey A (2018) Assessment of bearing capacity and failure mechanism of single and interfering strip footings on sloping ground. Int J Geotech Eng. https://doi.org/10.1080/19386362.2018.1540099
11. Rajani BB, Robertson PK, Morgenstern NR (1995) Simplified designs for method pipelines subjected to transverse and longitudinal soil movement. Can Geotech J 32:309–323

Determination of Lateral Load Carrying Capacity of Pile Group Located Near Contaminated Sand Slope Using Plaxis 3D

Chaitanya Kumar Singh, S. Biswas, and A. K. Sinha

Abstract With the rapid growth of industrialization and urbanization, the occurrences of war and natural disaster soil get contaminated due to spilling of oil. Contaminated soil with oil is unfit for civil engineering structures as the shear strength of the soil is reduced to a high extent which further impacts heavily loaded structures. The investigation on sand slope consists of numerical studies on two groups of piles (i.e., 2 and 4) with a different slenderness ratio of 16.67 and 21 at various spacing of 3.5 D_P, 4.5 D_P and 5.5 D_P. Oil content varies as 0 (i.e., clean sand), 2, 4 and 6% to understand the impact of oil content on lateral load carrying capacity of the pile group in contaminated sand. The influence of oil content on lateral load carrying capacity of group of pile decreases as the slenderness ratio of piles rises and slenderness ratio have greater influence on the loading capacity of group of the pile as compared to spacing in reducing the effect of oil content. Further, sand slope affects the lateral loading capacity of the pile to a great extent and it can be replaced by a pile group in most adverse conditions.

Keywords Contaminated soil · Slenderness ratio · Piles · Lateral capacity ratio

1 Introduction

Several civil engineering structures like tall buildings, culverts, bridges, irrigation structures, etc., are constructed beside the sloping sand using deep foundation. Lateral load due to wind influence the loading capacity of pile. In the past, many studies were investigated either experimentally or using the software which relate the effect of ground slope and lateral or axial loading capacity of pile. Mezazigh and Levacher [1] experimentally investigated the impact of sloping sandy surface using aluminum pile and concluded that the distance at which the pile is free of influence was 12 times the

C. K. Singh (✉) · S. Biswas · A. K. Sinha
Department of Civil Engineering, National Institute of Technology, Jamshedpur, India

S. Biswas
e-mail: sbiswas.ce@nitjsr.ac.in

A. K. Choudhary et al. (eds.), *Advances in Geo-Science and Geo-Structures*,
Lecture Notes in Civil Engineering 154,
https://doi.org/10.1007/978-981-16-1993-9_24

diameter of pile at a fixed slope of 2V:3H. Dhatrak et al. [2] experimentally examine the behavior of single pile in oil contaminated slope with an axial compressive load using steel pile in model tank, and concluded that pile capacity is maximum when ratio of length of contaminated soil and length of embedded depth of pile (L_c/L_p) is equal to 0.5. Aldelhalim et al. [3] perform an experiment and validate with numerical approach to understand the behavior of single pile having 21 mm diameter embedded in contaminated sand slope to examine the loading capacity, lateral displacement and bending moment of pile. They concluded the decrease in lateral loading capacity of pile located beside contaminated sand slope and also came to conclusion as the distance from the crest of slope increases the loading capacity of pile increases.

Moreover, due to rapid growth of industrialization and several human activities like deforestation and wars, areas get polluted resulting in contamination of soil causing land pollution. In the USA and reported that 20–25% of crude oil directly discharged to the soil and contaminating the soil. In India, there are about 600 million tons reserve of crude oil. In India, 82% are imported causing several problem of leakage while transportation. According to survey report, 10% of the oil reserves in India get wasted due to human error and leakages per year. Khamechiyan et al. [4] carried out an investigation to understand the behavior of soil in cohesive and cohesion less soil at different percent of oil content and concluded decrease in friction angle and permeability and of soil with increase in oil content in soil. Further, Nasr [5] investigated using FEM and calculated the decrease in bearing capacity factor (N_γ) in strip footing with increase in oil content. Nasr and Krishna Rao [6] carried out many triaxial experiments for computing angle of friction in contaminated soil. Results show comparatively less value of friction angle in sandy soil. Thus, the bearing capacity of soil falls due to the pollution of soil. There are very limited research and study involved in this type of condition. Research work should be done to analyze the capacity of a group of pile loaded laterally in these conditions having a sloping surface of sand and oil contaminated soil.

The prime objective of the current study and research is to numerically investigate and compare the behavior of groups of piles loaded in lateral direction beside the sand slope both in contaminated and non-contaminated sand by using PLAXIS 3D [7] software.

2 Model Preparation

PLAXIS Model in the Current Investigation

The numerical model equation can be solved using discretized approximate. Thus, the finite element method solves these approximate. In FEM, elements are divided into several numbers of nodes (stress and deformation nodes). In an element, deformation value (u) is obtained by discrete nodal value in vector (v) where the matrix (N) gets assembled:

$$u = Nv \tag{1}$$

Kinematic equation is given by Eq. 2:

$$\varepsilon = L_u \tag{2}$$

ε Strain component in six different direction.
L Differential operator.

Replacing u as N_v.

$$\varepsilon = LN_v = B_v \tag{3}$$

where, B = Strain interpolation matrix.
The basic equation for calculating deformation used in PLAXIS 3D software is:

$$\int \mathrm{BT}\Delta\sigma \mathrm{dV} = \int \mathrm{NTbidV} = \int \mathrm{NTtidS} = \int \mathrm{BT}\sigma_i - 1\mathrm{dV} \tag{4}$$

The dimension of the soil model as shown in Fig. 1 is considered as 1000 mm × 500 mm × 600 mm to analyze the lateral load carrying capacity of the pile beside contaminated sand slope with sand slope at the slope of 1V:2H. The contamination depth of oil in sand is 200 mm from top of the ground surface and four different percentages of oil content were used, i.e., 0 (clean sand), 2, 4 and 6%. The distance of the first pile from crest (X) is also kept constant (i.e., 4 D_P). The configuration of pile group is taken in model as 2 × 1 and 2 × 2. The spacing between the piles are taken as 3.5, 4.5 and 5.5 D_P. Two different length of piles were taken, i.e., 16.67 and 21 D_P. Plate acts like pile cap and depth are considered to be 5 mm and each side of plate is 215.5 mm length. Schematic diagram of model prepared in PLAXIS 3D is shown in Fig. 1.

Different properties of contaminated and uncontaminated soil, embedded beam and pile cap used in the PLAXIS MODEL are tabulated in Tables 1, 2 and 3, respectively.

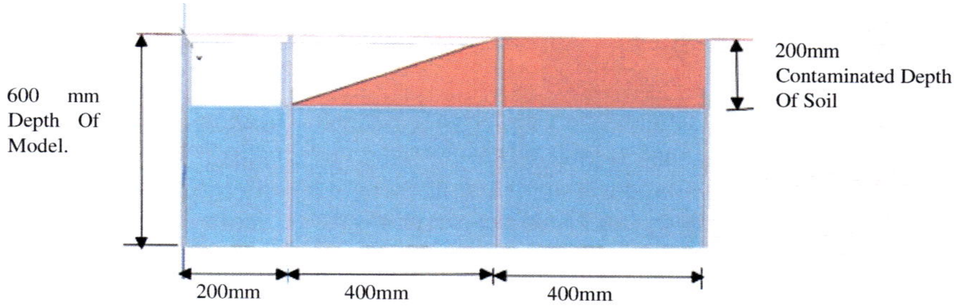

Fig. 1 Schematic diagram of model

Table 1 Properties of contaminated and uncontaminated

Parameters	OC = 0%	OC = 2%	OC = 4%	OC = 6%
Young's Modulus E_{ref} (kN/m^2)	15,000	12,500	11,000	10,000
Poisson's ratio (υ)	0.3	0.3	0.3	0.3
Friction angle (Φ) (°)	37	34.4	32.5	31.3
Dilatancy angle (Ψ)	7	4.4	2.5	1.3
Interface reduction (R_{inter})	0.75	0.7	0.7	0.7

Table 2 Properties of embedded beam used in the model

Parameters	Properties	Parameters	Properties
Young's Modulus (E_{ref}) (kN/m^2)	21.00E7	Diameter (mm)	21
Unit weight (γ) (kN/m^3)	78.50	Thickness (mm)	1.8
Poisson's ratio (υ)	0.2	–	–

Table 3 Properties of pile cap used in the model

Depth (m)	0.01
Unit weight (γ) (kN/m^3)	78.50
Young's Modulus (E) (kN/m^2)	21E7

The result obtained from the numerical analysis validate and give a good result with Aldelhalim et al. [3] of the single pile beside the contaminated sand slope.

3 Results and Discussion

In order to examine the lateral load carrying capacity of the pile beside contaminated sand slope, different variables parameters are taken and the effect of these parameters are analyzed using load-deflection curve numerically by PLAXIS 3D.

3.1 Impact of Oil Content and L/D *Ratio*

Effect of oil content and slenderness ratio on lateral loading capacity of pile group is shown in Figs. 2 and 3. Table 4 illustrates the impact of oil content on the deflection of pile when the force was applied in the lateral direction. To analyze the impact of oil content on the lateral loading capacity of the pile, variation of oil content. There is always decreasing curve trend as shown in Fig. 2 in lateral load capacity of pile group with an increase in oil content obtained in the analysis as due to the increment

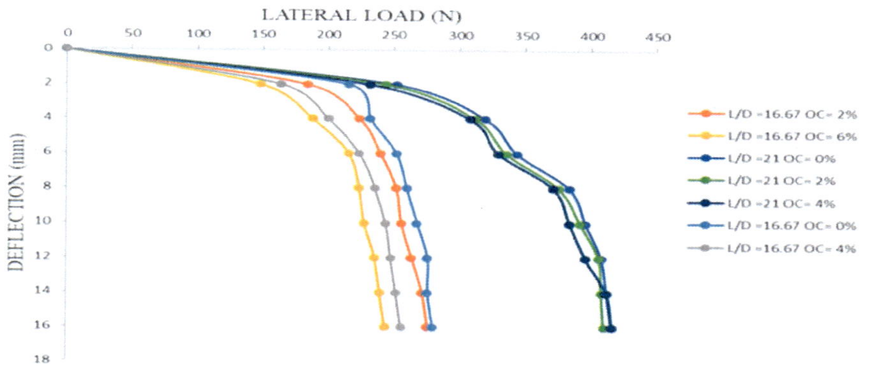

Fig. 2 Lateral load carrying capacity piles (2 × 1) with $L/D = 21$ and $L/D = 16.67$ for piles located at a spacing of 3.5D with different oil content

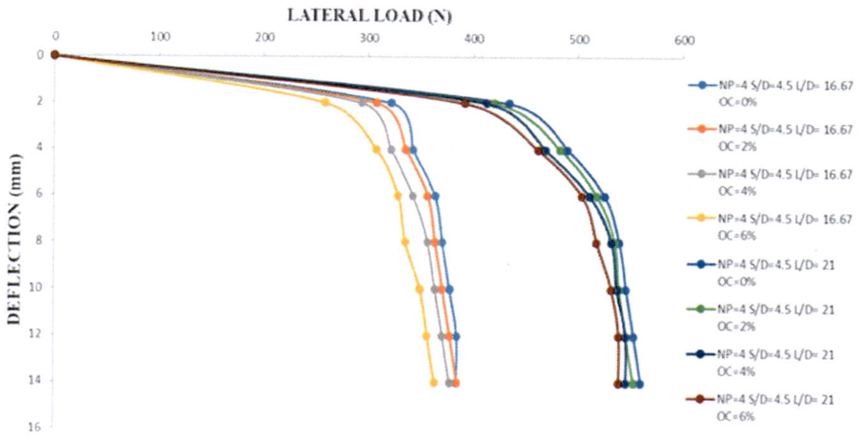

Fig. 3 Lateral load carrying capacity piles (2 × 2) with $L/D = 21$ and $L/D = 16.67$ for piles located at spacing of 4.5D with different oil content

in oil content, lubricates the soil particles which further decreases the friction angle leads to decrease in shear strength and ultimately lateral load decreases.

The increase in the lateral loading capacity due to an increase in slenderness ratio as given in 4 is due to the increase in the mobilization of passive resistance due to more embedment depth.

It can be seen from Table 5 that variation with higher *L*/*D* ratio of the pile in between the load is less as compared to variation in the lower *L*/*D* ratio. The conclusion can be drawn that group of piles with intermediate size are significant and effective in load carrying capacity than that with smaller size pile group.

Table 4 Percentage decrease in load carrying capacity of pile group (2 × 1) with the rise in oil content at *L*/*D* = 16.67 and 21

Oil content increment in percent	2%	4%	6%	The maximum difference from 0 to 6%
At spacing 3.5D, *L*/*D* = 16.67	4.76%	6.67%	3.57%	14.28%
At spacing 3.5D, *L*/*D* = 21	2.08%	1.064%	3.125%	4.166%
At spacing 4.5D, *L*/*D* = 16.67	20.83%	5.26%	3.88%	27.77%
At spacing 4.5D, *L*/*D* = 21	1.162%	3.53%	3.67%	8.54%
At spacing 5.5D, *L*/*D* = 16.67	2.56%	5.33%	5.41%	10.256%
At spacing 5.5D, *L*/*D* = 21	0.87%	1.755%	4.46%	6.95%

Table 5 Percent increase in load carrying capacity of pile group (2 × 1) with increase in *L*/*D* = 16.67–21

Sand with oil content (%)	0%	2%	4%	6%
At spacing of 3.5	16.67%	32.60%	41.46%	56.75%
At spacing of 4.5	19.44%	49.12%	51.85%	51.9%
At spacing of 5.5	47.43%	50%	51.35%	52.85%

Table 6 represents the effect of oil content in the lateral loading capacity of pile group configured to be 2 × 2. The percentage decrease in lateral loading capacity of pile group 2 × 2 is comparatively less as compare to 2 × 1 which is tabulated in Table 6. Thus, the effect of oil content in lateral loading capacity of pile group in 2 × 2 configuration is less as compared to 2 × 1 pile group. Further, Table 7 illustrates that the increment of pile in pile group leads to increase in the lateral loading capacity of the pile. But it is not so economical in adverse conditions.

Table 6 Percentage decrease in lateral load carrying capacity of pile group (configuration as 2 × 2) with the rise in oil content at slenderness ratio 16.67 and 21, respectively

Oil content increment in percent	2%	4%	6%	The difference from 0 to 6%
At spacing 3.5D, *L*/*D* = 16.67	2.22%	4.54%	2.4%	8.88%
At spacing 3.5D, *L*/*D* = 21	1.5%	1.52%	1.54%	4.47%
At spacing 4.5D, *L*/*D* = 16.67	1.92%	3.92%	4.1%	9.62%
At spacing 4.5D, *L*/*D* = 21	1.33%	1.36%	1.37%	4%
At spacing 5.5D, *L*/*D* = 16.67	1.78%	1.82%	3.7%	7.143%
At spacing 5.5D, *L*/*D* = 21	1.23%	2.5%	2.564%	6.17%

Table 7 Percent increase in load carrying capacity of pile group (2 × 2) with increase in *L*/*D* from 16.67 to 21

Sand with oil content (%)	0%	2%	4%	6%
At spacing of 3.5	48.88%	50%	54.76%	56.1%
At spacing of 4.5	44.23%	45.1%	48.98%	53.2%
At spacing of 5.5	44.64%	45.45%	44.44%	54.72%

3.2 *Impact of Spacing*

The impact of spacing on the pile group capacity loaded in lateral direction for a model is shown in Fig. 4 (Table 8).

Thus, percentage increase in lateral load as shown in Table 5 in pile group (2 × 2) becomes almost constant when there is increase in spacing at slenderness ratio of

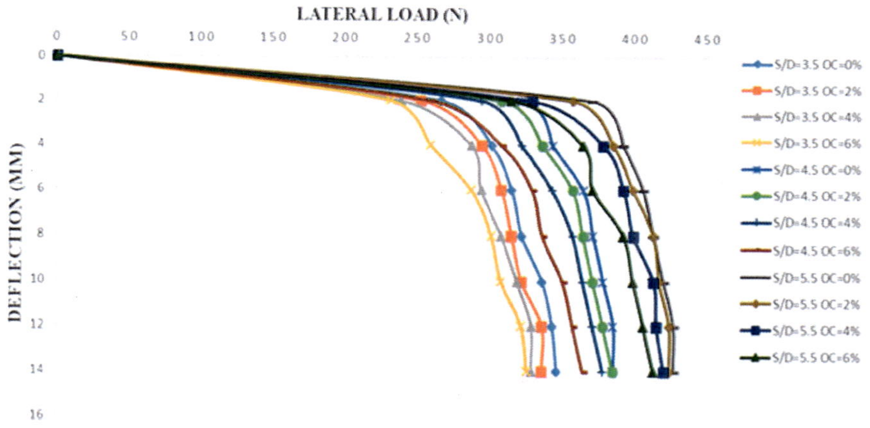

Fig. 4 Behavior of lateral deflection curve at *L*/*D* = 16.67 with different spacing and oil content of group (4) of piles

Table 8 Increase in load due to the rise in spacing at constant *L*/*D* (16.67 and 21) and at different oil content

Oil Content (%)	Spacing increases from 3.5D to 4.5D, *L*/*D* = 16.67 (%)	Spacing increases from 4.5D to 5.5D, *L*/*D* = 16.67(%)	Spacing increases from 3.5D to 4.5D, *L*/*D* = 21(%)	Spacing increases from 4.5D to 5.5D, *L*/*D* = 21(%)
0	15.55	11.53	11.9	12
2	15.9	11.76	12.12	12.16
4	16.67	14.28	12.3	12.3
6	14.63	12.76	12.5	12.5

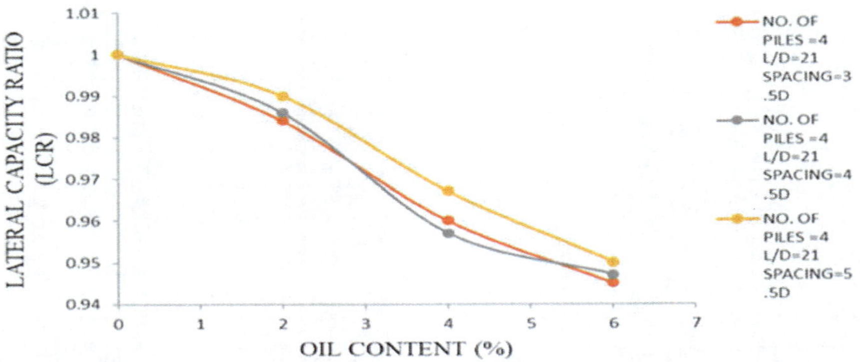

Fig. 5 Variation of lateral capacity ratio and oil content for short and long pile at different spacing for pile group (4)

21. This is due to effect in the pressure bulb overlapping. Due to increase in spacing, overlapping in the pressure bulb is least.

3.3 *Impact on LCR*

Figure 5 is the curve plotted between oil content and lateral capacity ratio for short and intermediate pile group at different spacing. The curve obtained is in decreasing order. As the oil content increases, LCR decreases with the least value at 6%. The decreasing curve of LCR with oil denotes the effect of oil in the sand as the oil increases the lubrication between sand particles and decreases its angle of friction. The decrease in angle of friction leads to a decrease in shear strength of the soil. Hence, as the oil content increases, more easily sand can be sheared with increment in load.

Thus, it can be concluded that the most critical value of the lateral load carrying capacity of the pile group is at 6% oil content for constant slenderness ratio and spacing as the sand can get easily get sheared at this percent on the application of load.

3.4 *Impact of Sand Slope*

Decrease in percentages of load carrying capacity of single pile due to sand slope is about 40.9% and 41.17% for short and intermediate pile, respectively. The most critical condition for sand slope is when oil content is 6% due to the consequence of a decrease in shear strength of soil. Considering the economic benefit, pile group (2)

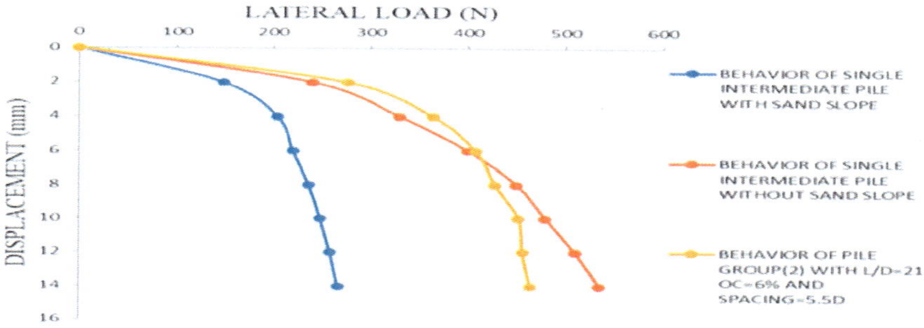

Fig. 6 Variation in the behavior of single intermediate pile with and without sand slope and behavior of pile group (2 × 1) at oil content 6% and spacing 5.5D

spacing 5.5 D_p with oil content 6% shows same behavior as that behavior of single pile without sand slope for both short and intermediate pile (Fig. 6).

The percentage increase in load carrying capacity for pile group (2) with oil content 6% is about 1.64 and 9.34% for short and intermediate as compare with a single pile without sand slope. Single pile without sand slope behaves almost in same manner as pile group (2) with contaminated sand having oil content 6% at spacing 5.5 D_p. Hence, in most adverse condition (sand slope with oil content 6%), single pile can be replaced by pile group (2) at spacing of 5.5 D_p.

4 Conclusions

(a) Group of piles subjected to lateral loading on the sand slope, with the increment of oil content in sand, the lateral load carrying capacity of group of pile reduces due to decrease in shear strength.

(b) As the L/D_P ratio increases, lateral load carrying capacity of the group pile increases. Due to the increment of the depth of embedment, it leads to an increase in the mobilization of passive resistance.

(c) The lateral loading capacity of the group does not get much affected due to an increase in the percent of oil content in the intermediate pile. Hence, it can be concluded that with an increase in the slenderness ratio effect of oil content reduces on lateral load carrying capacity of group.

(d) In group of pile, subjected to lateral loading beside sand slope, with increase in spacing lateral loading capacity also increases. Thus, pile with L/D_P=21 and spacing 5.5 D_P is having maximum lateral loading capacity adjacent to sand slope.

(e) The percentage change in lateral loading capacity decreases as oil content and spacing both increase. But, the influence of L/D_P ratio in reducing the effect of oil content on lateral load carrying capacity of a group of the pile beside contaminated sand slope is higher as compare to increment of spacing.

(f) Increase in oil content lead to decrease in lateral capacity ratio (LCR). This is due to an increase in lubrication between soil particles which decreases the friction angle of sand. Thus, shear strength of soil mass gets reduced as there is an increase in oil content.

(g) The efficiency of the pile group reduces due to the sand slope. Single pile without sand slope can be replaced by pile group at a spacing of 5.5 D_p with most adverse condition, i.e., contaminated sand slope with oil content of 6% for both short and intermediate pile.

References

1. Mezazigh S, Levacher D (2011) Laterally loaded piles in sand: slope effect on P–Y reaction curves. Can Geotech J 35(3):433–441
2. Dhatrak AI, Sabale SV, Thakare SW (2017) Performance of single pile in oil contaminated sand under axial loading. Int J Eng Sci Res Technol 6(6):574–582
3. Aldelhalim RA, Sawwaf ME, Nasr AM, Farouk A (2020) Experimental and numerical studies of laterally loaded piles located near oil-contaminated sand slope. Eng Sci Technol Int J
4. Khamehchiyan M, Charkhabi AH, Tajik M (2006) Effects of crude oil contamination on geotechnical properties of clay and sandy soils. Elsevier Eng Geol 89:220–229
5. Nasr AM (2009) Experimental and theoretical studies for the behavior of strip footing on oil contaminated sand. J Geotech Eng 135(12):1814–1822
6. Nasr A, Krishna Rao S (2016) Behaviour of laterally loaded pile groups embedded in oil-contaminated sand. Géotechnique 66(1):58–70
7. PLAXIS 3D (2020), PLAXIS 3D foundation user manual. Version 1.6 Plaxis Inc., Delft, The Netherlands
8. IS 2911-1-1 (2010) Design and construction of pile foundation—code of practice, Part 1: concrete piles. Section 1: driven cast in-situ concrete piles. CED 43: Soil and Foundation Engineering
9. IS 2911 (PART III)- 1980, Code for practice of design and construction of pile foundations. PART III Under reamed pile
10. Al-Sanad HA, Eid WK, Ismael NF (1995) Geotechnical properties of oil contaminated Kuwati sand. Am Soc Civ En (ASCE) J Geotech Eng (121):407–412
11. Dhatrak AI, Bhagat G (2016) Behaviour of single pile in reinforced slope subjected to inclined load. Int J Eng Res Appl 6(5):43–49. ISSN: 2248-9622, Issue 9

Use of Waste Plastic Bottle as a Geocell for Ground Improvement: A Review

Rohan Deshmukh, Gopal Patil, Urvi Bhatt, Ritik Shingote, and Tejas Patil

Abstract This study reviews the use of waste plastic bottles as a geocell for ground improvement. Characterization and properties of waste plastic bottle geocell (WPBG) for effective utilization for ground improvement have been discussed briefly. The effective aspect ratio of WPBG, the influence of infill material, has been well explained in this study. This paper reviews small-scale laboratory tests and numerical analysis of geocell mattresses made of plastic bottles carried out by many researchers. This review indicated that the inclusion of geocell mattresses made of plastic can produce a sevenfold enhance in the footing pressure; the 12-fold improvement was observed in the footing capacity while encased fly ash column geocell mattress composite system was introduced within the clay. It also indicated by numerical analysis that the cellular reinforced fly ash slopes that have higher ratios of (Lr/H) have an increment in load-carrying capacity because of the enhanced stiffness where cellular reinforcement material is made up of post-consumer waste plastic water bottles. However, the results encourage the utilization of waste plastic bottles for ground improvement, which successively may solve the disposal issue of plastic waste to a particular extent.

Keywords Geocell · Plastic bottle · Reinforcement · Stiffness · Confinement

1 Introduction

The biggest problem faced by the human race in the present era is dealing with the waste materials being generated at an alarming rate. Due to its production at an alarming rate, it has become a headache for mankind. The major waste material being used is waste plastic bottles. Their disposal is such a huge mess that it has become very difficult to find a solution. Nowadays, there is a shortage of land for the disposal of waste. Through the use of waste products, this problem can be countered effectively. This will reduce land pollution which is the effect of excessive dumping

R. Deshmukh (✉) · G. Patil · U. Bhatt · R. Shingote · T. Patil
Department of Civil Engineering, Terna Engineering College, Nerul, Navi Mumbai 400706, India

A. K. Choudhary et al. (eds.), *Advances in Geo-Science and Geo-Structures*,
Lecture Notes in Civil Engineering 154,
https://doi.org/10.1007/978-981-16-1993-9_25

Fig. 1 Photograph of plastic bottle geocell mattress [7]

of waste. The present study reviews the best solution available for this problem. At the same time, there is another issue of low-bearing capacity of soft soil and loose sand. There is a need for ground improvement to make structures safe resting on such soil. A new sustainable method to improve the quality of soil can be developed with the help of waste plastic bottles. Because of this, several past researchers have attempted to use these waste plastic bottles as geocell reinforcement for ground improvement. Geocell made from a waste plastic bottle is shown in Fig. 1.

2 Characterization of Material

All the previous researchers have used plastic bottle made from polyethylene terephthalate (PET or PETE) those are mostly used for drinking water, juice, and soft drinks. Sushovan and Mandal [7] in their study use bottles having a thickness of 0.3 mm and diameter ranging from 50 to 200 mm. The most important characterization for any reinforcement is its tensile stress–strain behaviour. In Fig. 2, plastic bottle strip of thickness 0.3 mm, width 50 mm, and 150 mm long was tested to understand tensile stress–strain curves. Properties of plastic bottles used by previous researchers for making cellular reinforcement are given in Table 1. Using plastic tie wire, discrete cells are connected to form a cellular mattress.

3 Experimental Studies on Conventional Geocell

The experimental study has confirmed that reinforced geocell lessens the settlement of foundation resting on soft soil by improving its bearing capacity [4, 32]. Marto et al. [23] have prepared a tentative test to uncover the effect of placement depth and width of the geocell reinforcement bearing capacity of sandy soil. Results of the experimental study have proven that if the geocell placed at $0.05B$ where B is the width of the foundation below the base of footing and optimum width of $b/B = 5$

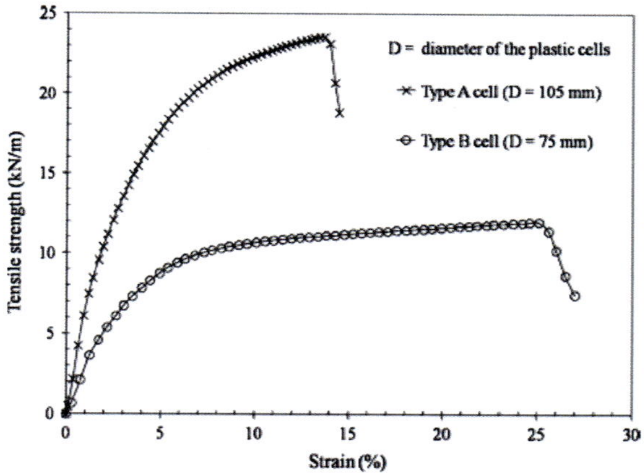

Fig. 2 Tensile strength-strain curves of plastic bottle specimens as per ASTM D5035-11 [10]

in the sand performing better in terms of improved bearing capacity. Also, Tafreshi Moghaddas and Dawson [29] have discovered the optimal value for deepness and width of the geocells 0.1*B* and 0.3*B*, respectively, for the footing subjected cyclic loading. "Geocell with inferior infill material like reclaimed asphalt pavement (RAP) has been effectively used as a base layer over a weak subgrade in the pavement application" [31]. The soil can be confined, and foundation can be stabilized using geocell within the limit to prevent the instability under loading [10]. Improvement of CBR values on expansive soil was noticed due to the influence of an increase in rubber strips, as the percentage of rubber, strips increased the CBR value, has also increased but only up to an optimum per cent, after which the CBR value has decreased. So, the increment in the CBR value of the soil was by 88% due to 5% rubber strips as optimal, Koteswararao et al. [15]. Chowdhary et al. [3] have considered the use of geosynthetics in improving the CBR values on the expansive soil and concluded that the swelling can be controlled significantly by inserting the reinforcement inside the soil subgrade.

However, the per cent of drop-in swell potential depends on the number and kind of reinforcement used. Increment in the number of reinforcing layers also increased the CBR value of soil. "The experimental study was carried out on the mitigation of traffic induced vibration using geocell inclusions. According to the results from the experimental results, the optimum depth for the effective mitigation of traffic vibration was suggested as 0.1*B*. The elasticity of the subgrade was improved by 96%. The amplitude versus frequency responses are also showed good agreement with the experimental results which predicted from the field and numerical study" Hegde and Venkateswarlu [12]. Han et al. [10] "carried out an experimental study on conventional geocell. The result indicated that by using geocells (3D cellular structures), the bearing capacity of the soil increases and settlements decrease. It is also used to stabilize foundations". Kumawat and Tiwari [16] "conducted a series of laboratory model tests on geocell reinforcement on the behaviour of fly ash beds as a foundation. According to the test results, the geocell-reinforced fly ash beds improved

Table 1 Properties of plastic bottles used by previous researchers

Researcher	Properties	Value
Nadaf and Mandal (2017)	Tensile strength	10 kN/m
	Tensile modulus	50 kN/m
	Height	10, 20, and 30 mm
	Diameter	50 mm
Dutta and Mandal (2017)	Tensile strength	24 kN/m
	Tensile modulus	177 kN/m
	Height	50, 100, and 150 mm
	Diameter	105 mm
Dutta and Mandal (2016)	Tensile strength	24 and 12 kN/m
	Tensile modulus	177 and 50 kN/m
	Height	50, 100, 150 and 200 mm
	Diameter	105 and 75 mm
Ram RathanLal and Mandal (2014)	Tensile strength	9 and 15 kN/m
	Tensile modulus	197 and 232 kN/m
	Height	15 and 30 mm
	Diameter	50 and 70 mm
Rao Nageshwar and Kumar (2019)	Tensile strength	450 MPa
	Tensile modulus	3100 MPa
	Height	–
	Diameter	80 mm
Adhana and Mandal (2011)	Tensile strength	12 kN/m
	Tensile modulus	–
	Height	10 mm
	Diameter	50 mm

the bearing capacity 3–4 times than the unreinforced fly ash beds with much smaller permanent deformations than the unreinforced fly ash beds". Lal and Mandal [22] "carried out the laboratory model tests to study the behaviour of a cellular reinforced prepared from wastewater bottles and fly ash which is used as a backfill exposed to strip loading. The test result showed that when the backfill is reinforced with cellular reinforcement may be feasible to use an Lr/H ratio less than the typical value of 0.7

3.1 Experimental Studies on WPBG

Karminder et al. [28] "studied the performance of the strip footing resting on sand bed reinforced with plastic bottle geocells. Results indicate that different heights had positive effects but beyond $h/B = 0.1$, not much effect was observed and suggested that optimum placement depth was $u/B = 0.25$. Significant improvement was observed in terms of CBR value for the sand reinforced with plastic bottle geocell mattress". Sushovan and Mandal [7] "considered a replacement quite encasement which was installed to limit or to stay the fly ash penetrated completely by using post-consumer waste plastic water bottles. The encased fly ash columns, geocell composite systems within the soft clay, have produced better footing capacity of about 5-fold, 8.5 times, and 12-fold, respectively, during a series of systematic model tests when it is compared with untreated clay bed. It has been noticed that the contribution from the encased column has decreased when the height of the mattress over the encased fly ash column increased".

Nageshwar and Kumar [27] "performed experimental study on geocell made up of polyethylene terephthalate (PET) and on soil subgrade of fibre reinforced under static and repetitive load. Improvement was observed in CBR, resilient modulus (Mr), and UCS values when soil is reinforced with basalt fibre. Lateral confinement is achieved in the soil by introducing PET bottles in the form of geocell. When a PET bottle of 80 mm diameter is provided at $1/3H$, from the top increment is observed in CBR and M_R". Adhana and Mandal [1] "carried out experimental study on fly ash steep slopes on soft foundation with and without reinforcement to check the stability of steep slope. Results indicated that both geocell strips and geogrid strips made of waste plastic bottle have better load-carrying capacity than the unreinforced soil".

Thakare and Sonule [30] analysed and carried out detailed experimental study on all the possible uses of plastic waste bottles for the improvement of soil bearing capacity. The geocells made of waste plastic bottles used for the experimental study were of diameter = 60 mm and height = 270 mm. Test set-up comprised of apparatus such as—steel tank of size of 0.5 m length, 0.5 m width, and 0.5 m height; loading frame; hydraulic jack; pumping unit and dial gauges to measure the footing settlement. The study observed that the bearing capacity ratio (BCR) was increased for spacing of $B/8$ between the layers. The maximum benefits were obtained at the space of $B/8$. The study concluded that the optimum results were obtained by providing L/B ratio of 2 with 3 reinforcing layers and with spacing of $B/8$. Lal et al. [18] carried out tests without and with three-dimensional circular geocell mattresses of height = 10 mm, diameter = 100 mm, and thickness = 0.4 mm and also on geocell strip with tensile strength as 12 kN/m. These geocells were made up of used waste plastic bottles. They observed the shape of failure wedges through perspex plate fixed on a side of tank. The observation suggested that the maximum load for failure was more for the geocell mattres as compared with the strip geocell when used as reinforcement. They concluded that for a particular surcharge value, the mattresses of geocell-reinforced wall show reduced normal panel displacement than the strip geocell. This indicated that the geocell mattresses reinforced wall performed better than geocell strip reinforced wall.

3.2 *Experimental and Numerical Studies on WPBG*

Nadaf Maheboobsab and Mandal [25] have conducted "numerical analyses using finite element analysis (FEA) PLAXIS 3D to confirm the experimental laboratory model test results of cellular reinforced fly ash slope. Properties of cellular reinforcement material made from waste plastic water bottles are shown in Table 1. Results from both experimental and numerical studies confirmed that the cellular reinforcement allow much better load distribution over a larger area. Lal and Mandal [19] studied the "performance of cellular reinforcement (geocell) in fly ash under the static triaxial test. With an increase in the height of cellular reinforcement for both the diameters of reinforcement, the peak deviator stress and shear strength parameters are found increasing. Similar results are also obtained with finite element simulation carried out by using PLAXIS 2D". Dutta and Mandal [7] reported that "cellular mattress fly ash composite bed improves seven times hooting capacity when compared with unreinforced fly ash bed. Finite element analysis in PLAXIS 3D is showing good agreement with experimental results". Dutta and Mandal [8] "carried out numerical analyses of cellular mattress-reinforced fly ash beds overlying soft clay. Results indicated an improvement in footing capacity by approximately 1.4 times larger over fly ash bed with the presence of a single geotextile separator". Dutta and Mandal [7] conducted "laboratory model tests on geocell mattress prepared from a waste plastic bottle and an infill material as fly ash overlying soft clay beds. According to them, when the height and the coverage ratio of cellular reinforcement are increased, the higher value of failure (q_f) will take place".

4 Behaviour of WPBG Infill with Marginal Material

Mohanty and Choudhary [24] studied the "behaviour of multilayered geocell-reinforced bottom ash in which geocell mattresses are made up of waste PET bottles. On increasing the aspect ratio of geocell mattress from 0.28 to 0.84, bottom ash has the highest ultimate bearing capacity at $h/d = 0.84$. The overall performance of the geocell material increases by increasing the number of geocell layers in the same mass". Dutta et al. [9] made an "overview on the use of waste plastic bottles and fly ash in civil engineering applications. With an increase in the edge distance from the slope crest, the ultimate bearing capacity (UBC) of the footing increases but current investigation UBC decreases at $2B$ edge distance. The difference in percentage for both $De/B = 1.5$ and 2.0 for CR5020 is nearly similar. All reinforcement lengths show the same improvement factor of $De/B = 1.5$ and 2.0. In the top layer of reinforcement, tension is more, and it is also subjected to bending".

5 Use of Natural Materials as Geocells

Balan [2] "carried out the experimental study of geocell made of coir which is placed on soft soil. Improvement in bearing capacity is seen by 2.6 times and 10.5 times with that of unreinforced sand cushion and unreinforced soft clay bed. More bearing capacity can be seen when woven planar coir geotextile is placed at the base of the coir geocell. Settlement observed is less in coir geocell than the synthetic geocell. But the bearing capacity of coir geocell is 2.5 times that of synthetic geocell". Amarnath and Sitharam [11] "carried out experimental study on bamboo geocells to check its performance in clay bed. Geocell and geogrid improve the ultimate bearing capacity of the clay bed by 6 times to that of unreinforced clay. It is observed that the ultimate bearing capacity of clay bed has been increased by 1.3 times by bomboocell and bamoogrid to that of geocell and geogrid". "An overview of natural materials as geocells and their performance evaluation for soil reinforcement was optimized. From the study, it is observed that the performance of geocells made of areca, coir, sisal is better than the geocell made of jute which has very less effect for the reinforcement. Comparing all the combinations of reinforcement, areca cell and areca grid are showing maximum bearing capacity. It gives 1.25 times greater bearing capacity than the geocell and geogrid", Kolathayar [14].

6 Conclusion

In this study, laboratory tests on as well as numerical analysis of geocells mattresses made of plastic bottles are carried out, and also the effective aspect ratio of WPBG and the influence of infill material have been well explained. The problem of effective use and disposal of waste plastic is countered with this study. Also, the hazard caused by the accumulation of plastic in the environment can be restricted by the use of plastic mattresses as geocells which also gives birth to a new recycling industry that produces plastic geocells. This will provide a major boost to the small-scale industries and sustainable development in our country. The main objective of the study is achieved as the geocells made of plastic have given the following conclusive observations which indicate the improvement in the bearing capacity of the soil structure. Geocell mattresses made up of plastic produce a sevenfold increment in the footing pressure. The fixing of encased fly ash column in geocell mattresses with a composite system in clay, the plastic geocell mattresses, shows a 12-fold increment in footing pressure. The slopes of cellular reinforced fly ash have a greater capacity to carry the load with a higher (Lr/H) ratio because of enhanced stiffness.

References

1. Adhana S, Mandal (2011) Reinforced fly ash slope using different geosynthetics. In: Proceedings of Indian geotechnical conference, 15–17 December 2011, Kochi (Paper No. J 194)
2. Balan K (2020) Behaviour of coir geocells over soft soil. In: Sitharam G et al (ed) Geocells. Springer Trans Civ Environ Eng
3. Chowdhary AK, Gill KS, Jha JN (2011) Improvement in CBR values of expansive soil subgrades using geosynthetics. In: Proceedings of Indian geotechnical conference
4. Swaraj C, Suman S (2015) A review of studies on geocell reinforced foundations. Res J Recent Sci 4
5. Dash SK, Krishnaswamy N, Rajagopal K (2001) Bearing capacity of strip footings supported on geocell-reinforced sand. Geotextiles and Geomembranes, 235–256
6. Dutta S, Mandal JN (2013) Feasibility study on waste plastic water bottles as encasements of stone columns for ground improvement. In: International symposium on design and practice of geosynthetic-reinforced soil structures. University of Bologna, Bologna, Italy, pp 379–388
7. Sushovan D, Mandal JN (2017) Model studies on encased fly ash column geocell composite systems in soft clay. J Hazard Toxic Radioactive Waste
8. Dutta S, Mandal JN (2015) Model studies on geocell reinforced fly ash bed overlying soft clay. J Mater Civ Eng ASCE
9. Dutta S, Nadaf M, Mandal JN (2016) An overview on the use of waste plastic bottles and fly ash in civil engineering applications. J Procedia Environ Sci 35(2016):681–769
10. Han J, Yang XM, Leshchinsky D, Parsons RL, Rosen A (2008) Numerical analysis for mechanisms of a geocell-reinforced base under a vertical load. In: Geosynthetics in civil and environmental engineering. Springer, Berlin, Heidelberg, pp 741–746
11. Amarnath H, Sitharam TG (2020) Performance of bamboo geocells in soft ground engineering applications, geocells. Springer Trans Civ Environ Eng
12. Amarnath H, Hasthi V (2019) Mitigation of traffic induced vibration using geocell inclusions. Front Built Environ 5:136
13. Husna H, Binil G (2016) A study on the improvement of CBR using waste plastic mat as geocell. Int J Eng Res Technol (IJERT) 05(09). ISSN: 2278-0181
14. Sreevalsa K, Renuka G, Sitharam TG (2020) Overview of natural materials as geocells and their performance evaluation for soil reinforcement, Geocells. Springer Trans Civ Environ Eng
15. Koteswara Rao D, Shilpa Devi GNVVSSL, Pranav PRT (2012) A laboratory study on the influence of rubber strips on the improvement of CBR values of expansive soil. IJESAT 2
16. Kumawat NK, Tiwari SK (2017) Bearing capacity of square footing on geocell reinforced fly ash beds. Mater Today: Proc, 10570–10580
17. Lal BRR, Mandal JN (2012) Feasibility study on fly ash as backfill material in cellular reinforced walls. Electr J Geotech Eng 17(J):1637–1658
18. Lal BRR, Deshmukh VB, Mandal JN (2013) Study of cellular reinforced fly ash under triaxial loading conditions. Int J Geotech Eng 7(1)
19. Lal BRR, Mandal JN (2013) Study of cellular reinforced fly ash under triaxial loading conditions. Int J Geotech Eng 7(1):91–104
20. Lal BRR, Mandal JN (2014a) Behavior of cellular-reinforced fly-ash walls under strip loading. J Hazard Toxic Radioactive Waste Manage ASCE 18(1):45–55
21. Lal BRR, Mandal JN (2014) Model tests on geocell walls under strip loading. Geotech Test J ASTM 37(3):477–548
22. Lal BRR, Mandal JN (2015) Model studies on cellular reinforced fly ash walls. Int J Geotech Eng 9(3):289–297
23. Aminaton M, Mohsen O, Amin E (2013) Effect of geocell reinforcement in sand and its effect on the bearing capacity with experimental test: a review. EJGE 18
24. Madhumita M, Kumar CA (2018) A study on the behaviour of multilayered geocell reinforced bottom ash. In: Indian Geotechnical Conference (IGC 2018)

25. Nadaf Maheboobsab B, Mandal JN (2019) Numerical simulation of loading strip footing resting on cellular mattress and strips: reinforced fly ash slopes. Int J Geosynth Ground Eng 3(2017):26
26. Rajakumar C, Meenambal T, Arumairaj PD (2014) CBR of expansive subgrade stabilized with waste materials. IJASGE 3
27. Nageshwar R, Ramesh A, Kumar M (2019) Experimental study on geocell and of fibre reinforced soil sub-grade under static and repetitive load. In: Sundaram R et al. (ed) Geotechnics for transportation infrastructure, Lecture Notes in Civil Engineering. Springer Nature Singapore Pte Ltd., p 29
28. Karminder S, Gurdeepak S, Kaur GG (2019) Behaviour of strip footing resting on sand bed reinforced with plastic bottle geocells. Int J Multi 04
29. Moghaddas Tafreshi SN, Khalaj O, Dawson A (2014) Repeated loading of soil containing granulated rubber and multiple geocell layers. J Geotext Geomembr 42(1)
30. Thakare SW, Sonule SK (2013) Performance of plastic bottle reinforced soil. Int J Eng Innovation Res 2(3). ISSN: 2277-5668
31. Thakur et al (2013) Creep behavior of geocell-reinforced recycled asphalt pavement bases. J Mater Civ Eng ASCE 25(10):1533–1542
32. Manish Y, Kumar AA, Akash P, Gaurav D (2014) Application of geocells in reinforcement of soil: a review. J Civ Eng Environ Technol

Studies on the Piled Raft Foundation for a High-Rise Building Using Finite Element Modeling

Pratish Kannaujiya, Sagar Jaiswal, and Vinay Bhushan Chauhan

Abstract Piled raft foundations (PRFs) are essential for improving the load-bearing capacity of the foundation and for controlling the total and differential settlements. For an efficient design of the foundation for high-rise buildings, components of PRF such as length, diameter, and raft thickness are important factors. This study aims to examine the interaction behavior of pile–soil foundations for various parameters, namely the diameter and the length of the pile and also the raft thickness, which plays an important role in improving the behavior of piled raft foundation. The important aspects that must be taken into account to achieve a reliable design and strategies for the optimized design of PRF that is subjected to load–settlement with regard to the distribution of shear force and bending moment are also analyzed. In order to understand the behavior of piled raft foundation subjected to uniform loading, numerical simulations are carried out in this study by using a numerical tool based on the finite element method (FEM), ELPLA. This study reveals that proper pile arrangement can result in a significant reduction in the total and differential settlements, as well as induced shear force and bending moments on the raft.

Keywords Piled raft foundation · Pile configuration · Numerical analysis · Pile settlement · FEM

1 Introduction

High-rise buildings are often built on piled raft foundations (PRF) that must withstand vertical, lateral, and overturning loads. The consideration of wind and seismic loads is also important while designing high-rise buildings. In such buildings, the foundation impacts a strong axial load on the ground. In the case of such structures, it is therefore important that the foundation is designed in such a way that it can resist the load of the superstructure and can also adequately transfer these loads to the ground. In previous

P. Kannaujiya · S. Jaiswal · V. B. Chauhan (✉)
Department of Civil Engineering, Madan Mohan Malaviya University of Technology, Gorakhpur 273010, India

A. K. Choudhary et al. (eds.), *Advances in Geo-Science and Geo-Structures*, Lecture Notes in Civil Engineering 154,
https://doi.org/10.1007/978-981-16-1993-9_26

studies, conventional methods were used to design the pile group to withstand the heavy loads acting on the pile foundations [1–3].

In the PRF, the load is distributed between the piles and the raft, so that the load from the superstructure can be partially taken over by the piles and the raft [4]. The analysis of the PRF is also performed by a hybrid method to analyze the total and differential settlement [5]. The simplified analysis based on the stiffness of the pile group and raft stiffness had been presented by many researchers for analyzing the load–settlement behavior of the PRF [6, 7]. In addition, the behavior of piled rafts has been analyzed with a pile of different lengths subjected to horizontal and vertical loadings by using the finite layer method [8]. It is noted from the available literature that the earlier researchers had developed three broad classes of different methods to analyze the behavior of PRF, such as simplified calculation methods, approximate computational methods, and more rigorous computational methods [6, 9–12].

Horikoshi et al. [13] adopted a centrifuge model to examine the behavior of a combined PRF under static horizontal and vertical–horizontal load and concluded that different results were obtained when the response of a single isolated pile is compared with the PRF of the same size. Also, the researchers had reported that the proportion of vertical load supported by the piles in a piled raft remains unchanged, whereas in the case of pile foundation, the proportion of horizontal load supported by the pile upsurges as the horizontal displacement decreases.

Rabiei [14] had conducted a numerical study to examine the effect of loading type on pile configuration and PRF using the FEM. The results showed that the piled raft foundation is economical foundation systems. Moreover, the study suggested that the design philosophy of a pile raft configuration should be based on both ultimate load and settlement responses.

From the previous studies, it has been found that the provision of the piled raft foundation improves the overall stability of the structures. Besides, the stability is controlled by many important parameters such as raft thickness, pile length, and pile diameter. These factors not only govern the overall stability of the structure but also contribute to an optimum selection of the aforementioned parameters, which leads to an economical design of the PRF.

Given the above, a comprehensive numerical study has been carried out to investigate the influence of pile length, pile diameter, pile spacing, numbers of piles, pile configuration, and raft thickness on the performance of the PRF by using a numerical tool based on the finite element method (FEM). Moreover, the detailed design of PRF system with optimum pile raft design and pile configuration scheme for a building of 20 storeys having a height of 60 m is carried out, and the total load (133 MN) of the building is calculated manually by using the codes of Indian Standards [15–18].

2 Numerical Modeling and Analysis

In the present paper, the designing of the piled raft foundation is performed by considering two configurations and their factor of safety as well, by using the finite element computational tool ELPLA [19]. A parametric study is performed to evaluate the behavior of PRF and also to analyze the parameters of different types of subsoil models. The thickness of layers of soil that are available for piling in case of loose, medium dense, and dense soils is 0–4 m, 4–15 m, and 15–35 m, respectively. The raft is modeled as an elastic element, and the size of the piled raft is 20 × 20 m, with a uniform load applied over the raft to investigate the maximum settlement and bending moment of piled raft foundations. In order to create a better understanding, two configurations of pile raft are considered for designing of foundation with an optimum selection of the pile length (L_p), pile diameter (d_p), and the raft thickness (t_R) which leads to an economical design. The details of the model configuration that are used in this study are shown in Fig. 1. The width (B) of the raft and the spacing between piles (s) are shown in Fig. 1. The properties of the soil and the foundation that are considered in this study are presented in Table 1. A comprehensive numerical study has been carried out to investigate the effect of length, diameter, and spacing between the piles on a pile group, and the raft thickness on the behavior of PRF using curves obtained from the numerical analysis.

In this study, two pile configurations are considered, namely pile configuration A and B, respectively. The pile configuration A has 34 and B has 25 piles, respectively, out of which the 25 piles of configuration A are of type P1 and 9 piles are of type P2, whereas the pile configuration B has 16 piles of type P3 and 9 piles of type P4. The

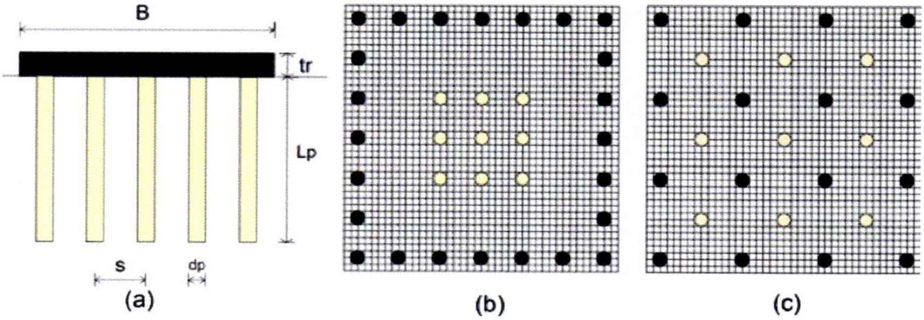

Fig. 1 Model configurations, **a** model condition, **b** pile configuration A, and **c** pile configuration B

Table 1 Properties used in the present study

Parameter	Soil type			Pile	Raft
	Loose	Medium dense	Dense		
Young's modulus, E (MPa)	10	18	22	3×10^4	2×10^4
Unit weight (kN/m^3)	18	19	20	25	25
Friction angle, ϕ(°)	22	31	38	–	–
Cohesion, c (kN/m^3)	1	2	2	–	–
Poisson's ratio, μ	0.3	0.3	0.3	0.18	0.25

pile configuration A having pile type P2 and B having pile type P4 has a constant length of 10 m, whereas the length of pile type P1 and P3 is varied to examine the behavior of pile length on the settlement of the foundation.

To analyze the behavior of pile raft, the length of pile type P1 and P3 is varied from 10 to 30 m, and the diameter of the pile and the raft thickness is kept constant at 1 m. Also, the embedded length of the piles (L_1/L_2) is varied from 1 to 3 in an interval of 0.5, for the selection of optimum depth of embedded pile length. Furthermore, for the ease of understanding, the length of pile and raft thickness are normalized with the pile diameter, i.e., L_p/d_p and t_R/d_p, and have been varied from 5.2 to 1.7 and from 0.5 to 1.5 in an interval of 0.2, respectively, to obtain the optimum dimensions.

3 Results and Discussion

To examine the behavior of the piled raft foundation, a parametric study is done for two configurations of PRF by varying the pile length, pile diameter, and raft thickness by using a finite element method-based numerical computational tool, ELPLA. The numerical results obtained from the present study are demonstrated in the form of curves and discussed in the following section.

Moreover, to validate the current numerical model, a numerical analysis is done for 9 identical piles placed at a spacing of 2 m vertical and 4 m horizontal concerning each other, and response of raft thickness versus the maximum bending moment and the differential settlement have been calculated (as shown in Fig. 2). The results obtained from the present study are in good comparison with those reported by Van Impe and Lungu [20]. Therefore, the current model can be described as well-validated and suitable for further analysis.

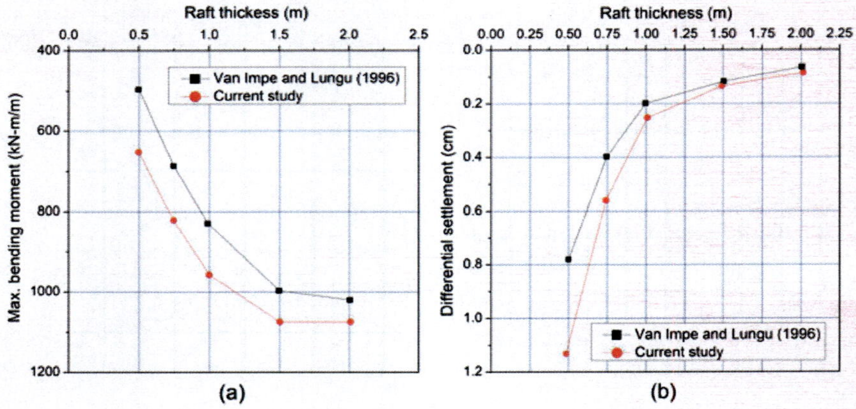

Fig. 2 Validation of the current numerical model

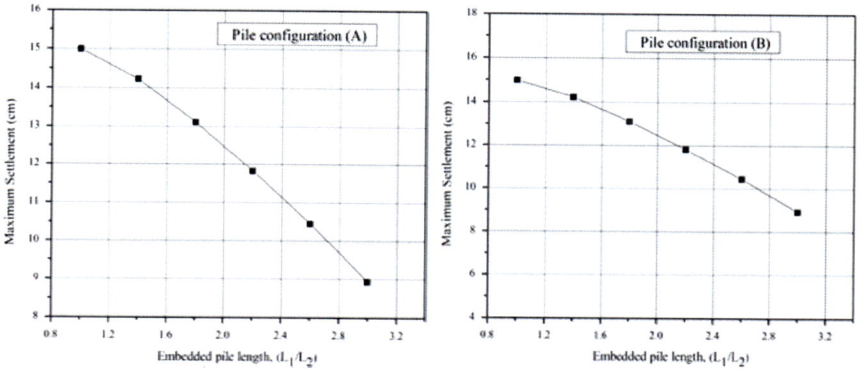

Fig. 3 Variation of maximum settlement with the embedded pile length (L_1/L_2) for pile configuration A and B

The variation of maximum settlement with the embedded pile length (L_1/L_2) is shown in Fig. 3. It can be noted from Fig. 3 that for the pile configuration (i.e., A and B), the settlement decreases with the increase in pile length. Moreover, the decrement in the settlement of configuration A and B is almost similar despite a greater number of piles present in A (i.e., 34). The settlement in both the configurations is reduced by 66.67%, as the embedded length is increased from 1 to 3. It is observed from both the configurations that the determinations of the proportional load carried by the piles are insensitive to the embedded length of the piles and there is no major difference in the settlement observed.

Furthermore, keeping constant the embedded pile length (L_1/L_2) as 2.6 and raft thickness 1 m, the diameter of the pile is varied for the procurement of optimum diameter which tends to enhance the load-bearing capacity and overall stability of the foundations. The interaction factor applied to the raft for a vertically loaded pile is 332 kN/m^2. The curve between maximum settlement versus normalized diameter of the pile is obtained on the application of load over the raft and shown in Fig. 4. It can be noted from Fig. 4, as the L_p/d_p ratio changes from 5.2 to 3.5, there is a sudden decrease of 6.6, and 13% in the maximum settlement is observed in configurations A and B, respectively. However, after $L_p/d_p = 2.8$, the change in the settlement is not very significant, and therefore, for keeping the design more economic, the diameter corresponding to $L_p/d_p = 2.8$ is kept constant for further analysis.

In a piled raft foundation, the design of the raft thickness is an important factor which also results in improving the ultimate load-bearing capacity and settlement behavior of the system. In the present study, the normalized raft thickness, i.e., (t_R/d_p) has been varied from 0.4 to 1.2 to obtain the optimum thickness of the raft. The variation of the maximum settlement versus normalized raft thickness is shown in Fig. 5.

The result indicates that the maximum settlement decreases linearly with the t_R/d_p ratio for both the configurations A and B, respectively. The maximum settlement at

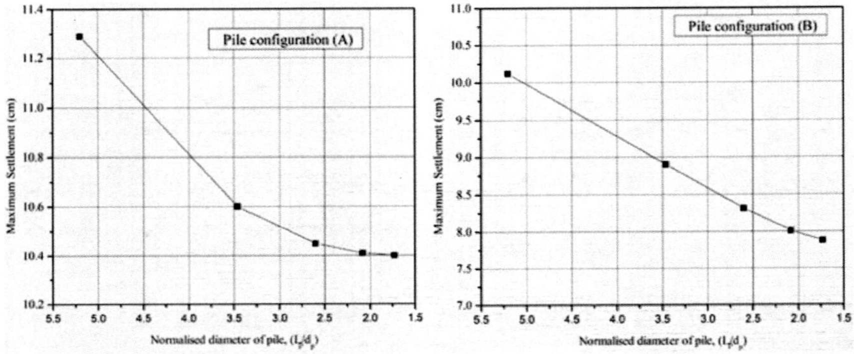

Fig. 4 Variation of maximum settlement with the normalized diameter of pile (L_p/d_p) in pile configuration A and B

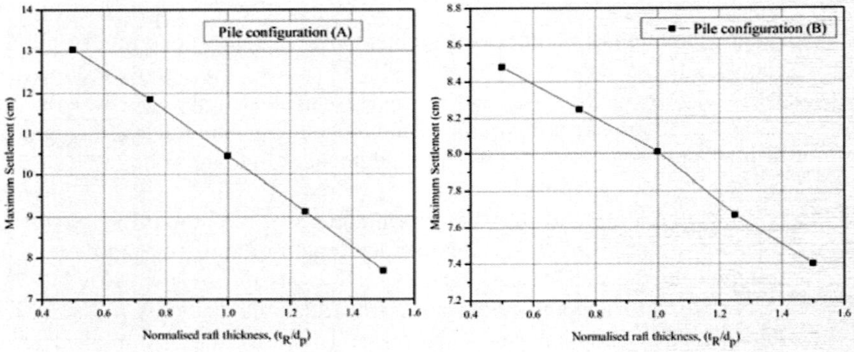

Fig. 5 Variation of maximum settlement with the normalized raft thickness (t_R/d_p) in pile configuration A and B

$t_R/d_p = 0.5$ is greater in case of configuration A than the configuration B, and this can be attributed to the fact that the arrangement of piles while the construction of a high-rise building affects the load-bearing capacity of the foundation immensely. For the same maximum settlement level, configuration B is far better than configuration A, despite having a greater number of piles.

Moreover, the variation between the maximum moment and the normalized thickness (t_R/d_p) is also investigated for the pile configuration (as shown in Fig. 6). The maximum moment in $x(M_x)$ and $y(M_y)$ direction is calculated, and also, their average values are evaluated for better accuracy. For both the pile configurations, it is noticed that the maximum moment resisted by the raft, in this case, increases with the increase in normalized thickness, and also, for all the corresponding values of t_R/d_p, the moment that the pile configuration A withstands is higher than the pile configuration B.

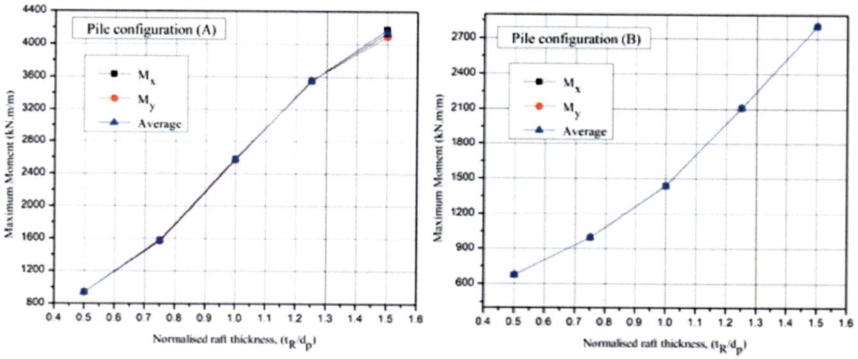

Fig. 6 Variation of the maximum moment with the normalized raft thickness (t_R/d_p) in pile configuration A and B

4 Conclusion

A numerical study is conducted to analyze the behavior of piled raft foundation by varying the length of the pile (L_p), pile diameter (d_p), and the raft thickness (t_R), by using a finite element numerical tool, ELPLA. This paper aims to determine the suitability of the PRF for the construction of high-rise building by examining the settlement behavior of the foundation. To analyze the effect of the number of piles on the settlement and maximum moment that the foundation can withstand, two different configurations of the piles are taken into account. The optimum length of piles, optimum diameter, and raft thickness have been calculated by performing a parametric study. The results showed that the arrangement of piles in the piled raft foundation plays a decisive role in improving the load-bearing capacity and also the maximum settlement of the foundation. The result obtained in the above numerical analysis is discussed below.

1. It would be a more efficient design strategy for improving the piled raft foundation. This can be observed that the embedded length of the pile ($L_1/L_2 = 2.6$) is a governing factor in improving the bearing capacity of the foundations and also improving the overall stability of the structures.
2. The pile configuration (B) is more effective than configuration (A) for the reduction of the overall settlement as it reduces 30% of the overall settlement at the corner and edges of the PRF.
3. It is observed that the raft thickness (t_R) has a great influence on the total and differential settlement of the foundation, and based on the present study, an optimum normalized raft thickness, t_R/d_p, of 1 is recommended.
4. Based on settlement and bearing capacity comparison, it is recommended to use the piled raft configuration B, as it is found as an economical configuration for the piled raft foundation.

References

1. Das BM (1984) Principles of foundation engineering, 7th edn
2. Tomlison M, Woodward J (1994) Pile design and construction practice, 4th edn. https://doi.org/10.1201/b12838
3. Bowles JE (1996) Foundation analysis and design, 5th edn. McGraw-Hill Companies Inc., Singapore
4. Poulos HG (2010) High-rise building foundations-a limit state design approach. Art Found Eng Pract. https://doi.org/10.1061/41093(372)25
5. Horikoshi K, Randolph M (1999) Estimation of overall settlement of piled rafts. Soils Found 39(2):59–68
6. Randolph MF (1994) Design methods for pile groups and piled rafts. In: Proceedings of the 13th international conference on soil mechanics and foundation engineering, vol 5. New Delhi, pp 61–82
7. Horikoshi K, Randolph M (1998) A contribution to optimum design of piled rafts. Geotech Int J Soil Mech 48(3):301–317. https://doi.org/10.1680/geot.1998.48.3.301
8. Chow HSW, Small JC (2005) Behaviour of piled rafts with piles of different lengths and diameters under vertical loading. Adv Deep Found. https://doi.org/10.1061/40778(157)20
9. Poulos HG (1997) Comparison of some methods for analysis of piled raft. In: Proceedings of 14th ICSMFE, vol 2. Hamburg, pp 1119–1124
10. Poulos HG, Davis EH (1980) Pile foundations analysis and design. Wiley
11. Butterfield R, Banerjee PK (1971) The elastic analysis of compressible piles and pile groups. Gétechnique 21(1):43–60. https://doi.org/10.1680/geot.1971.21.1.43
12. Burland JB (1995) Piles as settlement reducers. In: 18th Italian congress on soil mechanics. Pavia, Italy
13. Horikoshi K, Matsumoto T, Hashizume Y, Watanabe T, Fukuyama H (2003) Performance of piled raft foundations subjected to static horizontal loads. Int J Phys Model Geotech 3(2):37–50. https://doi.org/10.1680/ijpmg.2003.030204
14. Rabiei M (2010) Effect of pile configuration and load type on piled raft foundations performance. Deep Found Geotech Situ Test 34-41. https://doi.org/10.1061/41106(379)3
15. IS 875 Part 1 (1987) Code of practice for design loads (other than earthquake) for buildings and structures-dead loads-unit weights of building materials and stored materials. Bureau of Indian Standards, New Delhi, India
16. IS 875 Part 2 (1987) Code of practice for design loads (other than earthquake) for buildings and structures-imposed loads. Bureau of Indian Standards, New Delhi, India
17. IS 875 part 3 (2015) Design loads (other than earthquake) for buildings and structures-code of practice for wind loads. Bureau of Indian Standard, New Delhi, India
18. IS 1893 Part 1 (2016) Criteria for earthquake resistant design of structures: general provisions and buildings. Bureau of Indian Standards, New Delhi
19. El Gendy M, El Gendy A (2020) Analysis and design of raft and piled raft-program ELPLA. GEOTEC Software Inc., Calgary
20. Van Impe WF, Lungu I (1996) Technical report on settlement prediction methods for piled raft foundations. Ghent University, Belgium

Model Footing Tests on Sand Bed to Evaluate Efficiency of Tire Crumb as Infill Materials in Geocells

Sreevalsa Kolathayar and Rajesh Kumar Chitrachedu

Abstract This paper presents the results of model footing tests on sand bed reinforced with HDPE and coir geocells. The infill material influences the performance of geocells. The potential of sand tire crumb (STC) mixture as infill material is evaluated in the study in comparison with pure sand as infill materials. To select the appropriate blend of sand and tire crumb, the shear properties and density properties were analyzed by performing direct shear and density tests. The mix which has shown the highest shear strength and density properties has been selected, and plate load tests were carried to analyze the mechanism and effectiveness of tire crumb as infill material in geocell pockets. From the results, it is clear that the sand tire crumb mix is effective as infill materials, especially for coir geocells. The surface heave was found to reduce significantly with sand tire crumb mix as infill material.

Keywords Tire crumb · Geocell · Bearing pressure · Settlement · Footing

1 Introduction

Geocells are three-dimensional cellular confinement systems used to reinforce soil for wide applications in geotechnical engineering. Several researchers in the past have studied the reinforcement mechanisms of geocells in detail through experimental, analytical and numerical investigations [4–6, 12, 9, 10, 11, 20, 22, 23, 25–27, 29, 31], and others. Hegde [8] presented a detailed review of past research on geocells and gave directions for future studies. In recent years, there were efforts to develop geocells using natural materials such as bamboo, areca, coir and jute [1–3, 9–15, 21].

Soil stabilization is one of the most effective techniques to alter and enhance the physical properties of soils. A wide variety of cementitious and pozzolanic materials

S. Kolathayar (✉)
National Institute of Technology Karnataka, Surathkal, India
e-mail: sreevalsa@nitk.edu.in

R. K. Chitrachedu
Ashoka Institute of Engineering and Technology, Hyderabad, India

A. K. Choudhary et al. (eds.), *Advances in Geo-Science and Geo-Structures*,
Lecture Notes in Civil Engineering 154,
https://doi.org/10.1007/978-981-16-1993-9_27

like lime, cement and bitumen are popularly used to increase the shear strength and to improve the bearing capacity of the soil. But the chemical reactions between calcium-based additives and soils containing sulfates can cause heaving, disintegration, and finally lead to loss of strength. The pozzolanic reaction rate required to attain the strength will be reduced due to the presence of a high percentage of microbial biomass in organic soils. This indicates that there is a need for new stabilizing materials. In this paper, one of such material has been introduced. Finely shredded tire which is used for landfilling is taken as a stabilizing material. The infill materials used in the geocell pockets influence the overall performance of geocell reinforced soil significantly, and several researchers have used different infill materials like sand, gravel and seashell in geocell pockets. Hegde and Sitharam [12, 9, 10, 11], Pokharel et al. [24], Han [7], Venkateswarlu and Hegde [30], Kolathayar et al. [15, 31], Kolathayar and Kumar [17] discussed the potential of different infill material in geocells for soil retention systems. However, no studies have been performed on HDPE or coir geocells using tire crumb as infill material to reinforce the soil under the foundation. This study attempts to demonstrate the efficiency of sand tire crumb mix as infill material in geocell pockets of coir geocells through a series of model footing tests.

2 Materials

The sand was used as the foundation medium for the model footing tests. The sand is classified as poorly graded sand as per the unified soil classification system.

Two types of geocells were used in the present study, namely high-density polyethylene (HDPE) geocells and coir geocells. HDPE geocells were manufactured by ultrasonic welding of HDPE sheets. The HDPE geocells supplied by Geodukan were used for the study. The tensile strength of the geocell was determined as 11.5 kN/m from tensile strength test. The opening size of the honeycomb-shaped HDPE geocell used was 165 × 150 mm. The cell depth was 100 mm. The strip thickness was 1.74 mm.

Coir geocells were made from coir geotextile mat. The tensile strength of the coir mat was estimated as 47 kN/m. The tensile strength plot for coir mat is given in Fig. 2. The opening size of the honeycomb-shaped coir geocell used was 165 × 150 mm. The cell depth was 100 mm. The strip thickness was 7.4 mm.

Tire crumb. Tire crumb is obtained after crushing scrap tires into powdered form. Tire crumb passing through 4.75 mm sieve was mixed with sand to be used as infill material. The specific gravity, of the tire crumb, was determined as 1.14.

3 Experimental Program

The model footing tests were conducted in a steel tank of size 700 × 700 × 700 mm. The size of the footing was 150 × 150 × 12 mm. The tank was filled with sand using a constant height method maintaining a uniform relative density of 54%. The geocell layer was laid over the sandbed, and the geocell pockets were filled with infill material using the constant height of fall method. The model footing was placed at the center of the tank, and the loading was applied on the footing with an increment of 30 kg. The load was applied using a hydraulic jack, and the load was measured using a load cell. The settlement of the footing was measured at every increment of load using LVDTs. The surface heaves on the sides of the footing were also measured simultaneously.

The five models of testing with different combinations of geocell and infill material are listed below.

1. Unreinforced soil layer (without geocell).
2. Soil reinforced with HDPE geocell with sand as infill material.
3. Soil reinforced with coir geocell with sand as infill material.
4. Soil reinforced with HDPE geocell with sand tire crumb mixture as infill material.
5. Soil reinforced with coir geocell with sand tire crumb mixture as infill material.

To find out the optimum percentage of tie crumb to be added to sand for use as infill materials, different trials were done with varying percentage of tire crumb. Considering the internal friction angle and dry unit weight, 20% of tire crumb was chosen as optimum.

4 Results and Discussion

Figure 1 presents the bearing pressure settlement behavior of unreinforced soil, soil reinforced with HDPE geocell and soil reinforced with coir geocell with sand as the infill material. From the plot, it is evident that there considerable increase in bearing capacity of the soil when reinforced with geocell. The HDPE geocell performs better than coir geocell. However, HDPE geocell reinforced soil fails suddenly, whereas coir geocell reinforced soil fails gradually.

Figure 2 shows the bearing pressure settlement behavior of soil reinforced with HDPE geocells with different infill materials. It can be observed that the HDPE geocell infilled with sand performs better than that with sand tire crumb (STC) mixture as infill material. However, the failure is gradual when STC is used as infill material, unlike sand as infill material where failure is relatively sudden.

Figure 3 shows bearing pressure settlement behavior of soil reinforced with coir geocells with different infill materials. It is evident from the plot that coir geocell with sand as infill material performs better for initial loads, but for higher loads and high strain, the coir geocells with STC as infill material perform better. This is an interesting observation for the natural geocell, unlike HDPE geocell, where the sand infill was found to be effective.

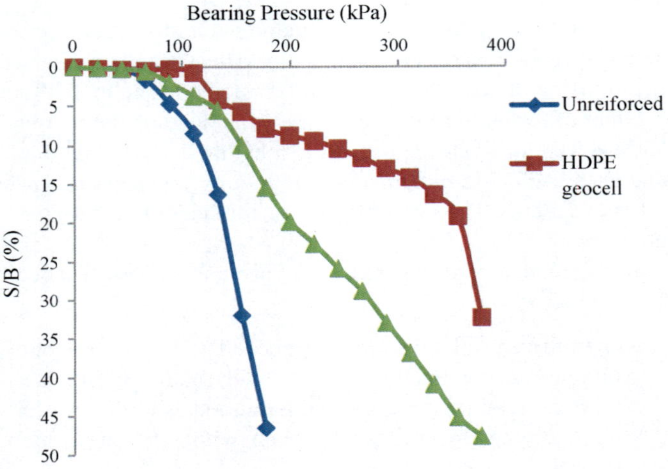

Fig. 1 Bearing pressure settlement behavior of unreinforced soil and soil reinforced with HDPE and coir geocells with sand as infill materials

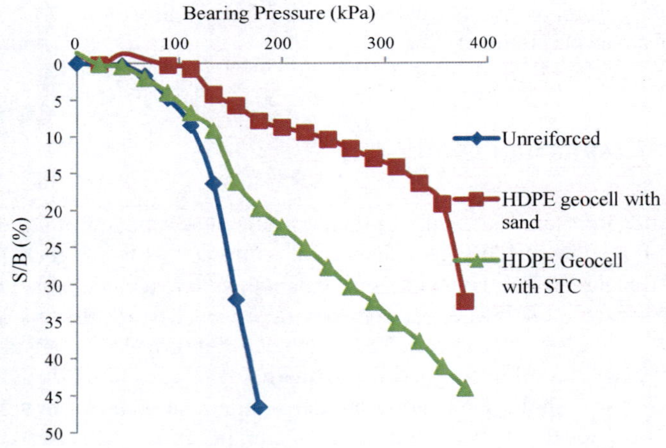

Fig. 2 Bearing pressure settlement behavior of unreinforced soil and soil reinforced with HDPE geocells with sand and STC as infill materials

Figure 4 presents the variation of surface heave with footing settlement for soil reinforced with coir geocells with different infill materials. From the figure, it is clear that the surface heave is much less when STC is used as infill material.

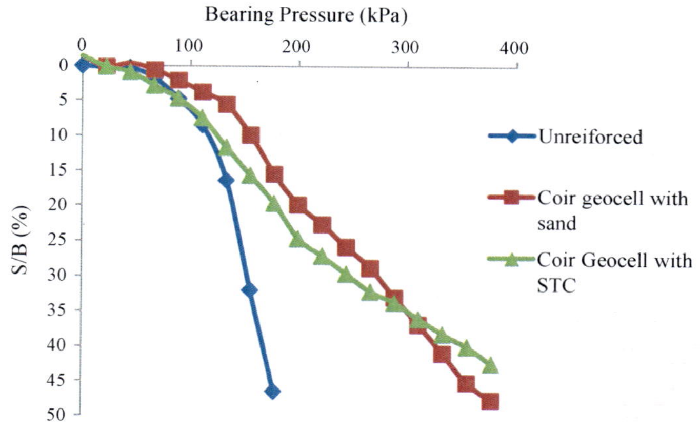

Fig. 3 Bearing pressure settlement behavior of unreinforced soil and soil reinforced with coir geocells with sand and STC as infill materials

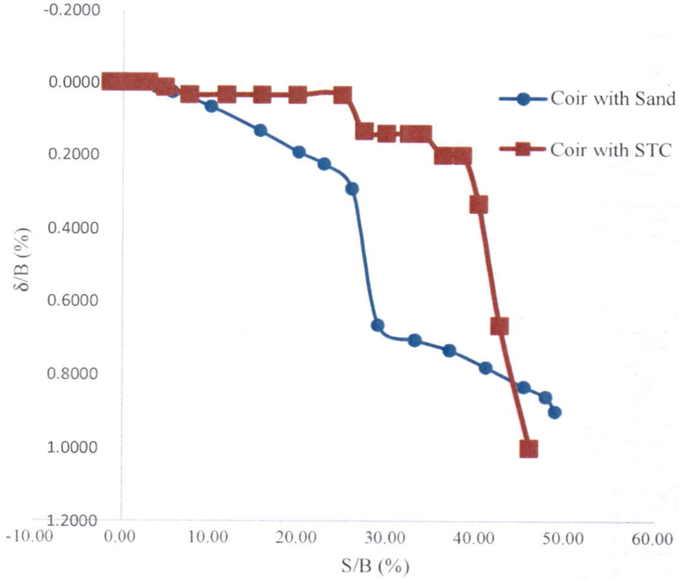

Fig. 4 Surface heave with the settlement, for coir geocells with sand and STC as infill materials

5 Conclusions

This study presented the performance evaluation of HDPE and coir geocells in reinforcing the soil as a foundation medium using sand and STC as infill materials, through a series of model footing tests. Major findings from the study are outlined below.

1. The soil reinforced with HDPE geocell is observed to increase the bearing capacity of the soil up to three times.
2. A two-fold increase in bearing capacity was attained when coir geocell was used to reinforce the soil. In addition, the failure was gradual, unlike HDPE geocell reinforced soil where the failure was sudden.
3. For HDPE geocell reinforcement, sand was found to be a better infill material.
4. For coir geocell, sand and STC both were found to be performing almost similarly when used as infill materials. However, it was found that for initial loads, sand was a better infill material but for higher load and high strain, STC as infill material was found to be more effective in resisting the load and settlement.
5. The surface heave was considerably reduced when STC was used as infill material in coir geocells.

References

1. Balan K, Sreelekha B (2016) Effect of coir geocell and fill material on bearing capacity improvement of soft clay-an experimental study. Indian J Geosynth Ground Improv 5(1):31–37
2. Balan K (2020) Coir geocells. In: Geocells. Springer, Singapore, pp 451–471
3. Chitrachedu RK, Kolathayar S (2020) Performance evaluation of coir geocells as soil retention system under dry and wet conditions. Geotech Geol Eng 1–14
4. Dash SK (2012) Effect of geocell type on load-carrying mechanisms of geocell-reinforced sand foundations. Int J Geomech 12(5):537–548
5. Dash SK, Krishnaswamy NR, Rajagopal K (2001) Bearing capacity of strip footings supported on geocell-reinforced sand. Geotext Geomembr 19(4):235–256
6. Dash SK, Sireesh S, Sitharam TG (2003) Model studies on circular footing supported on geocell reinforced sand underlain by soft clay. Geotext Geomembr 21(4):197–219
7. Han J, Pokharel SK, Parsons RL, Leshchinsky D, Halahmi I (2010) Effect of infill material on the performance of geocell-reinforced bases. In: Proceedings of 9th International conference on geosynthetics. International Geosynthetics Society, São Paulo, Brazil, pp 1503–1506
8. Hegde A (2017) Geocell reinforced foundation beds-past findings, present trends and future prospects: A state-of-the-art review. Constr Build Mater 154:658–674
9. Hegde AM, Sitharam TG (2015) Three-dimensional numerical analysis of geocell-reinforced soft clay beds by considering the actual geometry of geocell pockets. Can Geotech J 52(9):1396–1407
10. Hegde A, Sitharam TG (2015) 3-Dimensional numerical modelling of geocell reinforced sand beds. Geotext Geomembr 43(2):171–181
11. Hegde A, Sitharam TG (2015) Experimental and analytical studies on soft clay beds reinforced with bamboo cells and geocells. Int J Geosyn Ground Eng 1(2):13
12. Hegde AM, Sitharam TG (2015) Effect of infill materials on the performance of geocell reinforced soft clay beds. Geomech Geoeng 10(3):163–173

13. Hegde AM, Sitharam TG (2020) Performance of bamboo geocells in soft ground engineering applications. In: Geocells. Springer, Singapore, pp 429–449
14. Kolathayar S, Narasimhan S, Kamaludeen R, Sitharam TG (2020) Performance of footing on clay bed reinforced with coir cell networks. Int J Geomech 20(8):04020106
15. Kolathayar S, Suja P, Nair V, Krishna S, Tamilarasi G (2019) Performance evaluation of seashell and sand as infill materials in HDPE and coir geocells. Innov Infrastruct Solutions 4(1):17
16. Kolathayar S (2018). Vibration isolation of foundation using HDPE and natural geocells-a review. In: International congress and exhibition" sustainable civil infrastructures: innovative infrastructure geotechnology". Springer, Cham, pp 75–86
17. Kolathayar S, Kumar CR (2019) Potential of tire waste as infill material in geocells for soil retention systems. In: Geo-Congress 2019: geoenvironmental engineering and sustainability. American Society of Civil Engineers, Reston, VA, pp 193–201
18. Kolathayar S, Gadekari RS, Sitharam TG (2020) An overview of natural materials as geocells and their performance evaluation for soil reinforcement. In: Geocells. Springer, Singapore, pp 413–427
19. Kolathayar S, Sowmya S, Priyanka E (2020) Comparative study for performance of soil bed reinforced with jute and sisal geocells as alternatives to HDPE geocells. Int J Geosynth Ground Eng 6(4):1–8
20. Krishnaswamy NR, Rajagopal K, Latha GM (2000) Model studies on geocell supported embankments constructed over a soft clay foundation. Geotech Test J 23(1):45–54
21. Lal D, Sankar N, Chandrakaran S (2017) Behaviour of square footing on sand reinforced with coir geocell. Arab J Geosci 10(15):345
22. Latha GM, Dash SK, Rajagopal K (2009) Numerical simulation of the behavior of geocell reinforced sand in foundations. Int J Geomech 9(4):143–152
23. Oliaei M, Kouzegaran S (2017) Efficiency of cellular geosynthetics for foundation reinforcement. Geotext Geomembr 45(2):11–22
24. Pokharel SK, Han J, Leshchinsky D, Parsons RL, Halahmi I (2010) Investigation of factors influencing behavior of single geocell-reinforced bases under static loading. Geotext Geomembr 28(6):570–578
25. Rajagopal K, Krishnaswamy NR, Latha GM (1999) Behaviour of sand confined with single and multiple geocells. Geotext Geomembr 17(3):171–184
26. Sitharam TG, Sireesh S, Dash SK (2005) Model studies of a circular footing supported on geocell-reinforced clay. Can Geotech J 42(2):693–703
27. Sitharam TG, Saride S, Dash SK (2007) Performance of surface footing on geocell-reinforced soft clay beds. Geotech Geol Eng 25(5):509
28. Tafreshi SM, Dawson AR (2012) A comparison of static and cyclic loading responses of foundations on geocell-reinforced sand. Geotext Geomembr 32:55–68
29. Venkateswarlu H, Ujjawal KN, Hegde A (2018) Laboratory and numerical investigation of machine foundations reinforced with geogrids and geocells. Geotext Geomembr 46(6):882–896
30. Venkateswarlu H, Hegde AM (2019) Effect of infill materials on vibration isolation efficacy of geocell reinforced soil beds. Can Geotech J
31. Zhang L, Zhao M, Shi C, Zhao H (2010) Bearing capacity of geocell reinforcement in embankment engineering. Geotext Geomembr 28(5):475–482

Experimental Study on Black Cotton Soil Stabilization Using GGBS

Deena Nath Gautam, Md Azhar, and A. K. Sinha

Abstract About 20% of the land mass in India consist of black cotton soil (BCS), especially in the central part of country. It is the most problematic soils as it shows major volume change due to alteration in the moisture content. Soil is an integral part of civil engineering construction. If the stability of soil is not adequate for supporting the structural load, the properties of soil should be improved by soil stabilization technique. In present situation as the industrialization and urbanization are taking place, it generates many wastes. Hence, in this study, utilization of waste, i.e. ground granulated blast furnace slag (GGBS), for improving the soil properties is made. Addition of GGBS reduces free swell, plasticity index and maximum dry density. Unconfined compressive strength is observed to be increased with addition of GGBS. This facilitates the effective use of industrial waste for improving the strength of soil and also as a cheap alternative of additive.

Keywords Black cotton soil · Industrial waste · Soil stabilization · Swelling characteristics · UCS

1 Introduction

Utilization of industrial waste materials in the improvement of problematic soils is a cost-efficient and environmental friendly method. The present study evaluates the potential of GGBS generated from Tata Steel Jamshedpur in stabilizing the black cotton soil. In developing country like India due to remarkable development in basic infrastructure facilities, soil stabilization has become the major issue in construction activity. Stabilization is unavoidable for the purpose of building and road construction. Soil stabilization can be explained as the alteration of soil properties by chemical and physical means in order to enhance the engineering quality of the soil. The main objectives of the soil stabilization are to increase bearing capacity of the soil. Stabilization is a broad sense for the various methods employed in modifying the

D. N. Gautam · M. Azhar (✉) · A. K. Sinha
Department of Civil Engineering, National Institute of Technology, Jamshedpur, Jharkhand 831014, India

A. K. Choudhary et al. (eds.), *Advances in Geo-Science and Geo-Structures*,
Lecture Notes in Civil Engineering 154,
https://doi.org/10.1007/978-981-16-1993-9_28

properties of a soil to improve its engineering performance and used for variety of engineering works. In this study, emphasis is definitely placed upon the effective utilization of wastes products like ground granulated blast furnace slag GGBS, with view to decrease the construction cost. GGBS and BC soil mixtures are thoroughly prepared by varying the content of GGBS up to 40% of dry weight of soil. Black cotton soils are inorganic clays of high and medium compressibility and are characterized by high shrinkage and swelling properties. The black cotton soil has very low-bearing capacity. This poses serious problems as regards to subsequent performance of the proposed structure. In this context, there are several researchers who have made an attempt to stabilize the black cotton soil with different industrial waste. Madurwar et al. [1] have conducted a comparative study to stabilized BCS using RBI 81 and sodium silicate and find that both material enhance the engineering properties of BCS. However, RBI 81 has more efficient as compared to sodium silicate. Moreover, some of the researchers used lime as a binding material and thoroughly mixed with the BCS for subgrade improvement [2]. Recently, Ikeagwuani et al. [3] have conducted series of laboratory CBR on BCS for subgrade improvement, which is mixed with sawdust ash and lime. However, fly ash is also a major waste material which is used to stabilize the engineering properties of BCS [4, 5]. Fly ash can be attributed to pozzolanic activity forming stable compounds and control the swelling characteristics of soil [6]. Furthermore, several researchers have performed the laboratory tests to strengthen the BCS with reinforcing material such as geotextile [7–9] and plastic waste [10]. However, quarry dust, tier waste and squanders plastic also utilized to improve the CBR and SPT value for black cotton soil subgrade [11].

In this study, BCS and GGBS both are problematic in nature. BC soil is problematic because of its peculiar characteristics of swelling and shrinkage, and GGBS is problematic because of its unnecessary production in large quantity. Utilization of industrial waste materials such as GGBS in improvement of problematic soil is cost-efficient and environmental friendly method. It helps in disposal problems caused by the various industrial wastes. Present study is also an attempt to utilize GGBS in constructive way for improving the properties of black cotton soil. In this study, various tests have been performed with different proportion of 10, 20, 30, 40% replacement of soil with GGBS so that its effect on the black cotton soil can be observed.

2 Materials and Methods

2.1 Ground Granulated Blast Furnace Slag (GGBS)

GGBS is obtained from quenching of molten iron slag. It is a highly cementitious product. It can also improve the strength and durability of concrete. The basic properties of GGBS are provided by Tata Steel, Jamshedpur, which is listed in Table 1.

Table 1 Chemical composition of GGBS

Chemicals	Percentage
CaO	34.60
SiO_2	33.52
Al_2O_3	18.96
MgO	10.20
TiO_2	0.94
SO_3	0.47
MnO	0.62
Fe_2O_3	0.41
Slag basicity	1.90

Table 2 Properties of black cotton soil

Properties	Value
Colour	Black
Specific gravity	2.68
Free swell index (%)	100
Liquid limit (%)	72.8
Plastic limit (%)	31.57
Classification as per IS 1498-1970	CH
OMC (%)	28
Maximum dry density (kN/m^3)	14.72
Cohesion intercept, C (kN/m^2)	45
Angle of internal friction, Ø (°)	40.81

2.2 *Black Cotton Soil*

In this study, soil was black cotton soil which was brought from Ahwalwadi, Pune, (Maharashtra). It was collected at a depth of 15 cm below the natural ground level by removing the top cover of soil containing vegetation and foreign materials. The soil is dried and pulverized to perform the various experimental studies. Basic geotechnical properties of the black cotton soil as observed during testing are listed below in Table 2.

2.3 *Methodology*

In this study, an attempt is made to analyse the variation in the properties of black cotton soil on addition of varying percentage of GGBS. Methodology includes the methods and procedures followed during testing. The series of laboratory tests was

Table 3 Details of composite mixture

Mixture designation	Materials in %	
	BCS	GGBS
100% BCS + 0% GGBS	100	0
90% BCS + 10% GGBS	90	10
80% BCS + 20% GGBS	80	20
70% BCS + 30% GGBS	70	30
60% BCS + 40% GGBS	60	40

carried out on the virgin BCS, and BCS-GGBS mixture includes Atterberg's limits, free swell test, compaction test and unconfined compressive strength test as per IS provision. The sample was prepared by mixing of the BCS with varying percentage of GGBS up to 40%. Both BCS and GGBS were first oven dried at 105 °C for 24 h, and BCS is then pulverized and passed through 4.75 mm IS sieve to mix it with powdered GGBS. The amount of GGBS in the composite sample is the percentage of dry weight of GGBS in the total dry weight of composite sample. The curing period is 0, 7 and 28 days for UCS test. Table 3 shows the details of the composite mixture.

3 Results and Discussion

3.1 Free Swell Index

Free swelling index (FSI) tests were carried out in laboratory to determine the swelling potential of the black cotton soil under observation. FSI has been observed for both untreated BCS and treated BCS with GGBS, so that variation in FSI can be analysed with different percentages of GGBS. The results were obtained and presented in Fig. 1.

After analysis of results, it can be opined that the FSI is gradually decreasing for treated BCS with GGBS. From Fig. 1, degree of expansiveness for virgin BCS was 100% which is very high, and it significantly decreased to 30% for treated BCS with 40% of GGBS. Hence, degree of expansiveness reduced around 70% and belong to a moderate condition. The reduction in swelling with GGBS may be attributed by the calcium silicate mineral formation.

3.2 Atterberg's Limit

In this study, Atterberg's limit has been performed to understand the effect of GGBS with varying proportion on the plasticity characteristics of BCS. BCS is a highly plastic soil and contains very fine grains. The tests are conducted on BCS and obtained

Fig. 1 Variation of FSI of BCS with GGBS

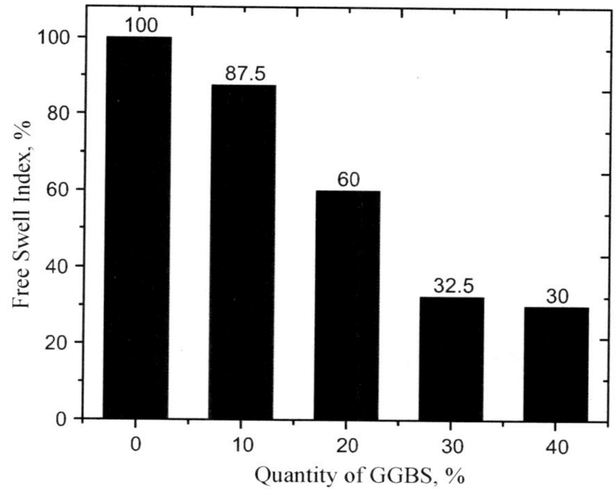

that liquid limit is ranging 70–75% and plastic limit for same ranges 30–35%. Thus, plasticity index is also on higher side. As such BCS has poor bearing capacity and rich in swelling characteristics. Figure 2 shows the variation in the Atterberg's limits with different percentages of GGBS.

After treating with GGBS, liquid limit of BCS was gradually decreased, and plastic limit was slightly increased. For virgin BCS, plasticity index was also on higher side and represents high plasticity clay on plasticity chat which is not permissible for civil engineering construction. After treatment with GGBS, plasticity index reduced to moderate condition and represent medium plastic silt on plasticity chart.

Fig. 2 Variation of Atterberg's limit with GGBS

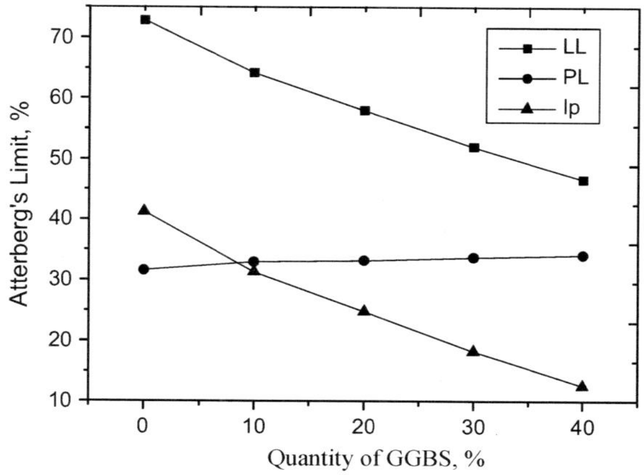

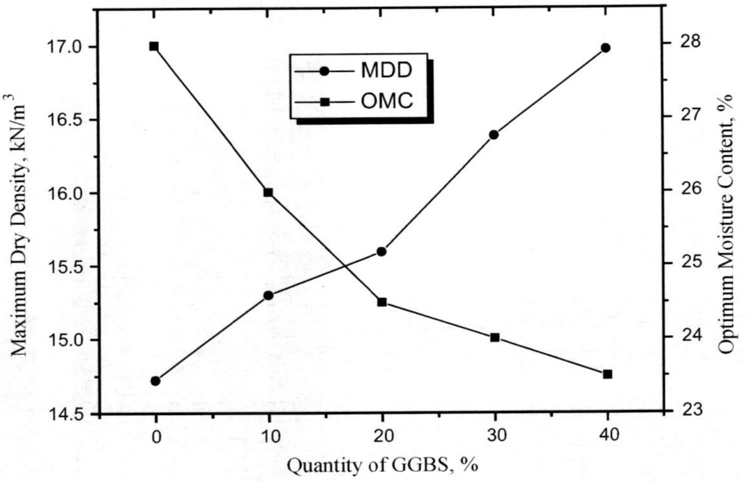

Fig. 3 Effect of GGBS on MDD and OMC of BCS

3.3 Compaction Test

Compaction is a process in which densification as air is displaced due to application of stress. Light compaction tests have been carried for BCS and mixture of BCS with varying proportion of GGBS to find out the effect of the GGBS on the optimum moisture contents and maximum dry densities. The obtained result is shown below in Fig. 3.

This study shows that there exists a relationship between MDD, OMC and GGBS content. From Fig. 3, it can be clearly concluded that as the quantity of GGBS increases from 0 to 40%, the maximum dry density increases from 14.7 to 17 kN/m^3, and optimum moisture content decreases from 28 to 23.5% due to hydration process. The optimum value of MDD is obtained as 17 kN/m^3 for BCS treated with 40% GGBS and OMC for same is 23.5%.

3.4 Unconfined Compressive Strength

The main objective of this test is to determine the unconfined compressive strength, which is used to determine the unconsolidated undrained shear strength of the clay soil under unconfined conditions. In this test, the unconfined compressive strength is taken as the maximum load attained per unit area, or the load per unit area at 20% axial strain, whichever occurs first during the performance of a test. The UCS tests were carried out on virgin BCS and treated BCS with GGBS at different curing

Fig. 4 Effect of GGBS on UCS at different curing period

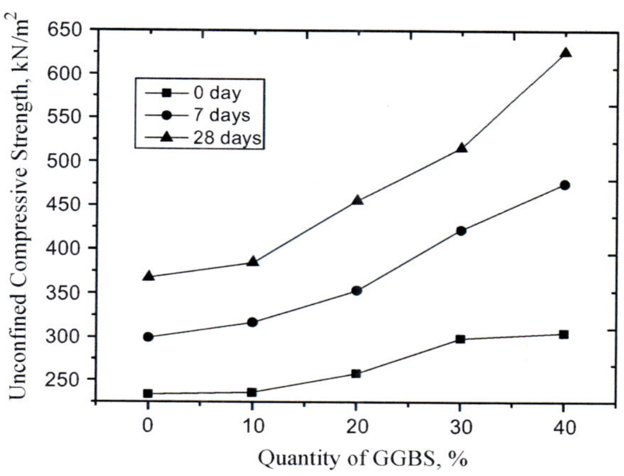

period (i.e. 0, 7 and 28 days). Figure 4 represents the effect of GGBS on UCS of BCS at corresponding curing period.

From Fig. 4, it can be clearly shown that the UCS of BCS increases with addition of GGBS. The effect of curing period is also increased the UCS of BCS irrespective of GGBS quantity. The significant increase in immediate UCS of BCS may be attributed to compacted packing between BCS and GGBS and able to sustain higher load. The significant change of UCS of BCS over curing period may be associated with pazzolanic reaction between aluminium and silicate of BCS and calcium of GGBS due to presence of sufficient moisture. The UCS value of virgin BCS is 233.7 kN/m^2 at 0 day, and it increases to 626 kN/m^2 for 40% GGBS at 28 days.

4 Conclusion

In this study, an attempt is taken to improve the expansive soil with GGBS. The black cotton soil is treated with the varying proportion of GGBS and analyses the effect of GGBS on the black cotton soil. The following observations are concluded on the behaviour of untreated soil and treated soil with GGBS of different amount.

- Free swelling index has been reduced with increase in the amount of GGBS and reduced around 70% with incorporation with 40% of GGBS. It eventually reduced the expansiveness of soil.
- Utilization of GGBS with BCS significantly decreases the plastic characteristics. The plasticity of BCS reduces from highly plastic to medium plastic soil.
- In light compaction test, it shows that by the use of GGBS, OMC decreases about 16% and MDD increases up to 15% as compared with untreated soil.
- As a result of increase in MDD, the void ratio and porosity have also gone down.

- The unconfined compressive strength is also increasing with increase in GGBS. On comparing the UCS of BCS with treated BCS with 40% GGBS, the overall increase in UCS after 28 days of curing is observed to be 70%. Increase in UCS value assured that stability of the soil increases with GGBS content.

References

1. Madurwar KV, Dahale PP, Burile AN (2013) Comparative study of black cotton soil stabilization with RBI Grade 81 and sodium silicate. Int J Innov Res Sci Eng Technol 2(2):493–499
2. Mehta KS, Sonecha RJ, Daxini PD, Ratanpara PB, Gaikwad KS (2014) Analysis of engineering properties of black cotton soil & stabilization using by lime. J Eng Res Appl 4(5):25–32
3. Ikeagwuani CC, Obeta IN, Agunwamba JC (2019) Stabilization of black cotton soil subgrade using sawdust ash and lime. Soils Found 59(1):162–175
4. Tastan EO, Edil TB, Benson CH, Aydilek AH (2011) Stabilization of organic soils with fly ash. J Geotech Geoenviron Eng 137(9):819–833. https://doi.org/10.1061/(asce)gt.1943-5606.0000502
5. Brooks RM (2009) Soil stabilization with fly ash and rice husk ash. Int J Res Rev Appl Sci 1(3):209–217
6. Chauhan KS, Rajesh J (2015) Effect of fly ash and fibre on index properties of black cotton soil. Int J Sci Technol Eng 2(01)
7. Naeini SA, Mirzakhanlari M (2008) The effect of geo-textile and grading on the bearing ratio of granular soils. Electron J Geotech Eng 13:1–10
8. Senthilkumar P, Rajkumar R (2012) Effect of geo-textile on CBR strength of unpaved road with soft subgrade. Electron J Geotech Eng 17:1355–1363
9. Truptimalapattnaik, Binayakbidyasagar, Biplabkesharisamal (2016) Application of geotextiles in pavement. Int J Eng Sci Res Technol (IJESRT), 5(4). ISSN: 2277-9655
10. Nsaif MH (2013) Behavior of soils strengthened by plastic waste materials. J Eng Sustain Dev 17(4):182–194
11. Poweth MJ, George S, Paul J (2013) Study on use of plastic waste in road construction. Int J Innov Res Sci Eng Technol 2(3):633–638

BIM and Transportation Geotechnics: An Outline

Lovnesh Kumar Goyal, Hardeep Singh Rai, and Raman Kumar

Abstract Building information modeling (BIM) integrates the various stages of building and infrastructure construction like design, construction, maintenance, and demolition. As a result, BIM can improve the performance, efficiency, mitigate risks, increase safety, decrease the waste, costs, and carbon footprint of the construction projects. Civil engineering is though late, but adopting the use of digital modeling for design, construction, and operations. The sub-branch geotechnical engineering of civil needs to design and calculate its solutions based on these technologies. All aspects like roads, railways, or highways that are related to geotechnical engineering are referred to as transportation geotechnics. One major challenge in transportation geotechnics is the substantial influence of the external environment on the life, performance, and serviceability of the pavement structures. Smart societies will be using BIM more judiciously as it can revolutionize the way to plan, design, and manage infrastructures. The present paper outlines the concept of BIM in the field of transportation geotechnics.

Keywords Transportation geotechnics · Building information modeling (BIM) · Geographic information system (GIS) · Geotechnical engineering

1 Introduction

The concept of building information modeling (BIM) has been around since the 1970s, but the technological constraints of that era marred its usability. BIM involves

L. K. Goyal (✉)
I. K. Gujral Punjab Technical University, Kapurthala, Punjab 144603, India

H. S. Rai
Civil Engineering Department, Guru Nanak Dev Engineering College, Ludhiana, Punjab 141006, India
e-mail: hsrai@gndec.ac.in

R. Kumar
Mechanical Engineering Department, Guru Nanak Dev Engineering College, Ludhiana, Punjab 141006, India

A. K. Choudhary et al. (eds.), *Advances in Geo-Science and Geo-Structures*, Lecture Notes in Civil Engineering 154,
https://doi.org/10.1007/978-981-16-1993-9_29

people, processes, and of course, technology. It is not a software but a virtual design and construction model of the project constructed and displayed on the computers. BIM gives an advantage of team collaboration that can detect any conflict in the design process in the early stages of the project. The architects and engineers can exchange their design ideas in more transparent ways. Similarly, the construction manager can provide engineering reports and sequencing. The subcontractors and vendors can also collaborate through 3D models, even in the early stages of the design. BIM has the maximum impact on the design phase of the construction project since this stage has the highest ability to influence the total cost of the project. The new discipline in geotechnical engineering in focus now is the transportation geotechnics. It is addressing the design and constructions of highways, railways, airports, and the modern metro railways. Both transportation and environmental geotechnics are the trends in geotechnical engineering. For the last some decades, the attention to these geotechnics branches arises by the environmental concerns and the demands for highway infrastructure. The expertise of the transportation and geotechnical engineers is required for the design, construction, and maintenance of the road pavements. Consequently, the success of any transportation project requires the synergy as a pre-requisite between both of these engineering groups [1].

Many global challenges have surfaced recently and require geotechnical inputs. These challenges include climate changes, urban sustainability, energy efficiency, material selection, raw resources management, and water harvesting management. The finding of solutions to these problems will require fundamental advances in estimating the properties of soil and rock along with advances in monitoring, sensing, and digital modeling of geo-systems. Furthermore, necessary research is required that can integrate big data into geotechnical engineering practices and improve the subsurface characterization of geo-materials [2].

The sites for new geotechnical projects like highways, railways, or airports are preferred at low-cost and low-utility lands like brownfield lands, landslides, old mining areas, or marshlands. At the same time, steeper slopes are opted for fill and cut by the contractors for economic considerations. A comparison and analysis need to be done between the reinforced and unreinforced slopes. The selection of such properties poses technical challenges to the engineers.

The transportation geotechnics has to deal with vast amounts of materials and their raw resources while constructing and maintaining the transportation infrastructure to work properly. Large quantities of fuel and a lot of energy are consumed in the process. Thus, there is an urgent need to address environmental concerns like energy efficiency and consumption, greenhouse gas emissions, material recycling, and carbon footprint [3]. BIM integrates the various stages of building and infrastructure construction like design, construction, maintenance, and demolition and hence can improve the performance and efficiency, mitigate risks, increase safety, decreasing the waste, cost, and carbon footprint of the construction project [4]. The use of BIM, geographic information system (GIS), and other digital engineering has grown a lot in recent times. Yet, civil engineering is late in adopting the use of digital modeling for design, construction, and operations as compared to other branches of knowledge.

Furthermore, it seems like the approach is to start the design considering ground as bottom and hence ignoring the geology and the subsurface strata for any analytical data [5]. The sub-branch geotechnical engineering needs to design and calculate its solutions based on these technologies [6]. It needs to harness the capability of the 6D and 7D levels of development of BIM that allows the optimal management of the project and infrastructure assets [7].

2 Building Information Modeling

In BIM, the objects are modeled parametrically and not as isolated elements. Hence, the changes in one object by any stakeholder reflect in all other related objects automatically, thus removing any redundancy of data creation. BIM realizes a reduction in design conflicts, order of changes, improvement in the quality of a project, and lowering of risk. BIM enables the rendering of a high-quality 3D construction model. BIM has thus a very diverse approach and is termed variously in different kinds of projects such as vertical or horizontal BIM, heavy, sustainable, green, or social BIM [8]. The project risks are contractually shared by all the BIM collaborators viz. the owners, designers, consultants, construction managers, contractors, and subcontractors. The project will need to be time-bound and within budget limits, for all the stakeholders to share the profit lest they all lose their fee. BIM can quickly simulate different engineering fields, sustainability, and other characteristics even before the actual construction has begun. Hence, the BIM model can compare various design configurations aiding most appropriate design decisions [9]. It can be observed that any engineering analysis based on BIM, be it energy, sustainability, structural, or daylight analyses, the evaluation can automate these processes, saving time, energy, and cost while improving the quality of design [10].

2.1 Transportation Geotechnics

All aspects like roads, railways, or highways that are related to geotechnical engineering are referred to as transportation geotechnics. The improved quality and enduring road life that are the features of the developed transportation infrastructure will contribute to the overall economic development of the country. One major challenge in transportation geotechnics is the substantial influence of the external environment on the life, performance, and serviceability of the pavement structures [11]. The properties of the soil subgrade play an essential role in improving the quality of the pavement structure as it bears the wheel loads of the vehicular traffic transmitted from upper pavement layers to it. Not all soils have desirable properties to be used in pavement construction. Some non-traditional material can be used to enhance the performance of such soils with appropriate reinforcement [12]. Geotechnical assets play an essential role in the efficient working of transportation corridors.

Geotechnical monitoring can be a vital tool to play an influential role in maintaining transportation assets effectively and safely [13].

2.2 Construction Material

The research of [14] is focused on the interaction between environmental concerns and the transport geotechnics. They emphasized the problem of using waste and reused or recycled materials and saving the subsoil from getting contaminated. Practical examples of utilizing the ash and waste from civil construction demolition were used. Potential contamination of the subsoil, either from the waste material or by transport management, is also discussed. The materials used in the road pavements, railways tracks, or airfields pose complex problems because they are heterogeneous, anisotropic properties depending on time and rate. It is the reason for finding an alternative to the mechanical process of modeling, analyzing, and then designing in artificial intelligence (AI) models and tools. However, geotechnical engineering has yet to realize the potential of these techniques and tools fully. Correia et al. [14] made attempts to increase collaboration between the AI and its efficiency as applied in the management of earthwork, jet grouting, pavement performance, etc. The mass concreting requires special techniques due to the emission of hydration heat that causes stress cracks by temperature variations. BIM technology was utilized to analyze the main characteristics of mass concrete construction. The process of using the temporary facilities, the selection of their location within the construction boundaries, construction site utilization planning (CSUP) is used to keep to a minimum the ineffective communication among various facilities by identifying spatial relationships, finding the best alternatives [15]. Deshpande and Whitman [16] provided a concise review of BIM, commonly used modeling tools, capabilities, weaknesses, and potential benefits.

2.3 Geotechnical Data Management

Correia et al. [14] opined that more integration of data mining, big data, and GIS tools would have a crucial role to play in the design, construction, and operation stages of transportation geotechnics. Most of the developed countries are facing their transportation assets to be either after their estimated life cycles or nearing it. Geotechnical asset management (GAM) applications are likely to be adopted as a standard by the transportation agencies of these countries [13]. The delays in completing the projects are caused by the unexpected conditions encountered underground and imprecise estimation in construction designing. BIM can cause revolutionary changes in transport geotechnics, but the required export and sharing of data is not the standard practice in the field [17]. Association of geotechnical and geo-environmental specialists (AGS) is the standard that is agreed upon by the UK and many other countries of the world for geotechnical data format.

Similarly, data interchange for geotechnical and geo-environmental specialists (DIGGS) form is emerging in the US as the preferred format. These formats permit the transfer of data within and among geotechnical and geo-environmental organizations [18]. The core of the BIM process is collaboration and data sharing. Precise specifications and standards can achieve BIM compliance of geotechnical projects on the delivery of AGS data [19]. Chandler et al. [20] reviewed the implementation of BIM in significant infrastructure projects to study the improvements in the decision-making process and the BIM model's usefulness by the inclusion of geotechnical data. Testing and sampling are made possible in the middle of an ongoing investigation by the real-time access of the BIM data.

2.4 *Integration of BIM and GIS*

GIS users want to harness the power of BIM; similarly, the GIS data is considered crucial for BIM processes to be integrated into geotechnical engineering. Its integration with geotechnical data can improve the BIM process. As most of the geotechnical projects do not incorporate the below-ground data information into account, Tawelian and Mickovski [21] are specifically concerned with the integration of geotechnical data and the BIM processes, its influence on time, and cost saving of the project. They considered and critically reviewed several methods for realizing the integration of geotechnical and BIM data and carried out the potential application as in transportation geotechnics for an optimal approach.

2.5 *City-GML*

Jayakody et al. [22] demonstrated the integration of BIM and GIS. The paper considered the development of Geo-BIM, an extension of City-GML for getting semantic IFC data in the context of GIS. The industry foundation classes (IFC) data model is used to exchange data among BIM applications working on a single construction. In contrast, City-GML models, the XML-based formats, represent the 3D models of entire cities and landscapes that are used to describe standard 3D features of objects found in the towns like roads, rivers, bridges, and vegetation. Both the modeling paradigms are used extensively in their domains. City-GML, the semantic information model developed by the open geospatial Consortium (OGC), is a format to exchange, store, disseminate, and do spatial analyses like solar potential, energy consumption, energy efficiency, and daylight analysis. It uses five different standards of details (LoDs) to represent 3D objects, allowing varying degrees of geometrical and thematic aspects of the city objects for purposes like data mining, simulations, facility management. As both of these format standards have structures so different, the exchange of information among models and tools based on them is tremendously challenging [23].

3 Geo-BIM

The integration of Geo and BIM data can prove to be a crucial step in confronting the multi-disciplinary technical and practical challenges of geotechnical engineering. The integration using substantial parts of both geospatial and BIM domains, termed Geo-BIM, has many uses from individual constructions to great cities. Noardo et al. [24] reported the second-phase work of the Euro-SDR Geo-BIM project that worked to make out the usage of Geo-BIM across Europe, focusing mainly on the perspective of national mapping and cadastral agency. A new concept was introduced for developing a hybrid of infrastructural information by IFC-SML, the building design data, and City-GML format 3D neighborhood models [23].

3.1 Sustainable Geotechnics

Sustainability refers to the contribution rendered by the geotechnical engineers in achieving ways to reduce adverse impacts on the environment while using limited resources. It includes some areas as energy consumption, recycling/reusing of the waste and material, carbon footprint, water, air, and soil pollution [3]. There are growing environmental concerns in response to global challenges of low-carbon economy in climate change, decreasing raw resources, and environmental management [25]. The sustainable culture that constitutes attitudinal and behavioral elements is the driving force for sustainability among the stakeholders in the construction industry [26]. Geotechnical engineering has a vital role to play in several areas of sustainability and construction resilience like underground infrastructure mapping, construction, operation, and development [2]. An essential aim of BIM is to show up in support of sustainable construction and architecture for a healthy, comfortable, and productive habitat [27]. Mahbub [28] considered the application of BIM necessary for sustainable construction projects that have long-term affordability, quality, and efficiency. The solid wastes in municipal landfills have construction, renovation, and demolition waste as its primary contributor. As some researches have come with the results that the construction-related waste has about 75% residual value, it has to be recycled or reused. The recycling of construction waste is as costly as going for new material purchases. It gives rise to the demand for going into sustainable construction options to maximize the reuse of construction waste and to optimize material usage. A real-time information system is needed to achieve both of these objectives [29]. Various environmental issues like pollution, poor air quality, and inclement climate changes prop up due to the rapidly growing construction activities and inappropriate use of energy resources. Many researchers foresee a healthy and safe environment with better use of land, transport, and strategic city planning, etc. [30]. BIM helps build more sustainable buildings to save precious resources. BIM performs various analyses in the early stages of design such as noise, comfort, daylight, visibility, energy efficiency, and consumption analyses. Different low-carbon technology solutions are compared in the construction phase of the building that can give a platform for the selection of various solutions to meet the requirement of green construction [31].

3.2 Road Safety

BIM can facilitate road optimization and safety by using digital simulation, visualization, and analysis while designing for efficiency and productivity [32]. Svatý et al. [33] tried to depict the implementation of the road safety aspects and BIM statistics, its applications, and strategies based on government resolutions in the Czech Republic. Besides, BIM-enabled virtual traffic lights can be used at road intersections to increase road safety [34].

3.3 Pavement

The way the world interacts with the built environment depends on the latest developments in information and communications technology (ICT). One can expect the future roads to be equipped with sensors, vehicles communicating with one another, with the help of the Internet of Things (IoT). Pavement inspection may be done by the robots that have already used self-healing construction material for its maintenance and stable condition [35]. As a measure against the pandemic COVID-19, the use of public transport is discouraged while providing more extensive cycling networks in the cities. Campisi et al. [36] focused on the design of cycling paths using infrastructural BIM (I-BIM). The use of slags as a sustainable alternative can have technical benefits in addition to the economic and environmental advantages in pavement construction [37].

3.4 Compaction

Though the compacted soils are unsaturated when the construction stage begins, eventually, the degree of saturation increases due to infiltration or loading, or it decreases due to evaporation, hence changing the properties of the soil considerably [39]. Prediction of characteristics of soil using any correlation equation has to be rational and global as the compaction characteristics are essential in geotechnical projects [38].

3.5 *Ground Improvement*

The selection and implementation of the most sustainable method that provides a comprehensive analysis of features such as building energy efficiency, water, health, carbon footprint, indoor air, life cycle, time, and cost can award higher sustainability ratings [3].

3.6 *Deep Mixing of Road Embankments*

The road construction methods in Nordic countries widely use deep mixing as an in situ ground improvement method. Either separate columns can be used for deep mixing or mass stabilization opted to mix the whole soil mass with a binder. Contractors have devised mechanical mass stabilization using mechatronics that controls the vertical positioning of pile columns to improve productivity [39]. BIM workflow and process are intelligent in such cases, based on the 3D model as against the traditional 2D methods. It provides efficiency, collaboration to all the project stakeholders beginning from planning, design, execution of the project, and operation or facility management of building assets and infrastructure. The partnership in the team is supported and enhanced by the BIM models by intelligent modelers. The information and documentation inherently shared in these models are used and collaborated throughout the lifecycle of the project [40].

3.7 *Reuse of Natural Geo-materials*

India, being a developing country, also has ambitious targets to expand its road and railway network. A problem facing transportation geotechnics is the immediate consumption of the natural aggregate resources that ensues the massive demand for pavement materials. On the contrary, the large quantity of fly ash is being generated by the thermal power plants and slag as the industrial waste by metal industries posing threats to ecology, public health, and landmass available for dumping them. The bulk utilization of waste material in geotechnical projects will result in preserving not only the natural resources of raw material but also ecology and precious landmasses. This likely usage in the construction of pavements or rail tracks, backfill in reinforced embankments, and soil structures need research for the most optimized process. It involves modern digital techniques, rational design methodologies, environmental controlling, instrumentation, and quality checking [41].

3.8 CO_2 Emissions in Earthworks Operations

The transportation geotechnics has a vital role in addressing the Green House Gas emissions by reducing deforestation by GIS mapping for route selection and alignments, reduced energy use in executing the projects [25].

3.9 Tunnels

An et al. [42] found that linked data is the most appropriate approach for exchanging geological information in the metro tunnel. Their paper focuses on the technical development process using linked data for BIM model extension. Hafner et al. [43] explain an integrated project delivery approach that involves collaboration among project stakeholders through a standard cloud-based IT media and the BIM workflow's digital transformation. It results in a new workflow that helps decision making and supporting optimization for underground traffic connections using urban underground space optimally.

3.10 Benefits of Incorporating Geotechnical Data into the BIM

The BIM can visualize and simulate various options at the design stage to minimize risks and uncertainty and increase safety in the construction process. The use of commonly accepted standards for digital data interchange has several benefits to the stakeholders. It can improve and streamline the internal processes, data quality, and also reduce legal exposure. All aspects of underground investigations can be enhanced by integrating them [44].

3.11 BIM Workflow

BIM workflow and process are intelligent, based on the 3D model as against the traditional 2D methods. Figure 1 shows the BIM workflow. It provides efficiency, collaboration to all the project stakeholders beginning from planning, design, execution of the project, and operation or facility management of building assets and infrastructure. The partnership in the team is supported and enhanced by the BIM models made by intelligent modelers. The information and documentation inherently shared in these models are used and collaborated throughout the lifecycle of the project [40].

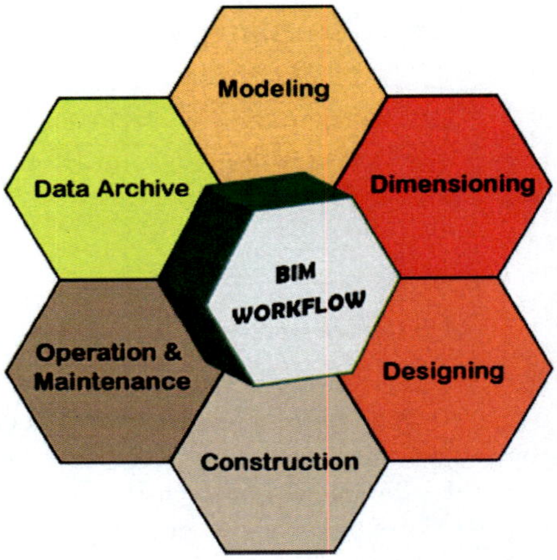

Fig. 1 BIM workflow

4 Conclusions

The present paper outlined the concept of BIM in the field of transportation geotechnics. BIM can manage construction projects that are large and complex. Digital data interchange has become a vital part of the geotechnical engineering to keep it abreast of the relevant technological advances. The geotechnical organizations and AI are creating enormous data repositories but data mining and other data analytic technologies are there to help analyze the data and visualize the alternatives in ways unimaginable today. The responsible agencies can be proactive in protecting the transportation infrastructure by predicting rockfalls, failure of steep slopes, pavement structure performance, or any other geotechnical hazard. Smart societies will be using BIM more judiciously as it can revolutionize the way to plan, design, and manage infrastructures. Using BIM tools in the projects has many associated benefits. Because of parameterization, the design time is reduced while alternative solution variants can be increased to select the most optimized one.

Acknowledgements The authors gratefully acknowledge the support provided by I.K. Gujral Punjab Technical University, Kapurthala, Punjab, India, and colleagues. The authors are also grateful to the anonymous reviewers of the paper for their careful reading and evaluation.

References

1. Karpurapu R, Gnanendran C, Saride S (2015) Special issue of Indian geotechnical journal: transportation geotechnics. Indian Geotech J 45(4):369–370. https://doi.org/10.1007/s40098-015-0169-z

2. Lu N, Mitchell JK (2019) Geotechnical fundamentals for addressing new world challenges. Springer International Publishing. https://doi.org/10.1007/978-3-030-06249-1
3. Gomes Correia A, Winter MG, Puppala AJ (2016) A review of sustainable approaches in transport infrastructure geotechnics. Transp Geotech 7:21–28. https://doi.org/10.1016/j.trgeo.2016.03.003
4. Dong RR, Martin A (2017) Research on barriers and government driving force in technological innovation of architecture based on BIM. Eurasia J Math Sci Technol Educ 13(8):5757–5763. https://doi.org/10.12973/eurasia.2017.01025a
5. Morin G et al (2017) Silvertown tunnel, London, England—a case study applying BIM principles to the geotechnical process. Geotech Special Publ (GSP 277) 587–595. https://doi.org/10.1061/9780784480441.061
6. Lacasse S (2016) GeoSuite–a modular system for geotechnical design. In: Proceedings of the 17th Nordic geotechnical meeting
7. Biancardo SA et al (2020) BIM approach for modeling airports terminal expansion. Infrastructures 5(5):1–14. https://doi.org/10.3390/infrastructures5050041
8. Rahman A et al (2013) Diverse approach of Bim in AEC industry: a study on current knowledge and practice. In: Proceedings of the 30th CIB W78 International Conference, pp 9–12. Beijing, China, 9–12 October 2013
9. Valinejadshoubi M, Bagchi A, Moselhi O (2016) Managing structural health monitoring data using building information modelling, pp 22–24. Available at: http://www.smar-conferences.org/smar/SMAR_2017_Proceedings/papers/197.pdf
10. Won J, Cheng JCP (2017) Identifying potential opportunities of building information modeling for construction and demolition waste management and minimization. Autom Constr 79:3–18. https://doi.org/10.1016/j.autcon.2017.02.002
11. Vanapalli, Sai K, ZH (2019) Addressing transportation geotechnics challenges using mechanics of unsaturated soils. In: Geotechnical design and practice. Springer Singapore, pp 91–104. https://doi.org/10.1007/978-981-13-0505-4
12. Ramesh A, Nageshwar Rao C, Kumar M (2019) Experimental study on geocell and of fibre reinforced soil sub-grade under static and repetitive load. In: Lecture notes in civil engineering. Springer, Singapore, pp 139–149. https://doi.org/10.1007/978-981-13-6713-7_11
13. Mazzanti P (2017) Toward transportation asset management: what is the role of geotechnical monitoring? J Civ Struct Health Monit 7(5):645–656. https://doi.org/10.1007/s13349-017-0249-0
14. Gomes Correia A et al (2013) Artificial intelligence applications in transportation geotechnics. Geotech Geol Eng 31(3):861–879. https://doi.org/10.1007/s10706-012-9585-3
15. Beide G, Jianhua Z (2019) Construction technology and quality control of mass concrete in super high-rise buildings based on BIM. In: 5th International conference on economics, management and humanities science, (Ecomhs). https://doi.org/10.25236/ecomhs.2019.133
16. Deshpande A, Whitman JB (2014) Evaluation of the use of BIM tools for construction site utilization planning. In: 50th ASC annual international conference proceedings, (Elbeltagi 2008). Available at: http://ascpro.ascweb.org/chair/welcome.php
17. Berdigylyjov M, Popa H (2019) The implementation and role of geotechnical data in BIM process. E3S Web Conf 85(October 2015):1–8. https://doi.org/10.1051/e3sconf/20198508009
18. Wang H et al (2020) Study of AI based methods for characterization of geotechnical site investigation data (No. FHWA/OH-2020–3). Ohio Department of Transportation, Office of Statewide Planning & Research, (135785)
19. Morin G, Hassall S, Chandler R (2014) Case study—the real life benefits of geotechnical building information modelling. Information technology in geo-engineering. In: Proceedings of the 2nd international conference (ICITG). Durham, UK, pp 95–102. https://doi.org/10.3233/978-1-61499-417-6-95
20. Chandler RJ, Mcgregor ID, Morin GR (2012) The role of geotechnical data in building information modelling. In: Australian–New Zealand conference on geomechanics (ANZ 2012), 15th–18th July 2012, vol 44, pp 511–516. Available at: http://www.keynetix.com/white-papers/the-role-of-geotechnical-data-in-building-information-modelling/

21. Tawelian LR, Mickovski SB (2016) The implementation of geotechnical data into the BIM process. Procedia Eng 143(Ictg):734–741. https://doi.org/10.1016/j.proeng.2016.06.115
22. Jayakody A et al. (2013) The development of the CityGML GeoBIM extension for real-time assessable model (integration of BIM and GIS). The development of the CityGML–PNCTM 2, 2(January), pp 109–113. https://doi.org/10.13140/2.1.1495.7443
23. Zadeh PA et al (2019) BIM-CityGML data integration for modern urban challenges. J Inf Technol Constr 24(May):318–340. https://doi.org/10.36680/j.itcon.2019.017
24. Noardo F et al (2019) EuRoSDR GeoBIM project a study in Europe on how to use the potentials of BIM and GEO data in practice. Int Arch Photogrammetry, Remote Sens Spatial Inf Sci ISPRS Arch 42(4/W15):53–60. https://doi.org/10.5194/isprs-archives-XLII-4-W15-53-2019
25. Hamza O, Bellis A (2016) Sustainable geotechnics. In: International conference on sustainability in energy and buildings, pp 11–12. https://doi.org/10.13140/RG.2.2.25353.44640/1
26. Yip Robin CP, Poon CS (2009) Cultural shift towards sustainability in the construction industry of Hong Kong. J Environ Manage 90(11):3616–3628. https://doi.org/10.1016/j.jenvman.2009.06.017
27. Akin Ö (2015) Necessity of cognitive modeling in BIM's future. In: Building information modeling, pp 17–27. https://doi.org/10.1002/9781119174752.ch2
28. Mahbub R (2014) Effective teaching of technology: building information modelling (BIM) module for built environment students. Recent Adv Educ Educ Tech 61–64
29. Hewage K, Porwal A (2011) Sustainable construction: an information modelling approach for waste reduction. Ucalgary CA, p 12. Available at: http://iiirr.ucalgary.ca/files/iiirr/148.pdf
30. Zavadskas EK et al (2017) MCDM assessment of a healthy and safe built environment according to sustainable development principles: a practical neighborhood approach in vilnius. Sustainability (Switzerland) 9(5). https://doi.org/10.3390/su9050702
31. Yuqi H, Jiajia, Y (2018) Benefit evaluation research of BIM application. In: 2018 7th International conference on social science, education and humanities research (SSEHR 2018), 2(SSEHR), pp 867–871. https://doi.org/10.25236/ssehr.2018.174
32. Strafaci A (2008) What does BIM mean for civil engineers? Road and highway project can benefit from design using building information modeling. CE News, Tranportation, pp 5–9
33. **Svatý Z, Kocián K, Mičunek**, T (2019) Integration of safety assessment in bim for transportation infrastructure. Int Arch Photogrammetry Remote Sens Spatial Inf Sci - ISPRS Arch 42(5/W3):143–148. https://doi.org/10.5194/isprs-archives-XLII-5-W3-143-2019
34. Andrisano O et al (2018) The need of multi-disciplinary approaches and engineering tools for the development and implementation of the smart city paradigm. Proc IEEE 106(4):738–760. https://doi.org/10.1109/JPROC.2018.2812836
35. Costin A et al (2018) Building information modeling (BIM) for transportation infrastructure—literature review, applications, challenges, and recommendations. Autom Constr 94(June):257–281. https://doi.org/10.1016/j.autcon.2018.07.001
36. Campisi T et al (2020) Cycling master plans in Italy: the I-BIM feasibility tool for cost and safety assessments. Sustainability (Switzerland) 12(11). https://doi.org/10.3390/su12114723
37. Gökalp İ et al (2018) Technical and environmental evaluation of metallurgical slags as aggregate for sustainable pavement layer applications. Transp Geotech 14(2018):61–69. https://doi.org/10.1016/j.trgeo.2017.10.003
38. Nagaraj HB et al (2015) Correlation of compaction characteristics of natural soils with modified plastic limit. Transp Geotech 2:65–77. https://doi.org/10.1016/j.trgeo.2014.09.002
39. Korkiala-tanttu L (2017) Transportation geotechnics: Deep mixing of road embankments. In: 3rd International conference on new advances in civil engineering, pp 27–29
40. Dastbaz M, Gorse C, Moncaster A (2017) Building information modelling, building performance, design and smart construction. In: Building information modelling, building performance, design and smart construction, pp 1–326. https://doi.org/10.1007/978-3-319-50346-2
41. Shahu JT, Sivakumar Babu GL, Usmani A (2020) Developments in transportation geotechnics. Indian Geotech J 50(2):157–158. https://doi.org/10.1007/s40098-020-00436-0

42. An Y, Liu Y, Zheng H (2018) Geological data extension for metro tunnel BIM model with the linked data approach. In: Proceedings of GeoShanghai 2018 international conference: transportation geotechnics and pavement engineering, vol 1, pp 90–99. https://doi.org/10.1007/978-981-13-0011-0_10
43. Hafner M et al (2019) Artificial intelligence support for tunnel design in urban areas. In: Tunnels and underground cities: engineering and innovation meet archaeology, architecture and art—Proceedings of the WTC 2019 ITA-AITES World tunnel congress, pp 2196–2205. https://doi.org/10.1201/9780429424441-232
44. Deaton SL (2018) What are the benefits of geotechnical data interchange? In: 69th highway geology symposium (HGS), pp 1–18

Numerical Analysis of Geocell-Reinforced Embankment Using PLAXIS 3D

Udaya Pratap, Md. Azhar, and Awdhesh Kumar Choudhary

Abstract The construction of road embankment on soft soil encountered with various challenges in practice, such as bearing failure, slope instability, local failure, differential settlement and lateral movements. Horizontal geocell mattresses have been progressively applied as a reinforced material of road embankment on problematic soil such as soft soil. The main objective of this study is to determine the performance of geocell of different aperture size as reinforcing material in road embankment considering load versus settlement behavior. The stability analysis of road embankment has been evaluated by using finite element software PLAXIS 3D. In this study, two-basic modeling was analyzed. First, the performance of road embankment was analyzed without any reinforcing material. Second model of embankment was analyzed by using geocell of different height to width ratio (i.e., 0.8, 1, 1.2, 1.4 and 1.6) as reinforcement. The small settlement and deflected shape of geocell mattress acted as a raft foundation for the road embankment. The result of this study indicated that the stability of embankment was highly influenced by tensile strength of geocell.

Keywords Soft soil · Geocell · Embankment · Reinforcement · PLAXIS 3D

1 Introduction

Due to rapid growth of urbanization and industrialization, there is a huge scarcity of good engineering land space all over the world. This result force to construct on the soft soil which is not preferable for construction work. The stability of structure on expensive soil is a challenge due to low bearing capacity and heaving tendency. The method of reinforced earth is to enhance the properties of weak soil by application of synthetic material in to soil. Geosynthetic as reinforcing material is the most adopted technique to improve the stiffness of soil [1, 2], increase bearing capacity and reduce settlement [3]. Several available product of geosynthetic material, geocell

U. Pratap (✉) · Md. Azhar · A. K. Choudhary
Department of Civil Engineering, NIT Jamshedpur, Jamshedpur, Jharkhand 831014, India

A. K. Choudhary et al. (eds.), *Advances in Geo-Science and Geo-Structures*, Lecture Notes in Civil Engineering 154,
https://doi.org/10.1007/978-981-16-1993-9_30

is gather an immense attention for protection and stabilization product for weak soil [4]. Geocell provide a faster, feasible and environmental solution. It is a three dimensional, polymeric, honeycomb-like structural cell which is interconnected at joint. Geocell can provide better lateral confinement to the infill material as compare to planar geosynthetic [2, 5]. The application of geocell has a broad range such as foundation, road embankments, retaining wall and buried pipe lines, many studies have reported that geocell significantly improve the stiffness and elasticity of soil [6, 7]. Several experimental results have been observed to analyze the bearing capacity of soft soil-reinforced geocell using various methods [2, 3, 8]. In design the complex structures, numerical modeling would be useful for better understanding. In addition, experimental studies are often time consuming and costly as well. Several studies have also been performed on geogrid and geocell-reinforced soil to enhance the stiffness and elasticity of soil [4, 8, 9]. Latha et al. [10] have performed laboratory experiments to design the geocell-reinforced embankments. The results were illustrated the slope-stability analysis and the critical slip surfaces of embankments, which should be checked by the slope-stability methods. Moreover, Zhao et al. [11] investigated the geocell-reinforced layers for embankments and suggested that the geocell layer functions could be divided into three aspects: (a) lateral resistance effect, (b) vertical stress dispersion effect and (c) membrane effect.

In this study, PLAXIS 3D software has been adopted to analyze the load versus settlement behavior of geocell-reinforced soil. PLAXIS 3D is three-dimensional finite element program, which is helpful for deformation and stability analysis of the geotechnical problems. The software has a user-friendly and easy-to-use graphical user interface that makes it easy to quickly create model geometry. PLAXIS 3D has used implicit integration solution planning to solve the initial and boundary value problem. In the implicit method, the solution is determined by solving a differential equation in both the current and future state of the system.

In this paper, a road embankment has been constructed over soft soil and having the embankment fill material is sand. The road embankment is reinforced with geocell having different width to height ratio. The geometric model has been analyzed for load-settlement behavior using PLAXIS 3D. Hence, the main objective of this paper is to determine the influence of width-height ratio of geocell over load-settlement behavior.

2 Numerical Analysis

All the finite element models in this paper are analyzed by using PLAXIS 3D. Finite element analysis of embankment material was defined as hardening soil. The foundation soil was termed as soft soil. The entire model was analyzed for Mohr–Coulomb failure criteria. Geocell was modeled as square structure and define as geogrid type structure (refer Fig. 1). Mainly, the geogrid structure was used to uphold the membrane stress. Moreover, the joints of geocell were simulated as the rigid joint.

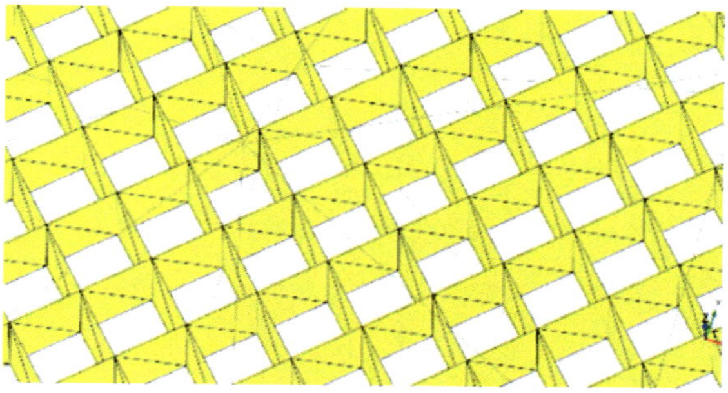

Fig. 1 Square model of geocell in PLAXIS 3D

Further, the interface between geocell and soil was linearly modeled with Mohr–Coulomb failure criteria. The thickness of geocell can resist the bending load a part from membrane structure. The mesh was generated with 10 node triangular element of modeled geometry (refer Fig. 2). The node and element for the model geometry meshing was 68,834 and 49,709, respectively.

In this study, the embankment of 3 m height was established over soft soil to analyze the load-settlement behavior using finite element methods. The embankment fill material is sand and maintained the relative density of fill around 60%. The crest of embankment is 10 m wide and side slop of embankment is 2H:1V (refer Fig. 3). The embankment was reinforced with geocell with varying height to width ratio (i.e., 0.8, 1, 1.2 1.4 and 1.6). The aperture size of reinforced geocell was adopted as 250 mm × 250 mm. Moreover, in this paper, the aperture size of geocell kept constant, but the height of geocell was varying; hence, the height-width ratio (*h*/*b*) was varying from 0.8 to 1.6. The numerical model is able to capture the actual response of geocell-reinforced embankment under 100 kN/m^2 surcharge pressure on

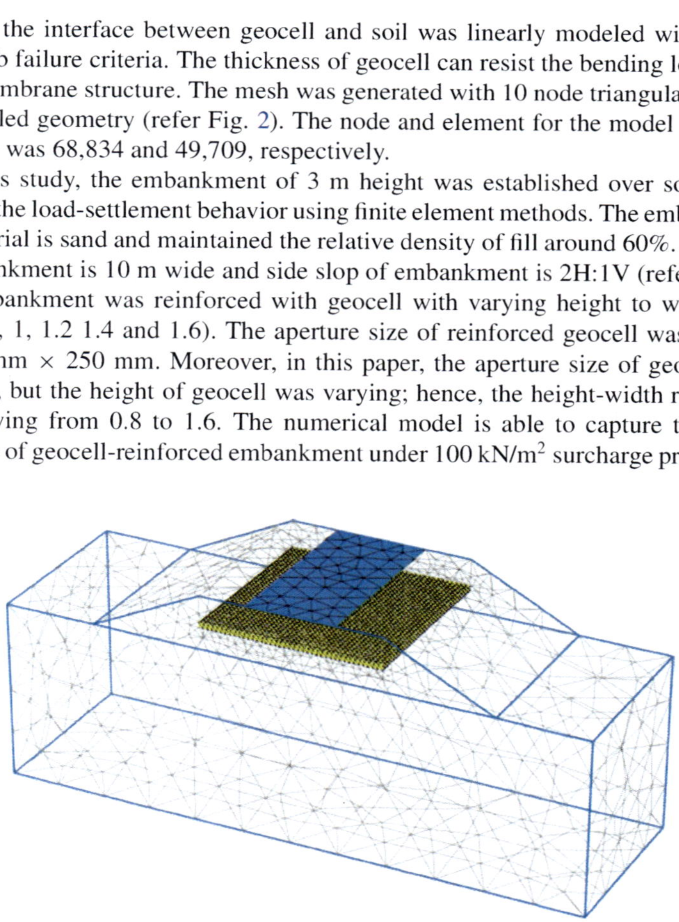

Fig. 2 Mesh generation of reinforced embankment model in PLAXIS 3D

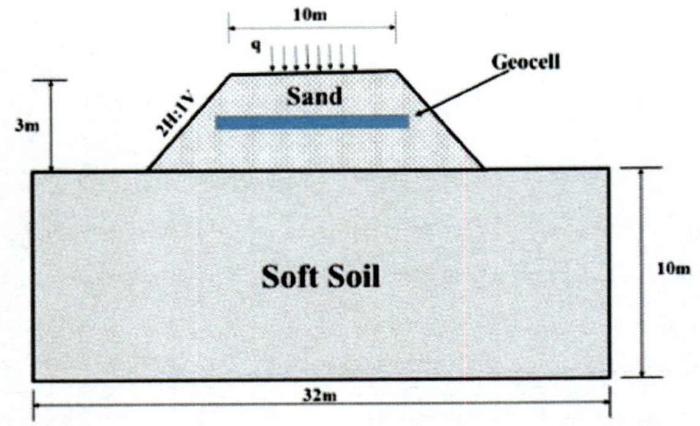

Fig. 3 Geometric description of embankment

the crest of embankment over an area of 5 m × 10 m. Geocell was reinforced at the depth of 1.5 m from the crest of embankment and cover an area of 10 m × 10 m. The objective of this study is to analyze the settlement behavior of geocell-reinforced embankment with varying height under constant static load.

2.1 Materials

For the numerical modeling, the embankment fill (i.e., sand) and foundation soil was simulated to determine the settlement behavior. Thereafter, the properties of sand and soft soil was adopted in PLAXIS 3D, as tabulated in Table 1. The reinforcing material was adopted to reduce the settlement characteristics of embankment. Hence, the properties of reinforcement material has been adopted from Hegde and Sitharam [4] and tabulated in Table 2.

Table 1 Properties of soil used in PLAXIS 3D

Properties	Soft soil	Sand
Specific gravity (G_S)	2.58	2.66
The angle of internal friction (φ)	14°	33°
Maximum dry density ($\gamma_{d,\ max}$) in kN/m^3	16.8	16.67
Minimum dry density ($\gamma_{d,\ min}$) in kN/m^3	–	14.5
OMC in %	19.5	–
Relative density	–	60%
Cohesion (C) in kN/m^2	40	0
Classification of soil (USCS)	CI	SP

Table 2 Properties of geocell

Properties	Quantity
Young's modulus, E (MPa)	275
Poisson's ratio, μ	0.45
Interface friction angle, φ_1 (°)	30
Cell size (mm)	250 × 250
No. of cell per unit area (per m^2)	16
Strip thickness (mm)	1.5
Density (g/cc)	0.95

3 Results and Discussion

Numerical models were investigated to determine the influence of geocell reinforcement on the embankment of soft soils. In addition, the geocell was reinforced with varying h/b ratios into the embankment at a depth of 1.5 m, where $b = 250$ mm, which is constant. The results were obtained from a numerical analysis for a reinforced geocell embankment and compared with the results for the unreinforced embankment. Figures 4, 5, 6, 7, 8 and 9 have described the vertical settlement for unreinforced and reinforced embankment using numerical analysis.

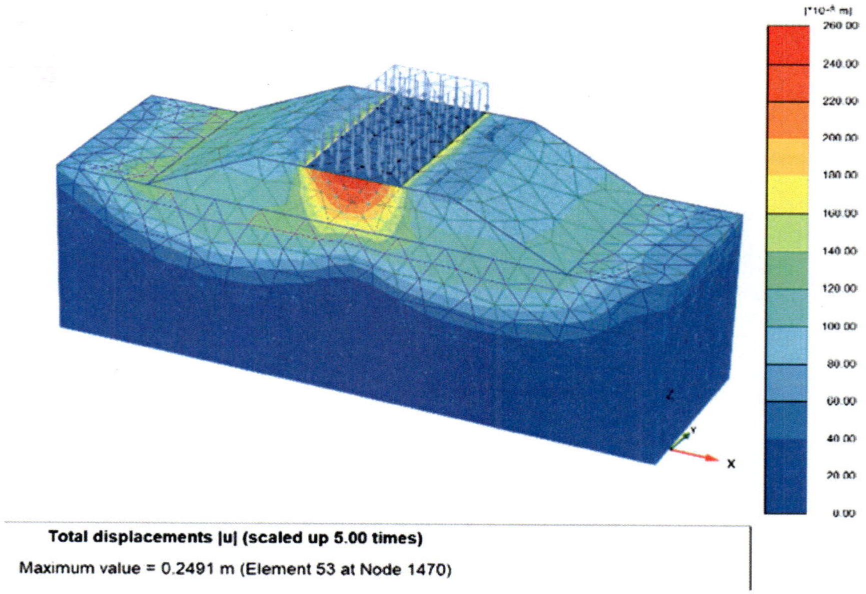

Fig. 4 Contour of vertical settlement for unreinforced embankment

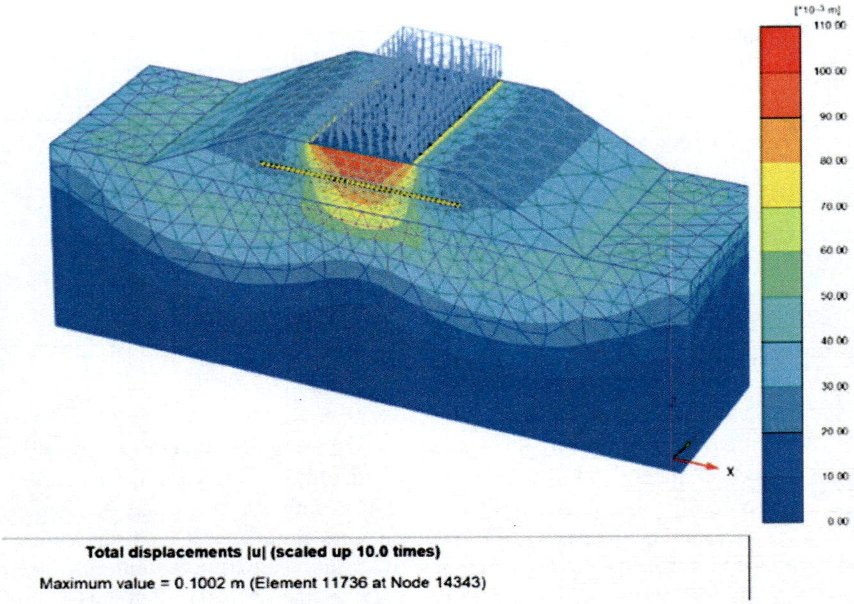

Fig. 5 Contour of vertical settlement for reinforced embankment ($h/b = 0.8$)

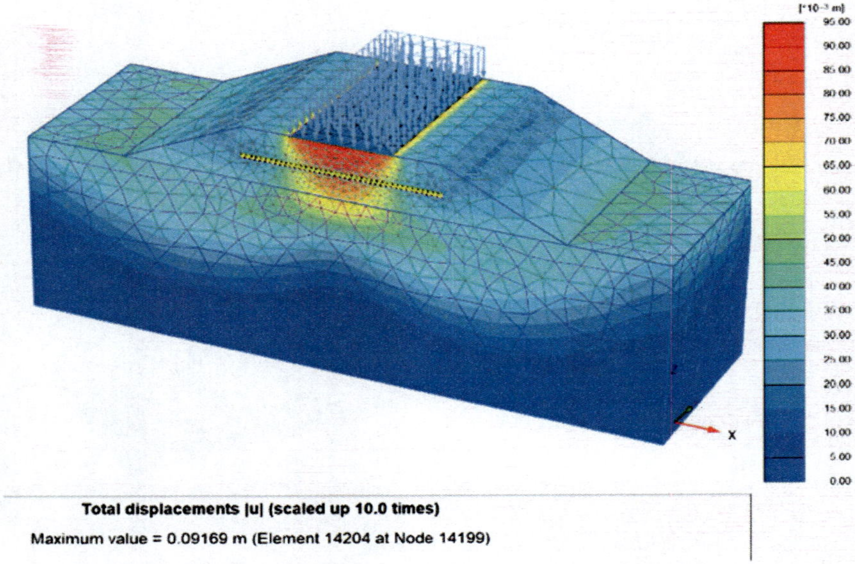

Fig. 6 Contour of vertical settlement for reinforced embankment ($h/b = 1.0$)

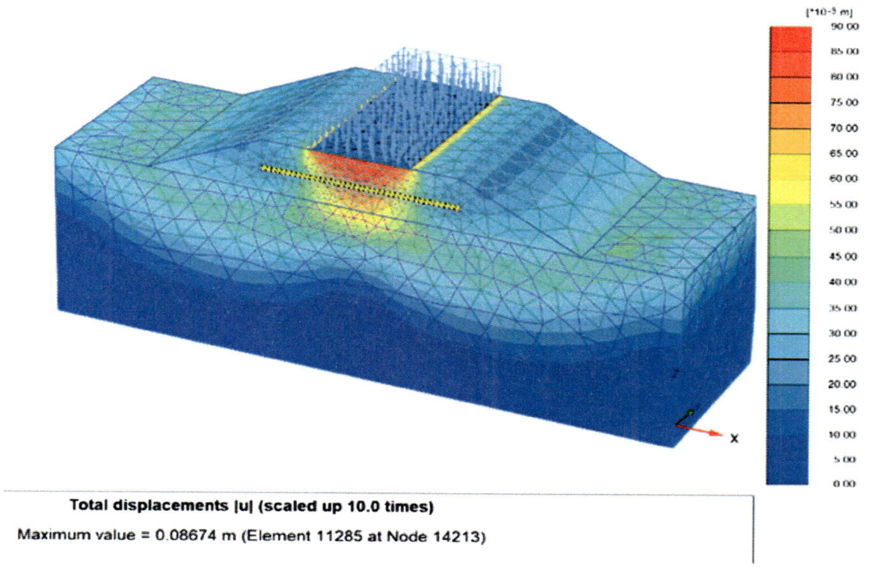

Fig. 7 Contour of vertical settlement for reinforced embankment ($h/b = 1.2$)

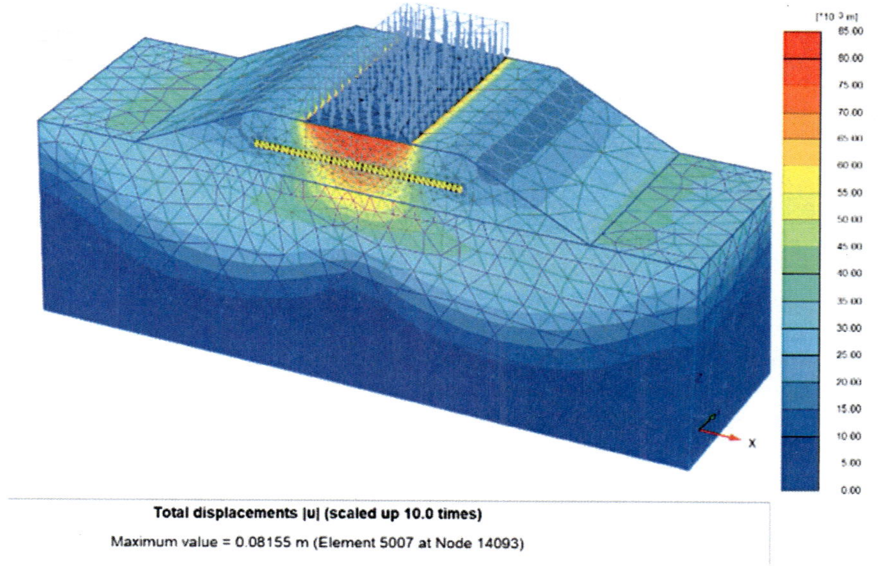

Fig. 8 Contour of vertical settlement for reinforced embankment ($h/b = 1.4$)

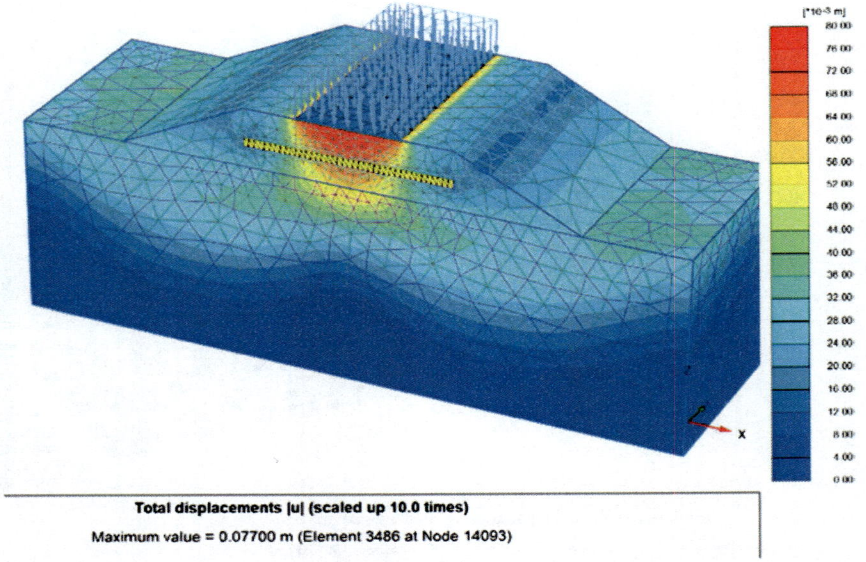

Fig. 9 Contour of vertical settlement for reinforced embankment ($h/b = 1.6$)

From the above contour figures, it can be noticed that the settlement of crest of the embankment significantly reduces within permissible limit for reinforced soil. The vertical displacement of unreinforced embankment is beyond the permissible limit. Thereafter, the reinforced geocell embankment increasingly enhanced the bearing capacity of soft soil as compare to unreinforced embankment. The load versus settlement behavior has been discussed for unreinforced and reinforced embankment (refer Fig. 10).

From the numerical analysis, the vertical settlements of embankment are 249 mm, 100 mm, 92 mm, 87 mm, 81.5 mm and 77 mm for unreinforced embankment and geocell-reinforced embankment with $h/b = 0.8$, 1, 1.2, 1.4 and 1.6, respectively. The reduction in vertical settlement of embankment was caused by using geocell as reinforcing material. From the obtained results, it can be concluded that as the height of geocell increases, and the settlement continuously reduces as well. The mechanism behind that geocell may be acted as the raft foundation below the crest of embankment. Hence, as the height of geocell was increased, the thickness of geocell raft also increased and due to that it could reduce the settlement and enhance the bearing capacity of soil. Moreover, the height of geocell also provided the confinement to infill material. Hence, as the height of geocell was increased, it provided the better lateral confinement to infill material as compare to geocell with less height. This may cause to reduce the settlement around 1/3rd of unreinforced settlement for reinforced soil.

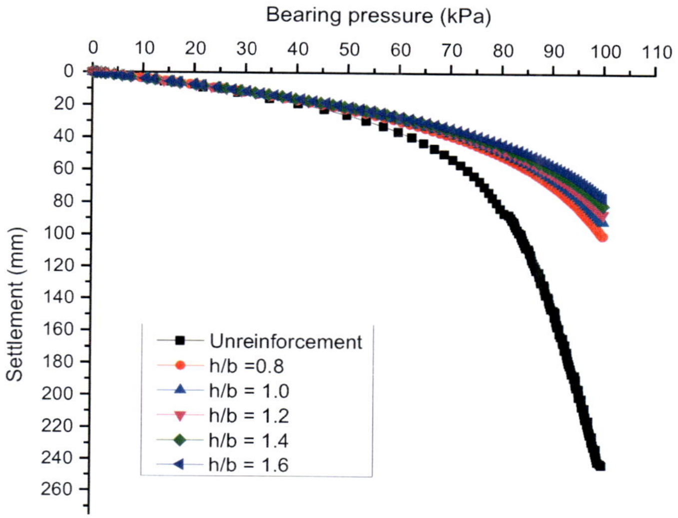

Fig. 10 Vertical settlement of embankment under static loading

4 Conclusion

In this paper, the different height of geocell has been analyzed to determine the settlement behavior of embankment under static load using PLAXIS 3D. The settlement of reinforced embankment also reduced with increasing the height of geocell. By increasing the height of geocell, the stiffness and elasticity of soil was also improved, which could improve the bearing capacity of soft soil. Furthermore, geocell with $h/b = 1.6$ provided more lateral confinement to infill material. In this paper, geocell acted as a virtual raft foundation to the embankment. The numerical result of PLAXIS showed a well-defined trend and consistency in settlement of embankment for unreinforced and reinforced, both.

Acknowledgements The authors sincerely acknowledged the financial support provided to the third author by Technical Education Quality Improvement Programme (TEQIP-III), India in the form of minor research (seed) grant (NITJSR/DCE/2019-20/023) at the Department of Civil Engineering, NIT Jamshedpur.

References

1. Dash SK, Rajagopal K, Krishnaswamy NR (2001) Strip footing on geocell reinforced sand beds with additional planar reinforcement. Geotext Geomembr 19(8):529–538
2. Hegde AM, Sitharam TG (2015) Effect of infill materials on the performance of geocell reinforced soft clay beds. Geomech Geoeng 10(3):163–173

3. Hejazia SM, Sheikhzadeha M, Mahdi SA, Zadhoush A (2012) A simple review of soil reinforcement by using natural and synthetic fibers. Constr Build Mater 30:100–116
4. Hegde A, Sitharam TG (2015) Experimental and numerical studies on protection of buried pipelines and underground utilities using geocells. Geotext Geomembr 43(5):372–381
5. Yang X (2010) Numerical analyses of geocell-reinforced granular soils under static and repeated loads. Ph.D. dissertation, University of Kansas, New Jersy,USA.
6. Sanjei C, De Silva LIN (2016) Numerical modelling of the behaviour of model shallow foundations on geocell reinforced sand. In: 2016 Moratuwa engineering research conference (MERCon). https://doi.org/10.1109/mercon.2016.7480142
7. Venkateswarlu H, Hegde A (2018) Numerical analysis of machine foundation resting on the geocell reinforced soil beds. Geotech Eng J SEAGS AGSSEA 49(4):55–62
8. Dash SK, Rajagopal K, Krishnaswamy NR (2007) Behaviour of geocell-reinforced sand beds under strip loading. Can Geotech 44:905–915
9. Dash SK, Bora MC (2013) Improved performance of soft clay foundations using stone columns and geocell-sand mattress. Geotext Geomembr 41:26–35
10. Latha GM, Rajagopal K, Krishnaswamy NR (2006) Experimental and theoretical investigations on geocell-supported embankments. Int J Geomech 6(1):30–35
11. Zhao MH, Zhang L, Zou XJ, Zhao H (2009) Research progress in two-direction composite foundation formed by geocell reinforced mattress and gravel piles. Chin J Highway Transp 22(1):1–10

Seismic Response of Soil-Like Material in MSW Landfill Using Equivalent Linear Approach

Kumar Shubham, Durgesh Prashad, and Subhadeep Metya

Abstract Soil-like material (SLM) is one of the most important and the largest proportion of reclaimed materials from the old municipal solid waste (MSW) landfills which usually constitute over 50% of the excavated materials. The engineering properties of SLM vary with the depth from which it has been obtained which shows the dependency on the age of the material. This variation in properties leads to varied response when subjected to various loads. So, it becomes necessary to study the behaviour of the SLM under different loading conditions so that it can be used safely as a filling material in construction works. In this paper, the seismic response of the SLM has been analysed by DEEPSOIL using an equivalent linear approach. The data available in the literature for SLM from landfills is used, and then, equivalent linear-frequency domain solution is being applied in pressure-dependent modified Kondner Zalesko soil model with the Chi-Chi input motion seismic vibrations, and the seismic responses of different layers are discussed. At last, based on the observed response of the SLM materials, effect of various factors like age of waste from MSW landfills and engineering properties is discussed.

Keywords Soil-like material (SLM) · DEEPSOIL · Equivalent linear approach

1 Introduction

Population explosion across the globe has led to the massive utilization of resources creating a large-scale production of wastes. Rapid urbanization and intensive lifestyle have resulted in a large amount of production of municipal solid waste (MSW). There is five hierarchy of waste management practices, namely source reduction, recycling of materials, combustion or incineration, landfills, and composting. Most of the developing nations like India rely greatly on the landfilling method for waste management.

K. Shubham (✉) · D. Prashad · S. Metya
Department of Civil Engineering, National Institute of Technology, Jamshedpur, India

K. Shubham
Department of Engineering, Arka Jain University, Jamshedpur, India

A. K. Choudhary et al. (eds.), *Advances in Geo-Science and Geo-Structures*, Lecture Notes in Civil Engineering 154,
https://doi.org/10.1007/978-981-16-1993-9_31

According to a study, around 79.3% of MSW in China and 70.6% of MSW in India are disposed of in the landfills. So, it becomes very necessary to look after the proper operations of landfills. The engineering properties of MSW landfills are complex, and thus, an application of external forces can make it unstable. Various research works are going on to determine the stability of MSW landfill. The stability of analysis of an MSW landfill site is a major field of research in which many researchers have contributed through their work. The stability of MSW depends on the behaviour of the foundation, engineering properties of the MSW landfill, the containment patterns, and the composition of the MSW [1–6]. It is also very important to analyse the stability from the geological point of view considering the material properties of the MSW landfills rather than only considering the engineering properties [7]. The ecological issues related to open dumps as of now demand prompt recovery. One of the choices is that open dumps can likewise be looked as a wellspring of materials like recovered secondary soils, refused derived fuels, and so forth. Landfill mining can be the possible alternatives to reduce accumulated aged waste at a dumpsite and using the site for other purposes or expanding landfill capacity [8–10]. Landfill mining can be defined as the process of excavating and processing the dumped solid wastes [11]. The constitutes found during the landfill mining are (1) oversized stones and debris; (2) material which can be used for composting; (3) combustible material; and (4) soil-like material. Soil-like material (SLM) is the major constituent [12, 13]. Apart from this, the stability of the MSW landfill must be checked based on seismic conditions and the local site conditions [14, 15]. Characterization of the soil-like fraction of waste is one of the areas on which researchers are working at present. One of the major findings obtained from previous studies is that the majority of the reclaimed material (50–70%) is soil like. The high soil-like proportion found in the reclaimed waste from landfill mining can be the placing of daily cover, dust collected from the sweeping of streets, silts from drains, and construction and demolition (C and D) waste along with the degradability of organic fractions in the MSW [16].

Many researchers have worked on the determining the seismic responses of MSW landfills and dynamic stability of new and expanded landfills using different methods, development of models to characterize the MSW properties based on seismic conditions and kinematics of waste bodies [17–19]. However, some researchers have worked on another aspect of MSW landfill stability by including the geological properties like the geological body, its structure, and geo-environment along with the engineering properties. It has also been found that the ageing of wastes in the MSW landfills leads to the change in engineering properties and hence can affect the stability [20]. So, this paper aims at determining the seismic response of the SLM reclaimed from the landfills using the equivalent linear approach as it is one of the necessary aspects of stability which we need to keep in mind while designing a sanitary MSW landfill.

Table 1 Engineering parameters for the SLM present in the soil profile

Parameters	SLM with the age of 706 days	SLM with the age of 1472 days	SLM with the age of 2182 days	SLM with the age of 2379 days
Thickness of layer (m)	7.5	7.5	7.5	7.5
Unit weight (kN/m^3)	11.5	11.8	12.6	12.8
Shear wave velocity (m/s)	150	200	300	400
Effective vertical stress (kN/m^2)	43.125	130.5	222	317.25

2 Methodology

The literature available for the SLM from the previous research on Santo Tirso landfill (north of Portugal) between 2001 and 2007 has been used to develop the soil profile diagram of the MSW landfill [4]. The data which has been used is placed according to their age of dumping to form a stratified profile of the landfill. Different parameters are then assigned to the materials to analyse the response when subjected to a seismic force.

Table 1 depicts the engineering parameters used for developing the model. This model is analysed using DEEPSOIL by an equivalent linear approach. This software has an in-built function to generate acceleration, velocity, displacement, and Arias intensity time histories, as well as the response spectrum and Fourier amplitude spectrum for the selected motion. Figure 1 shows the input seismic vibration which has been selected for the analysis. The geotechnical properties of the SLM are not linear/specific as it varies with depth and greatly depends on the age of dumped wastes and their composition. So, the equivalent linear approach is selected for analysis. The equivalent linear (EL) model uses an iterative method in the selection of the shear modulus and damping ratio soil properties. These properties can be defined by discrete points or by defining the soil parameters that define the backbone curve of one of the nonlinear models. The option of defining the soil curves using discrete points is only applicable for the equivalent linear analysis. For this option, the $G/G_{\max}$ and damping ratio (%) are defined as functions of shear strain (%).

3 Observations and Results

The soil profile is developed using the data of samples from boreholes collected Santo Tirso landfill (north of Portugal) between 2001 and 2007 and is arranged based on the age of the waste from top to bottom keeping the thickness of each layer as 7.5 m. Figure 2 shows the parametric variations in the soil profile. The introduced vibrations

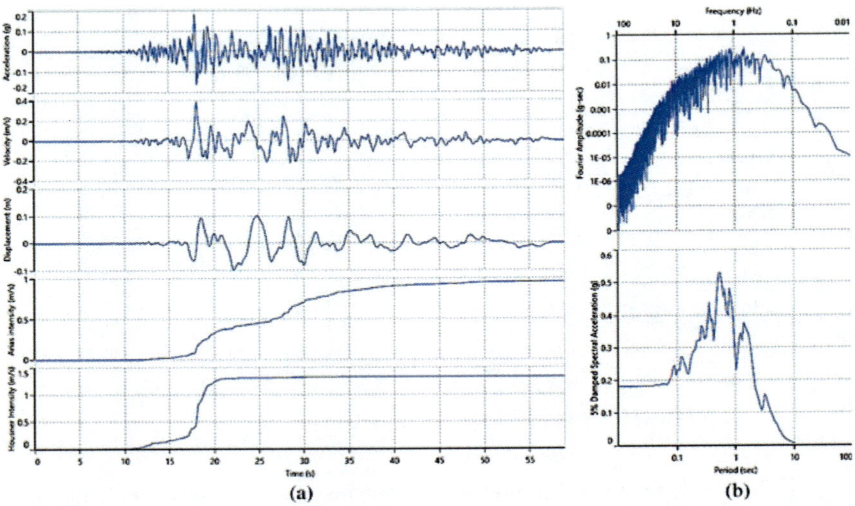

Fig. 1 **a** Graph representing the input vibration motion (Chi-Chi model) parameters like PGA, PGV, PGD, Arias intensity, Housner intensity. **b** Graph representing the Fourier amplitude and spectral acceleration corresponding to 5% damping with respect to period

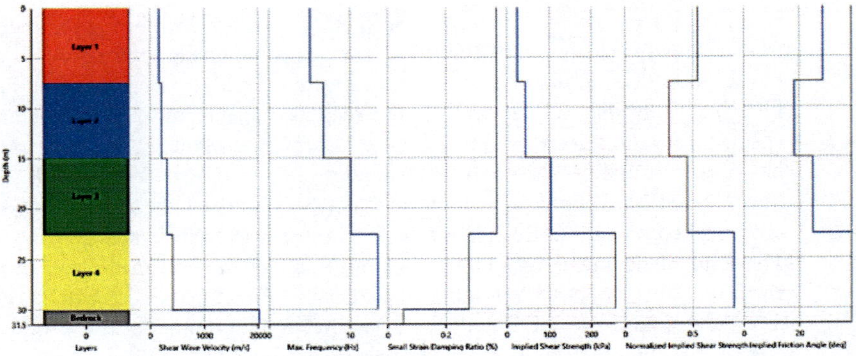

Fig. 2 Soil profile plot depicting variation in shear wave velocity, maximum frequency, small strain damping ratio, implied shear strength, and the implied friction angle

lead to the development of seismic responses which are unique for each layer. Figure 3 shows the time history plot of the seismic vibrations after being introduced in the soil profile. It can be inferred from Fig. 4 that the strain developed decreases with the depth. It has also been observed from Fig. 5 that the peak ground acceleration (PGA) and peak ground displacement (PGD) decreases with the increase in depth of the prepared soil profile while the maximum strain (%) increases up to the second layer and then decreases with the depth.

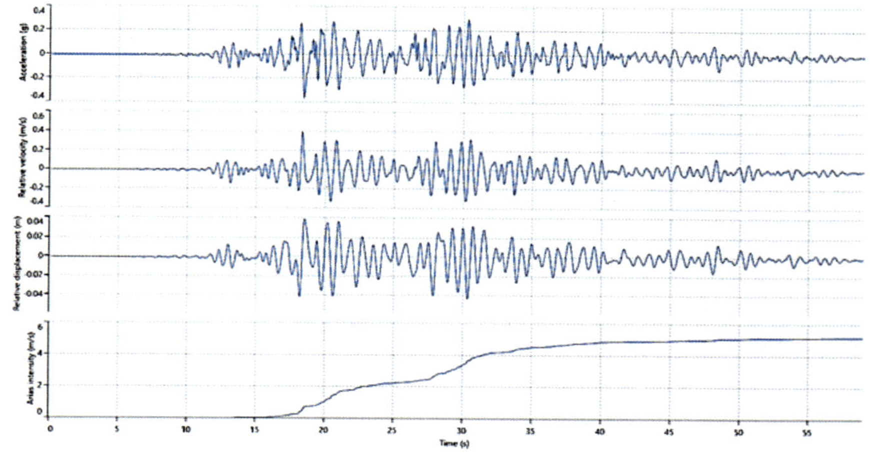

Fig. 3 Time history plot of the seismic vibrations after being introduced to the soil profile

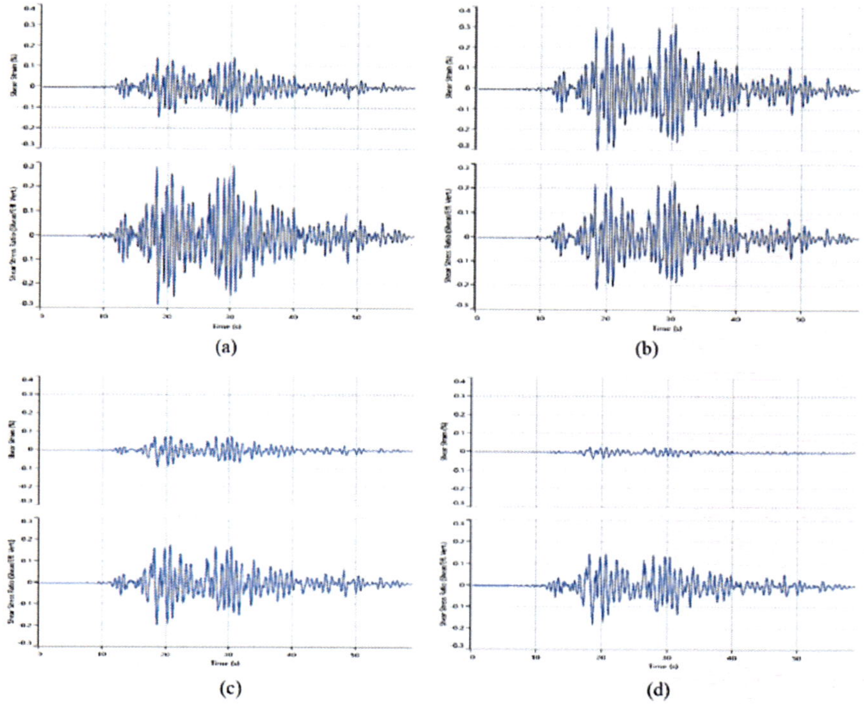

Fig. 4 **a** Stress–strain plot of SLM ageing 706 days. **b** Stress–strain plot of SLM ageing 1472 days. **c** Stress–strain plot of SLM ageing 2182 days. **d** Stress–strain plot of SLM ageing 2379 days

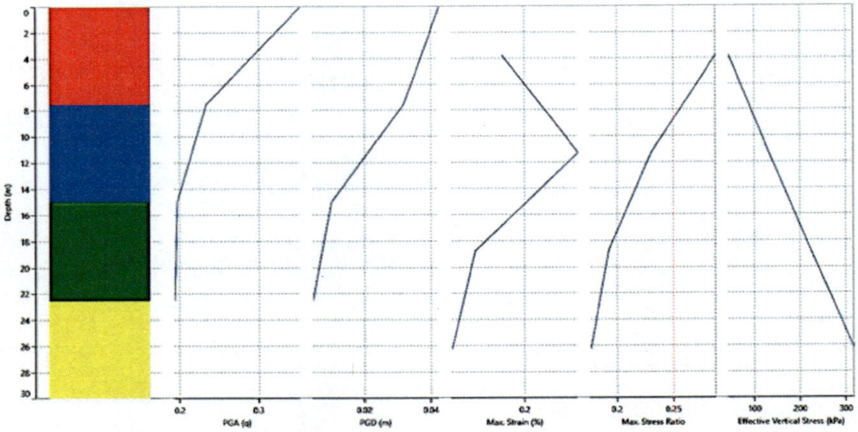

Fig. 5 Profile plot showing the variations in PGA, PGD, maximum strain (%), maximum stress ratio, and effective vertical stress

The final obtained seismic response spectrum as shown in Fig. 6 demarcates a clear variation and uniqueness in the response behaviour of the SLM present at different depths. It can be observed that the new wastes give higher values of response as compared to the older wastes present at the deeper layer of the MSW landfill. The similar trend can be observed in the tripartite chart shown in Fig. 7. So, the obtained graphs can be used to calculate the values of parameters which are required for the determination of seismic stability of any structure.

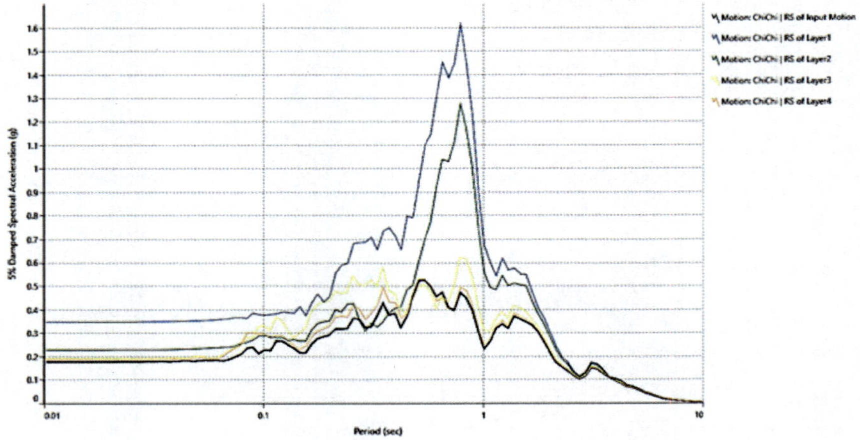

Fig. 6 Seismic response of the different layers with SLM ageing 706 days, 1472 days, 2182 days, and 2182 days, respectively, when subjected to Chi-Chi input motion

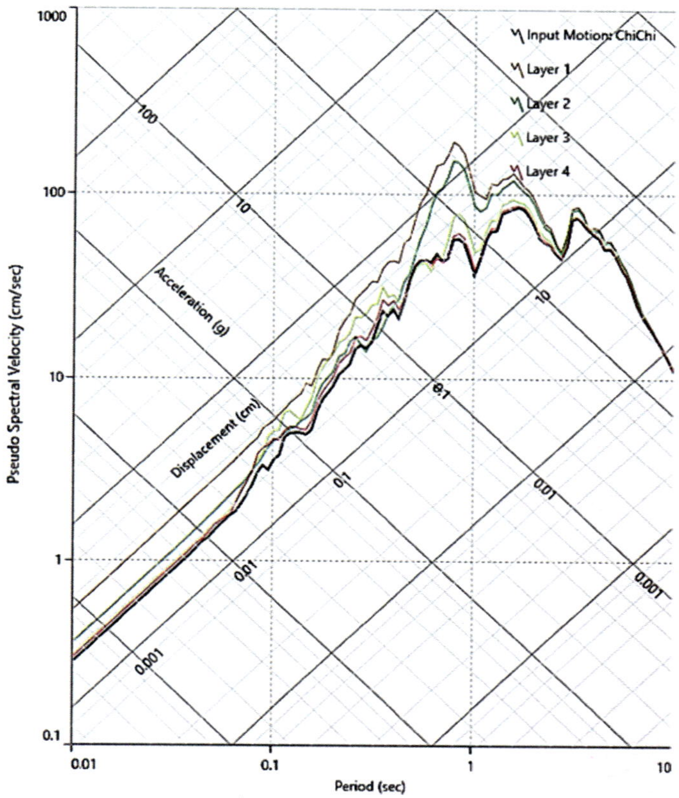

Fig. 7 Tripartite graph showing the variation of seismic response with pseudo-spectral velocity, time period, acceleration, and displacement

4 Conclusions

Some of the main conclusions which can be derived from this study based on the seismic responses are summed up as follows:

1. *Effect of shear behaviour and strength parameters*—The stress–strain curves shows there is a continuous increase in the shear strain when moving from the bottom layer to the top layer. So, it can be concluded that the shear strengths of SLM present at the greater depths are more as compared to those present at shallower depths. This typical upward curvature is due to the contribution of the fibrous waste components, where the tensile strength is progressively mobilized due to the intersection of the shear surface with the fibrous plane.
2. *Effect of waste composition*—The top layer of the MSW landfills usually contains a large amount of inert solid materials like plastics, papers, textiles, wood, metal, glass, etc., along with a considerable amount of moisture which

either gets evaporated or seeps to the bottom layers resulting in the increased values of acceleration of shear waves. Thus, the top layer becomes susceptible to failure and the observed seismic response is higher.

3. *Effect of ageing and decomposition*—Ageing and decomposition of wastes in MSW landfills lead to (i) an increase of the relative quantity of inert materials and fine fraction and in the fraction of granular materials; (ii) a decrease in particle size, the amount of organic matter, fabrics, and plastics, and the number of materials with sheets and wire forms With the amount of fibrous and granular elements increases, values of cohesion and friction angle increases, respectively. Waste ageing or degradation leads to a decrease in the overall resistance and the cohesive resistance but a slight increase in frictional resistance. So, being the most stable layer, it can be concluded that the oldest SLM provides the least response.

References

1. Annapareddy VR, Pain A (2020) Influence of foundation soil on the seismic factor of safety of geosynthetic-lined solid-waste landfills: equivalent linear approach. Nat Hazard Rev 21(3):04020027
2. Babu GS, Reddy KR, Srivastava A (2014) Influence of spatially variable geotechnical properties of MSW on stability of landfill slopes. J Hazard Toxic Radioactive Waste 18(1):27–37
3. Choudhury D, Savoikar P (2011) Seismic stability analysis of expanded MSW landfills using pseudo-static limit equilibrium method. Waste Manage Res 29(2):135–145
4. Gomes C, Lopes ML, Oliveira PJV (2013) Municipal solid waste shear strength parameters defined through laboratorial and in situ tests. J Air Waste Manag Assoc 63(11):1352–1368. https://doi.org/10.1080/10962247.2013.813876
5. Burlakovs J, Kaczala F, Vincevica-Gaile Z et al (2016) Mobility of metals and valorization of sorted fine fraction of waste after landfill excavation. Waste Biomass Valor 7:593–602. https://doi.org/10.1007/s12649-016-9478-4
6. Hölzle I (2019) Contaminant patterns in soils from landfill mining. Waste Manage 83:151–160. https://doi.org/10.1016/j.wasman.2018.11.013
7. Huang Y, Fan G (2016) Engineering geological analysis of municipal solid waste landfill stability. Nat Hazards 84(1):93–107
8. Hull RM, Krogmann U, Strom PF (2005) Composition and characteristics of excavated materials from a New Jersey landfill. J Environ Eng 131(3):478–490. https://doi.org/10.1061/(ASCE)0733-9372(2005)131:3(478)
9. Jahanfar A, Amirmojahedi M, Gharabaghi B, Dubey B, McBean E, Kumar D (2017) A novel risk assessment method for landfill slope failure: case study application for Bhalswa Dumpsite, India. Waste Manag Res 35(3):220–227
10. Kaartinen T, Sormunen K, Rintala J (2013) Case study on sampling, processing and characterization of landfilled municipal solid waste in the view of landfill mining. J Clean Prod 55:56–66. https://doi.org/10.1016/j.jclepro.2013.02.036
11. Krook J, Svensson N, Eklund M (2012) Landfill mining: a critical review of two decades of research. Waste Manage 32(3):513–520. https://doi.org/10.1016/j.wasman.2011.10.015
12. Kurian J, Esakku S, Palanivelu K, Selvam A (2003) Studies on landfill mining at solid waste dumpsites in India. In: Proceedings, Sardinia Margherita di Pula. CISA, Lucca, pp 248–255
13. Masi S, Caniani D, Grieco E, Lioi DS, Mancini IM (2014) Assessment of the possible reuse of MSW coming from landfill mining of old open dumpsites. Waste Manage 34(3):702–710. https://doi.org/10.1016/j.wasman.2013.12.013

14. Pain A, Annapareddy VR, Sarkar S (2018) Effect of local site condition on the seismic stability of municipal solid waste landfills. In: Geotechnical earthquake engineering and soil dynamics V: slope stability and landslides, laboratory testing, and in situ testing. American Society of Civil Engineers, Reston, pp 46–55
15. Pain A, Annapareddy VR, Sarkar S (2019) Seismic stability analysis of municipal solid waste landfills using strain dependent dynamic properties. Indian Geotech J 49(2):204–215
16. Parrodi JCH, Höllen D, Pomberger R (2018) Characterization of fine fractions from landfill mining: a review of previous investigations. Detritus 2:46–62. https://doi.org/10.31025/2611-4135/2018.13663
17. Qian X (2006) Translational failures of geosynthetic lined landfills under different leachate buildup conditions. In: Advances in unsaturated soil, seepage, and environmental geotechnics, pp 278–289
18. Savoikar P, Choudhury D (2010) Effect of cohesion and fill amplification on seismic stability of municipal solid waste landfills using limit equilibrium method. Waste Manage Res 28(12):1096–1113
19. Savoikar P, Choudhury D (2012) Translational seismic failure analysis of MSW landfills using pseudodynamic approach. Int J Geomech 12(2):136–146
20. Somani M, Datta M, Ramana GV, Sreekrishnan TR (2018) Investigations on fine fraction of aged municipal solid waste recovered through landfill mining: case study of three dumpsites from India. Waste Manage Res 36(8):744–755. https://doi.org/10.1177/0734242X18782393

Numerical Analysis on Load Carrying Mechanism of Single Pile Due to Twin Stacked Tunnelling

Arnab Choudhury and Awdhesh Kumar Choudhary

Abstract In recent years, the construction of twin tunnels in several populous cities increases significantly for better infrastructural development. In many situations, tunnel alignment must be laid below the high-rise buildings, bridges or important civil engineering structures. In such condition, the chances of detrimental effect on the foundations of these structures become very high. Hence, it is important to understand the tunnel–pile interaction mechanism for better safety and serviceability of these structures existing on the pile. In view of this series of numerical analysis are carried out on twin tunnel–pile interaction behaviour. It is observed that sequence of twin tunnel excavation gives significant effect on load carrying mechanism of pile. Two different tunnel excavation sequences are considered in the present study. In first sequence, first tunnel excavation is carried out near the mid-depth of pile and second tunnel excavation is carried out near the pile toe followed by the first one (ST). In second sequence, the excavation near the pile toe is carried out first and second tunnel excavation is carried out near the mid-depth of pile (TS). It is found that a sudden change in load bearing capacity is observed for first twin tunnels sequence (ST). Whereas, a smooth transition in load bearing capacity of pile is observed for second twin tunnel sequence (ST).

Keywords Tunnel · Pile · Tunnel excavation sequence · Load carrying capacity of pile

---

A. Choudhury (✉) · A. K. Choudhary
NIT Jamshedpur, Jamshedpur, Jharkhand, India

A. K. Choudhary
e-mail: awdhesh.ce@nitjsr.ac.in

A. K. Choudhary et al. (eds.), *Advances in Geo-Science and Geo-Structures*,
Lecture Notes in Civil Engineering 154,
https://doi.org/10.1007/978-981-16-1993-9_32

1 Introduction

With the expansion of cities, the need of sustainable flawless traffic system is also increasing rapidly. Underground rapid transit system comes out as a best representative of sustainable traffic system. Series of underground excavation or tunnel excavation takes place as a part of rapid transit system. Tunnel excavation causes stress relaxation of soil which ultimately leads to ground settlement. This tunnel-induced ground settlement imposes significant effect on load transferring mechanism of pile which becomes a major concern of safety and serviceability of existing buildings.

For this reason, several experimental, analytical and numerical studies have been carried out to analyse the tunnelling effect on adjacent pile behaviour. From these studies, it is found that the change in pile behaviour will be significant when distance between pile and tunnel is less than the tunnel diameter [1]. Pile head settlement, increase in axial load, change in shaft resistance, decrement in end bearing capacity of pile due to tunnel excavation were reported in the literature [2, 3]. The researchers also reported that the effect on load carrying mechanism of the pile depends on various tunnel positions with respect to pile [4–7]. For twin tunnels, the effect of tunnel on pile is more as compared to a single tunnel. A clear decrement in load carrying capacity of pile for twin tunnel excavation in sand was reported by Ng et al. [6]. Soomroo et al. [7] performed detailed numerical studies on twin tunnel-induced pile behaviour in clay. Most of the researchers performed analysis on tunnel–pile interaction behaviour when tunnels placed on two sides of pile. Ng and Lu [6] conducted centrifuge test on twin tunnel excavation sequence when both the tunnels are placed on same side of pile in sand. Soomro et al. [8] performed numerical study on excavation sequence of twin tunnel on same side of pile. Different settlement values were reported by authors due to various twin tunnel excavation sequences.

In view of this, the series of numerical studies have been carried out to analyse pile behaviour for various twin tunnel construction sequences. The change in load bearing capacity of pile for various twin tunnel sequences has been evaluated and discussed here.

2 Methodology

In the present study, numerical analysis has been carried out by using FLAC 2D software package. FLAC 2D based on finite difference techniques is capable to solve various complex geotechnical problem with varying boundary condition. Highly nonlinear and irreversible response of soil and rock can be analysed with the help of several constitutive models which are default in FLAC. The package also has several structural elements like pile, cable, strip, liner to simulate more accurate field conditions in during modelling. A typical layout of FLAC 2D model with the details of loading and the boundary conditions is shown in Fig. 1, where LP is the length of

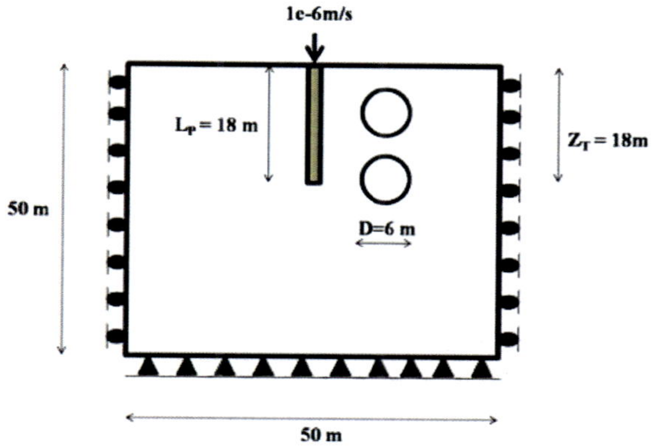

Fig. 1 Configuration of typical numerical model for twin stacked tunnel

pile and Z_T is the depth of second tunnel from soil surface. A downward displacement is applied at a rate of 1e−6 m/s.

The FLAC 2D model used in this study is presented in Fig. 2. The number of grids used in this study is 2500. The number of grids used in present model is sufficiently large to minimize the boundary effects during numerical simulation. The horizontal displacement along the side boundary is restricted, and vertical displacement for same boundary is allowed. Both horizontal and vertical displacements are restricted along the bottom boundary. Pile is installed at middle of the model. Predesigned pile elements of pile are used to simulate pile behaviour. Normal and shear spring coefficients are used to simulate the soil–pile interaction behaviour. Tunnel is excavated

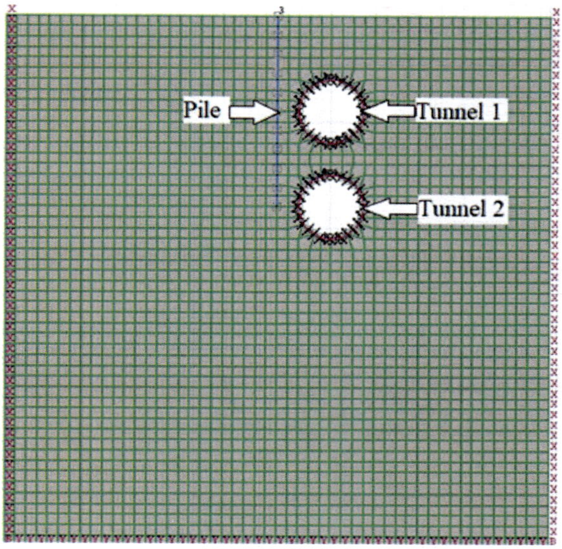

Fig. 2 Finite difference grids and boundary conditions of numerical model for ST

Table 1 Parameters used in the present study

Sl. No.	Properties	Value
1	Clay bulk modulus (Pa)	100e6
2	Clay shear modulus (Pa)	48e6
3	Friction angle (°)	10°
4	Density (kg/m^3)	1200
5	Stiffness of pile material (Pa)	35e9
6	Stiffness of liner material (Pa)	35e9

by using null command. Predefined liner elements are used to simulate concrete liner behaviour for present study. All parameters used in the present study are summarized in Table 1.

As per requirement of study, tunnels have been excavated in two sequences: (a) first tunnelling at mid-depth of pile (9 m) and second tunnelling near the pile toe (18 m) (ST) and (b) first tunnelling at toe of pile (18 m) and second tunnelling near mid-depth of pile (9 m) (TS). Tunnel volume loss of 1% is considered for present study.

3 Results and Discussion

3.1 Load Displacement Response of Pile for Various Tunnel Positions

Figure 3 represents load displacement response of pile due to tunnel position at toe of pile and mid-depth of pile. It is observed that ultimate bearing capacity of pile is

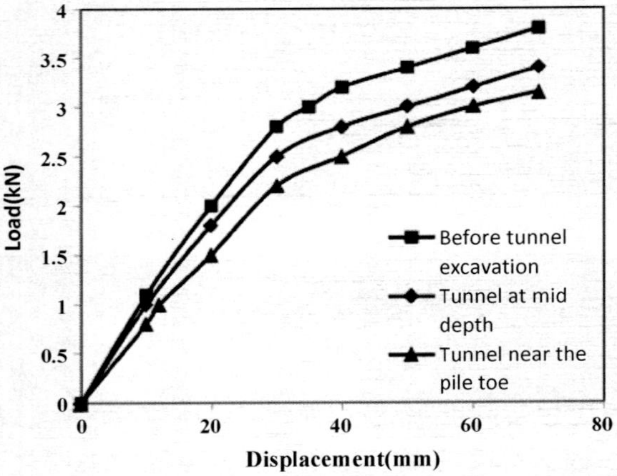

Fig. 3 Load–displacement curve of pile for different position of tunnel

minimum when tunnel is placed near the toe of the pile. The state of soil around pile after tunnelling operation is depicted in Figs. 4 and 5. From Fig. 4, it is observed that shear yielding takes place along the entire pile length. Shear yielding effects on load transfer mechanism of pile ultimately decreases the end bearing capacity and shaft resistance capacity of pile. Whereas, negative skin friction (tension in soil) is developed at upper portion of pile when tunnelling is carried out near the mid-depth of pile (presented in Fig. 5). Negative skin friction is generated due to downward soil movement after tunnel excavation. This clearly indicates that load transfer mechanism of pile changes due to tunnel excavation and depends on tunnel position with respect to pile.

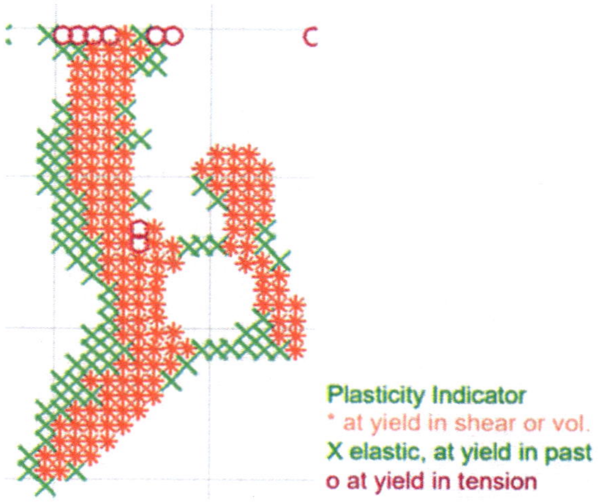

Fig. 4 Plastic zone of this model when tunnel excavated near the mid-depth of pile

Fig. 5 Plastic zone of this model when tunnel excavated near the toe of pile

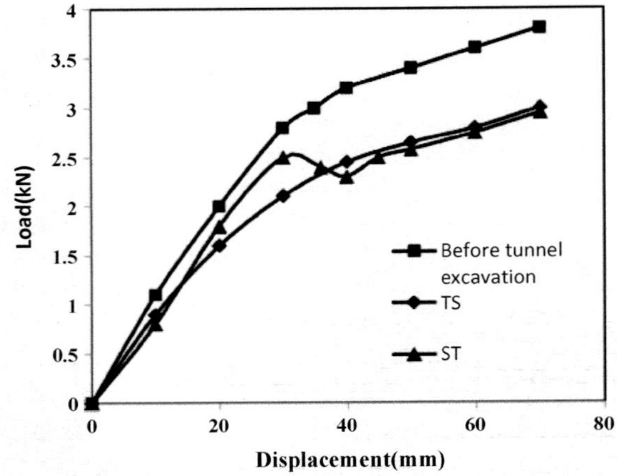

Fig. 6 Load–displacement curve of pile for different tunnel excavation sequence

3.2 Load Displacement Response of Pile for Two Different Sequences of Twin Tunnel Excavation

Figure 6 represents load displacement response of pile for first tunnel and second tunnel excavation in twin stacked tunnelling system ST and TS. In case of ST, first tunnel excavation is done near the pile mid-depth and second tunnel excavation is done near the pile toe followed by first excavation. A sudden jerk is found in load displacement response of pile for ST sequence. This is occurred because of sudden reduction in soil stiffness after second tunnel excavation near the pile toe. However, load–displacement curve with smooth transition is observed for TS excavation sequence. It clearly signifies load transferring mechanism of pile is affected by twin tunnels excavation sequences.

4 Conclusion

A series of numerical analysis is performed in this study to improve the understanding on effect of two different twin tunnels construction sequence on load bearing capacity of pile. Based on the present study following conclusions can be drawn.

1. The effect of tunnel excavation on load carrying capacity of pile depends on tunnel positions. Maximum reduction in load bearing capacity is obtained when tunnel is position at the pile toe level.
2. For twin stacked tunnel excavation, a sudden jerk in load–displacement curve of pile is observed for ST sequence. Whereas, a smooth transition in load–displacement curve is obtained in TS sequence.

References

1. Bezuijen A, Schrier JS (1994) The influence of a bored tunnel on pile foundations. Centrifuge 94:681–686
2. Lu H, Shi J, Wang Y, Wang R (2019) Centrifuge modeling of tunnelling-induced ground surface settlement in sand. Underground Space
3. Malhotra M, Sahu V, Srivastava A, Misra KA (2019) Impact of pile foundations adjacent to tunnels in sandy stratum. Sadhana 44:184
4. Jacobsz SW, Standing JR, Mair RJ, Hagiwara T, Sugiyama T (2004) Centrifuge modeling of tunnelling near driven piles. Soils Found 44:49–56
5. Loganathan N, Poulos HG, Stewart DP (2000) Centrifuge model testing of tunnelling induced ground and pile deformations. Geotechnique 50(3):283–294
6. Ng WWC, Lu H, Peng YS (2013) Three-dimensional centrifuge modelling of the effects of twin tunnelling on an existing pile. Tunnell Undergr Space Technol
7. Soomro MA, Ng CWW, Liu K, Memon NA (2017) Pile responses to side-by-side twin tunnelling in stiff clay: effects of different tunnel depths relative to pile. Comput Geotech 84:101-116S
8. Soomro MA, Kumar M, Xiong H, Mangnejo AD, Mangi N (2020) Investigation of effects of different construction sequences on settlement and load transfer mechanism of single pile due to twin stacked tunnelling. Tunnell Undergr Space Technol 96:103171

Recent Advances in Metallic and Geosynthetic Reinforced Soil Walls: A Review

K. Purna Vindhya and B. Giridhar Rajesh

Abstract Soil reinforcement is defined as a method to enhance the engineering characteristics of the soil. In the last few decades, the construction of reinforced earth walls in various scenarios has gained importance due to its flexibility, economy, and ease of construction. Various reinforcing elements were used since inception. Metallic reinforcements were used in the construction of early reinforced earth walls. Due to the corrosion of metal reinforcements over a period of time, geosynthetic reinforcements have gained importance as an alternative. Several analytical, numerical, and experimental research works and many case studies on both corrosion in metallic reinforced earth walls and the development toward geosynthetic reinforced earth walls were performed. In the present study, an attempt has been made to provide an insight into the recent developments and to present an overview of these advances. A critical literature review of corrosion in metallic reinforced earth walls and advancement toward geosynthetic retaining walls was conducted, focusing on compaction-induced stresses and geosynthetic reinforced soil bridge abutments. Gaps in research in both metallic and geosynthetic retaining walls were identified, and future scope for research is briefed.

Keywords Reinforced soil wall · Corrosion · Geosynthetics · Compaction-induced stresses · GRS bridge abutment

1 Introduction

Since the invention of reinforced earth walls by Henry Vidal in 1963, it has become an economical solution to build retaining structures and many types of reinforced soil structures. These are especially appreciated when the design conditions are difficult

---

K. Purna Vindhya (✉) · B. Giridhar Rajesh
Department of Civil Engineering, National Institute of Technology Andhra Pradesh, Tadepalligudem 534101, India

B. Giridhar Rajesh
e-mail: rajesh@nitandhra.ac.in

A. K. Choudhary et al. (eds.), *Advances in Geo-Science and Geo-Structures*, Lecture Notes in Civil Engineering 154,
https://doi.org/10.1007/978-981-16-1993-9_33

and the space available is restricted. Reinforced earth technology involves reinforcing the soil with a material which is strong in tension; thereby, the stability of the structure is ensured. These reinforcing materials may be metallic strips, bars, or geosynthetics. From Fig. 1, it is evident that strength of soil is improved with the inclusion of reinforcements in the soil.

Metal tensile elements are widely used as the soil reinforcement, owing to their high tensile strength, availability, and ease of installation. The service life of reinforced soil structures is about 50–75 years. This service life of embankments, walls, and slopes reinforced with metallic inclusions is affected by the deterioration of inclusions with time. Corrosion is an electrochemical process, which results in the flow of electrons. The phenomenon of corrosion is affected by the presence of moisture, type of reinforcement, chemicals such as dissolved sulfate and chloride ions, pH, temperature, microbiological activity, and soil resistivity. Equation (1) indicates the chemical loss that governs corrosion.

$$\mathrm{Fe} > \mathrm{Fe}^{2+} + 2e^{-} \tag{1}$$

Of lately, the most widely used reinforcements as an alternative to metal inclusions are geosynthetics. These are used along with soil in geotechnical constructions as a reinforcement. Natural products like coir, jute, hemp, etc., and artificial products like polymeric and metallic geotextiles, geogrids, geocells, geonets, geomembranes, etc., are used. Compaction may significantly affect the internal stresses of reinforced soil walls, and hence, it is necessary to model this phenomenon correctly. Field observations indicate that compaction-induced stresses (CIS) promote displacement during the construction period and reduce both settlement and horizontal displacement due to compaction-induced load application after construction. Compaction

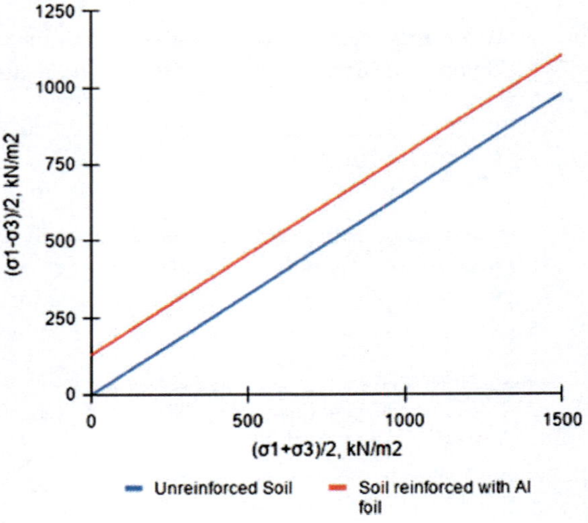

Fig. 1 Strength envelopes of reinforced and unreinforced soils. Adopted from Saran [1]

introduces an over-consolidation on the soil that promotes a stiffer behavior at the end of construction. Although there are several studies on reinforced earth walls centered on model scale tests and numerical simulations, to the best of authors' knowledge, there is no specific study focusing on the effect of corrosion of reinforcement, compaction-induced stresses, and geosynthetic reinforced soil bridge abutments. Hence, the objective of this present study is to present a summary of the past studies, current trends, and the scope for future research on corrosion of metal reinforced earth walls and geosynthetic reinforced earth walls, focusing mainly on compaction-induced stresses and geosynthetic reinforced soil bridge abutments.

2 Literature Review

2.1 *Studies on Corrosion in Metal Reinforced Earth Walls*

The literature related to the metallic reinforced earth walls is divided into analytical studies, field studies, and numerical studies. These studies are discussed in the following sub-sections.

2.1.1 Analytical Studies

Not many analytical studies of the corrosion in MSE were carried out by many researchers. Previous researchers performed tests to evaluate the performance of the MSE walls reinforced with metallic strips and observe the corrosion of the reinforcements with time. Romanoff [2] conducted a large study on mechanically stabilized earth (MSE) walls at 128 different sites consisting of 14 soil types for 17 years to monitor corrosion. The author proposed a power law, given by:

$$e = At^n \tag{2}$$

where e = average corrosion rate per side.

A = corrosion parameter, which is a constant between 150 and 180 μm for carbon steel, or a constant between 5 and 75 μm for galvanized steel,

t = time after burial,

n = constant varying between 0.5 and 0.6 for carbon steel.

Beckham et al. [3] monitored the development of corrosion in four MSE walls in Kentucky, USA. Equations (3) and (4) give the average and maximum value of metal loss for carbon steel, respectively, adopted by the authors.

$$x = 25t^{0.65} \tag{3}$$

$$x = 50t^{0.65} \tag{4}$$

2.1.2 Field Studies

Guilloux and Jailloux [4] carried out corrosion studies on a full-scale reinforced earth wall. Walls of height 6 and 5 m long and reinforced with cold-rolled steel reinforcements were built and brought to break by accelerated corrosion. The accelerated corrosion was brought by spraying brine. It was observed that the corrosion of reinforcements increased with the conductivity of the soil. The metal loss was found to vary along the length of the reinforcement. To account for the heterogeneity of soil, the authors concluded that the rate of strength loss in reinforcement is greater than that of the metal loss. Blight and Dane [5] studied the failure of a reinforced wall built on the coastline at Tweepad, South Africa. The backfill soil was sand, and the reinforcement used was galvanized steel. By comparing the corroded MSE wall with a non-corroded wall nearby, it was observed that the extent of corrosion was not spatially uniform. The authors conclude that only fill properties do not define the metal loss and rate of corrosion. The discrepancy in the variation of strip tension with the depth of the mechanically stabilized earth (MSE) wall is attributed to the construction procedure adopted. High-stress concentrations in the strips, which are usually observed when the strips are bent or out of shape, cause accelerated corrosion. Interestingly, the authors observed that the chloride content does not affect the rate of corrosion. Sagüé et al. [6] investigated corrosion in 10 MSE walls reinforced with galvanized steel at eight different locations in Florida, USA. It was observed that the red rust, which indicates the highest oxidation, was predominant in older MSE walls, and the zinc corroded completely. The increase in localized corrosion can be attributed to spots of stress concentrations and other irregularities in soil moisture, chemical, and oxygen content throughout. Corrosion in galvanized steel is found to be more uniform than that observed in plain steel. Elias et al. [7] conducted many open electrochemical tests on the existing MSE walls in the field and found that the resistivity of the soil strongly influenced the metal loss, as described by the corrosion mechanism.

2.1.3 Numerical Studies

Owing to the heterogeneity of corrosion in MSE walls, limited numerical studies were performed by researchers. Chau et al. [8] developed a finite element model to study the long-term effect of corrosion on reinforced earth walls by adopting the equation for metal loss, given by Eq. (2). Homogenization of the mechanical properties of the soil and reinforcement was done by adopting a multi-phase model. Parametric studies indicated that in cohesive soils, with an increase in cohesion of backfill soil, the life of the MSE wall decreases, contrary to predicted. This is because the cohesive soils trap moisture and air in pockets around reinforcement to accelerate pitting corrosion and increase the lateral deformation. Similar studies were performed by Bourgeois et al. [9] to analyze the long-term deformations of reinforced walls subject to four different corrosion scenarios. Interestingly, it was analyzed that the failure observed at the end of design life was due to the loss of stiffness, although the tensile stresses in the reinforcement did not reach the failure state.

2.2 Recent Advances in Geosynthetic Reinforced Earth Walls

The recent literature related to the GRS is divided into experimental studies and numerical studies. These studies are discussed in the following sub-sections.

2.2.1 Experimental Studies

Full-scale tests were conducted to study the field observations of GRS walls on a large scale. Helwany et al. [10] conducted a large-scale laboratory test on 144 GRS retaining walls of different reinforcement, backfill, and wall configurations. The authors concluded that the type of backfill had the most profound effect on the behavior of the GRS retaining wall. If the backfill was of lower stiffness and shear strength, the stiffness of the geosynthetic reinforcement had considerable effect. Bathurst et al. [11] studied the effect of GRS reinforcement stiffness on the CIS by testing four full-scale GRS walls with propylene geogrid, polyester geogrid, and welded wire mesh. The authors observed a decreasing horizontal displacement with increasing stiffness of reinforcement. At higher surcharge loads, the influence of stiffness on the lateral deformation of the GRS wall was more prominent. An important observation made by the authors was that with decreasing mobilized reinforcement stiffness and with increasing surcharge, the displacement of the wall increases non-linearly. Also, for a constant stiffness with increasing surcharge, the displacement of the wall increased linearly.

A majority of the previous researchers have used the reduced model tests to study the efficacy of the geosynthetics. The typical test setup consists of a tank connected to the hydraulic jack to apply the load via a steel plate. Various geosynthetic materials, viz. geocells, woven geosynthetics, polymer strips, geogrids, etc., were used. Ehlrich et al. [12] conducted a laboratory model study on the effect of CIS on a geogrid reinforced wall using both light and heavy compaction techniques. The authors concluded that 80% of the lateral movement of the GRS wall occurs before the surcharge, and the remaining 20% occurs in stages with increasing surcharge. The lateral movement due to surcharge is prominent, especially when the surcharge is greater than the CIS. With increasing compaction stresses, the sensitivity of the wall to lateral displacement decreases due to an over-consolidation of the backfill. Xiao et al. [13] conducted model tests of vertically loaded footings on GRS walls. The authors found that the ultimate bearing capacity of the footing depended on the offset distance of the footing to the wall facing, the footing size, the geogrid length, and the method of connection between geogrid and facing blocks. Chen and Chiu [14] studied the effect of height and the facing angle of the wall, structure type, surcharge loading area, and the position of the reinforcement. The authors observed that the introduction of reinforcement reduced the deformation of the wall because of more mobilization of frictional resistance. Also, it was suggested that the facing-type wall be reinforced throughout the height of the wall. Song et al. [15] observed that the failure surfaces predicted by the general limit equilibrium method by both reduced-scale and centrifuge models were very similar to [14]. Latha et al. [16] observed that the general limit equilibrium method, with an assumption of polygonal failure surface, is more accurate than Rankine's analysis with a planar failure surface.

Several studies on centrifuge model tests were conducted to ascertain the findings of the reduced-scale model tests and to check the extent to which the full-scale tests replicate the findings of the model studies. Song et al. [17] performed centrifuge studies to observe the influences of aspect ratios and wall-facing inclination angles. The authors described that the failure mechanisms of geocell reinforced retaining walls were similar to those of slopes through centrifuge model tests. For flexible walls, the internal failure surfaces were observed to pass through the toe of the GRS wall.

In addition to laboratory tests on MSE walls, many studies were conducted on triaxial tests to note the improvement in the strength of the soil due to geosynthetics. Rajagopal et al. [18] performed triaxial tests with four different types of geocells with a varying number of geocell pockets. Triaxial tests were performed at different confining pressures, and the authors conclude that the confinement in the geocells introduced an apparent cohesion even in cohesionless soils. Chen et al. [19] used different shapes of cells such as circular, rectangular, and hexagonal to construct geocells. They found out that the circular shape was most effective in increasing the apparent cohesion.

Wu [20] reported a field test of a pier in GRS in Virginia, USA. The pier was constructed on a three-layer geogrid reinforced foundation. It was observed that the lateral deformation of the wall facing increased with the applied footing load. The maximum strains in the geotextiles occurred in the area near the middle of the wall facing. Adams and Saunders [21] conducted field tests of a GRS wall supporting an integrated bridge system (IBS). These bridges shift from traditional bridges which comprise a single simply supported girder, with a pair of abutments via fixed (or hinged) and moveable bearings. The GRS-IBS had a wrapped-around facing and was of 1.5 m wall height. The differential settlement between the bridge and nearby road was found to be 13 mm, and no bump at the connection was observed. Adams et al. [22] presented detailed design guidelines for GRS-IBS. These guidelines, outlined by Federal Highway Administration (FHWA), were specifically developed for small bridges with GRS abutment that acts as a composite load-bearing support structure. Xu et al. [23] conducted 13 model tests to investigate the ultimate bearing capacity of the GRS walls. It was concluded that the ultimate bearing capacity depends on the reinforcement spacing and backfill friction angle.

2.2.2 Numerical Studies

The most convenient and widely adopted method of analysis is the numerical analysis of GRS walls. Various researchers studied GRS walls used in several applications to model geosynthetics and simulate their effect in MSE walls by modeling them numerically. The numerical studies will help in the thorough understanding of failure mechanisms and the impact of various parameters on the stability of the wall. Han and Leshchinsky [24] worked on the analysis of back-to-back mechanically stabilized earth walls. The finite difference method and limit equilibrium method (the Bishop simplified method) were adopted for this analysis. The authors concluded that the interaction between the walls changes the location and depth of the critical surface. Huang et al. [25] evaluated the performance of piles in GRS retaining walls using a 3D numerical software based on a finite difference method and generally predicted the lateral earth pressures. It was demonstrated that the linearly elastic-perfectly plastic soil model was sufficient for predicting the behavior of GRS retaining walls. Huang et al. [26] performed numerical modeling to simulate a laterally loaded pile in the GRS retaining wall using an advanced soil constitutive model, the Cap-Yield model. This soil model can characterize the compression and shear hardening/softening behavior of reinforced backfill and is capable of accounting for both shear and compression yielding. Chen et al. [27] performed numerical analysis on geocell reinforced walls. The numerical and analytical studies are based on an equivalent composite model for representing the strength and stiffness of geocell confined soil. The results of the numerical studies were similar to [14]. The authors observed that an increase in geocell length at certain depths can reduce the wall movement and settlement of the backfill drastically.

Pierson et al. [28–30] performed field tests to study the behavior of laterally loaded piles installed in a GRS retaining wall without isolation. The facing deflections, lateral earth pressures, and strains in geogrids were measured to evaluate the performance of the wall. The authors observed a reduction in the lateral load capacity of the group piles which resulted from the pile group effect. It was found from the study that the MSE wall could provide considerable lateral support for the piles. The authors observe that the laterally loaded piles only caused additional tension in the geogrid closely surrounding the shaft. This is explained as the uniaxial geogrid has weak strength in the direction along with the depth, which prevents the tension from being transmitted to a larger area. Hence, the movement of the shaft only leads to a strain increase in the geogrid in the vicinity of the pile.

The concrete key is currently being used primarily for the alignment of the wall-facing units in GRS walls. Kim and Bilgin [31] studied the effect of the length of the concrete key under MSE walls on the wall deformations. The authors concluded that the horizontal deformations of an MSE wall decrease as the length of the concrete key under the wall increases. Also, as the reinforcement length decreases, the wall deformations were observed to increase.

The use of GRS for bridge piers and abutments has gained increasing popularity due to ease of construction, shortened construction duration, and low maintenance during their service life. Zheng et al. [32] presented the numerical model of a realistic

MSE wall-supported bridge abutment using a finite difference program, to study the response of the MSE wall-supported bridge abutment. The authors concluded that as the load applied on the bridge increased the lateral earth pressures on the lower wall and vertical pressures at soil foundation level and under the abutment structure also increased significantly. Shen et al. [33] used both 2D and 3D numerical analyses to evaluate the performance of GRS piers. The parametric study concluded that the reinforcement stiffness did not influence the maximum tension in reinforcement. However, the reinforcement spacing influenced the performance significantly. Zheng et al. [34] represented a numerical study of maximum reinforcement tensile forces in GRS bridge abutments. The parametric study concluded that the reinforcement spacing and backfill friction angle influenced the tensile stresses the most. Khosrojerdi et al. [35] performed a numerical investigation to study the effect of constitutive soil on the behavior of GRS piers. The authors observed that the plastic-hardening model can best predict the behavior of GRS piers under service loads. Also, a parametric study was done and noted that as the load on the GRS pier increases, sliding initiates at the corners pier and progressively shift toward the center of the pier. Gebremariam et al. [36] aimed to evaluate the connection load in GRS-IBS structures under service loads based on field monitored GRS-IBS sites as well as design guidelines. It was found that the lateral pressure distribution was not triangular, rather follows a bin pressure distribution. Also, the reinforcement connection loads measured in the field were in agreement with those obtained from reinforcement strains and lateral stresses.

The effect of compaction-induced stresses on the behavior of the GRS wall was studied by various authors under different contexts. Mirmoradi and Ehrlich [37] performed a numerical analysis with two methods of application of CIS: one with CIS introduced at top of the compacted layer and the other with CIS acting both above and below the compacted soil layer. The authors observed that when CIS was applied, the maximum tension measured in the reinforcement did not correlate well with the values extracted from numerical modeling. This indicates a redistribution of stresses along with the depth of the wall. The authors concluded that if the CIS is assumed to act only on top of the layer, the tensile stresses are overestimated against the measured values. The maximum tension mobilized in the reinforcement depended only on the external surcharge acting on the GRS. Ambauen et al. [38] performed numerical modeling on a geogrid reinforced wall to study the effect of confinement due to the spacing of reinforcement under working stress. The authors observed that the CIS significantly increased the reinforcement tensions and deformations and that the numerical and laboratory data of lateral displacement were in agreement. Yu et al. [39] performed a numerical study on two existing polyethylene uniaxial geogrid reinforced walls in Seattle, Washington, using a finite difference program. The GRS walls were designed using the K-stiffness method. To simulate CIS, a temporary surcharge was placed on top of the layer. Each layer was added sequentially and upon reaching equilibrium, the compaction surcharge was removed. The authors considered the effect of CIS on soil parameters at the end construction, by removing the load at the end of the stage. The lateral displacement of the wall was found to decrease with a decrease in the compaction pressure for both the linear elastic and nonlinear elastic models. The authors observed that an increase in the compaction pressure increased the strain in the reinforcement and was most distinct at the facing. Scotland et al. [40] conducted a numerical study using a geogrid reinforced flexible wall. A strain-hardening soil with the Mohr–Coulomb failure model was used. The authors adopted a staged construction model and applied a load-unload cycle compaction stress above and below each layer. By considering a layer-by-layer construction technique, CIS was applied sequentially. In a full-height construction technique, CIS was applied and removed at the end of construction. It was observed that the wall deformation was sensitive to the method of construction.

3 Summary of Past Studies

The majority of past research focused on metallic reinforcements in MSE walls. The most significant problem posed to researchers was that of corrosion. Studies were performed to analyze the corrosion in MSE walls, but due to the unpredictable nature of corrosion, limited conclusions were drawn. Numerical analysis of corrosion in GRS was found to be difficult to model, and hence, very few researchers have focused on the long-term numerical analysis of corrosion. Due to the shortcomings of metal reinforced MSE walls, alternate reinforcements like geosynthetics, polymer strips, geogrids, etc., gained increasing importance in various applications of geotechnical engineering. Many studies were conducted on full-scale models, reduced-scale models, centrifuge models, field observations, and numerical investigation of different GRS walls. Most of these works focused on the parametric study to understand the influence of the reinforcement strength on the stability of the wall on the reinforcement strength, pullout resistance, lateral deformation, and connection stability of facing blocks of GRS walls. Most of these applications are limited to retaining walls, pavements, and foundations. Also, novel geosynthetics such as geocells do not have well-documented case histories for comparison and calibration. The applications of geosynthetics can be extended to high geosynthetic reinforced walls and bridge abutments.

4 Current Research Trends

In recent years, research focus has shifted toward the seismic performance of MSE walls. More studies are being carried out on the model tests and centrifuge tests to evaluate seismic and cyclic loadings on MSE walls. Geosynthetics are gaining popularity in applications such as underground utilities. Geosynthetics such as geotextiles and geogrids can be utilized to protect buried pipelines against deformation and damage. Interestingly, collapsed GRS walls are nowadays widely accepted to reconstruct embankments and traditional reinforced walls that collapsed by floods and earthquakes. The walls are reconstructed as embankments with steep slopes or stage constructed rigid facing. The FHWA has been promoting an approach based on using only geosynthetics to reinforce soil to construct bridge abutments. This approach blends the bridge superstructure with the integrated approach through a joint-less connection aiming to prevent the bump that may occur on the surface of the road at the end of bridges supported on deep foundations. More researchers are now involved with working on the different aspects of the performance of GRS-IBS.

5 Future Prospects

The following enlists the future scope of research in the area of reinforced earth walls and research gaps in the field of study.

- Majority of the past studies are large-scale field observations with little scope to predict and foresee corrosion in metallic reinforcements in MSE walls. Hence, more and more field tests with the proper instrumentation and monitoring arrangements shall be conducted. These tests will help to understand the long-term behavior of MSE walls and thereby developing the appropriate design considerations.
- Most of the studies did not consider the depth of the groundwater table. A better insight into the effect of the depth of the groundwater table on corrosion will help extend the applications of metal reinforcements in MSE walls.
- The scope of the effect of backfill is limited to short-term analysis. The knowledge about the long-term consolidation of soil and its effect on the reinforcement tension and wall deformation is unknown.
- There are limited studies on the influence of stiffness and bending of the geosynthetics in MSE walls, and this need area needs further investigation.
- The seismic studies focusing on corrosion, geosynthetic reinforced soil abutments, and compaction-induced stresses are very few and shall be investigated thoroughly.

6 Conclusions

The use of reinforcements in earth walls is gaining a lot of importance owing to various factors such as cost, flexibility, settlements, maximum height, constructability, the pace of construction, foundation requirements, seismic response, and variations to suit site conditions. The main objective of this paper was to present the reader with a summary of the past studies, current trends, and the scope for future research on metallic and geosynthetic reinforced earth walls. The paper has discussed numerous studies related to corrosion of metallic reinforcements, compaction-induced stresses in geosynthetic reinforced soil walls, and geosynthetic reinforced soil bridge abutment. It was observed from the past studies that there exist certain research gaps in the area of MSE walls as detailed in future prospects. If these gaps are addressed, it can improve the overall understanding of failure mechanisms and long-term behavior of MSE walls. In future, MSE walls will be more prominent in many novels in many sustainable applications of Civil Engineering.

References

1. Saran S (2011) Reinforced soil and its engineering applications. IK International Pvt Ltd
2. Romanoff M (1957) Underground corrosion. US Government Printing Office, Washington, DC
3. Beckham TL, Sun L, Hopkins TC (2005) Corrosion evaluation of mechanically stabilized earth walls
4. Guilloux A, Jailloux JM (1979) Full scale failure test on reinforced earth wall by accelerated corrosion. CR Coll Int Soil Reinf 503–508
5. Blight GE, Dane MSW (1989) Deterioration of a wall complex constructed of reinforced earth. Geotechnique 39(1):47–53
6. Sagüés AA, Scott R, Rossi J, Peña JA, Powers R (2000) Corrosion of galvanized strips in Florida reinforced earth walls. J Mater Civ Eng 12(3):220–227
7. Elias V, Fishman K, Christopher BR, Berg RR, Berg RR (2009) Corrosion/degradation of soil reinforcements for mechanically stabilized earth walls and reinforced soil slopes (No. FHWA-NHI-09–087). National Highway Institute (US)
8. Chau TL, Bourgeois E, Corfdir A (2012) Finite element analysis of the effect of corrosion on the behavior of reinforced earth walls. Int J Numer Anal Meth Geomech 36(15):1741–1756
9. Bourgeois E, Corfdir A, Chau TL (2013) Analysis of long-term deformations of MSE walls based on various corrosion scenarios. Soils Found 53(2):259–271
10. Helwany SMB, Reardon G, Wu JTH (1999) Effects of backfill on the performance of GRS retaining walls. Geotext Geomembr 17(1):1–16
11. Bathurst RJ, Nernheim A, Walters DL, Allen TM, Burgess P, Saunders DD (2009) Influence of reinforcement stiffness and compaction on the performance of four geosynthetic-reinforced soil walls. Geosynth Int 16(1):43–59
12. Ehrlich M, Mirmoradi SH, Saramago RP (2012) Evaluation of the effect of compaction on the behavior of geosynthetic-reinforced soil walls. Geotext Geomembr 34:108–115
13. Xiao C, Han J, Zhang Z (2016) Experimental study on performance of geosynthetic-reinforced soil model walls on rigid foundations subjected to static footing loading. Geotext Geomembr 44(1):81–94
14. Chen RH, Chiu YM (2008) Model tests of geocell retaining structures. Geotext Geomembr 26(1):56–70
15. Song F, Liu H, Chai H, Chen J (2017) Stability analysis of geocell-reinforced retaining walls. Geosynth Int 24(5), 442–450
16. Latha GM, Dash SK, Rajagopal K (2009) Numerical simulation of the behavior of geocell reinforced sand in foundations. Int J Geomech 9(4):143–152
17. Song F, Xie YL, Yang YF, Yang XH (2014) Analysis of failure of flexible geocell-reinforced retaining walls in the centrifuge. Geosynth Int 21(6):342–351
18. Rajagopal K, Krishnaswamy NR, Latha GM (1999) Behaviour of sand confined with single and multiple geocells. Geotext Geomembr 17(3):71–184
19. Chen RH, Huang YW, Huang FC (2013) Confinement effect of geocells on sand samples under triaxial compression. Geotext Geomembr 37:35–44
20. Wu JT (2001) Revising the AASHTO guidelines for design and construction of GRS walls (No. CIlOT-D1D-R-2001-16). United States. Federal Highway Administration
21. Adams MT, Saunders SA (2007) Upper Ouachita National Wildlife Refuge GRS Abutments for Replacement Bridges. Presentation, FHWA

22. Adams M, Nicks J, Stabile T, Wu JT, Schlatter W, Hartmann J (2011) Geosynthetic reinforced soil integrated bridge system, synthesis report (No. FHWA-HRT-11-027). United States. Federal Highway Administration
23. Xu C, Liang C, Shen P (2019) Experimental and theoretical studies on the ultimate bearing capacity of geogrid-reinforced sand. Geotext Geomembr 47(3):417–428
24. Han J, Leshchinsky D (2010) Analysis of back-to-back mechanically stabilized earth walls. Geotext Geomembr 28(3):262–267
25. Huang J, Parsons RL, Han J, Pierson M (2011) Numerical analysis of a laterally loaded shaft constructed within an MSE wall. Geotext Geomembr 29(3):233–241
26. Huang J, Han J, Parsons RL, Pierson MC (2013) Refined numerical modeling of a laterally-loaded drilled shaft in an MSE wall. Geotext Geomembr 37:61–73
27. Chen RH, Wu CP, Huang FC, Shen CW (2013) Numerical analysis of geocell-reinforced retaining structures. Geotext Geomembr 39:51–62
28. Pierson MC, Parsons RL, Han J, Brennan JJ (2009) Capacities and deflections of laterally loaded shafts behind mechanically stabilized earth wall. Transp Res Rec 2116(1):62–69
29. Pierson, M.C., Parsons, R.L., Han, J., Brown, D. and Thompson III, W.R., 2009. Capacity of laterally loaded shafts constructed behind the face of a mechanically stabilized earth block wall (No. K-TRAN: Ku-07–6).
30. Pierson MC, Parsons RL, Han J, Brennan JJ (2011) Laterally loaded shaft group capacities and deflections behind an MSE wall. J Geotech Geoenviron Eng 137(10):882–889
31. Kim HS, Bilgin Ö (2007) Studying the effect of concrete key size on mechanically stabilized earth wall deformations using finite element method. In: Computer applications in geotechnical engineering, pp 1–8
32. Zheng Y, Fox PJ, Benson Shing P (2014) Numerical simulations for response of MSE wall-supported bridge abutments to vertical load. In: 2014 GeoShanghai International Congress: Ground Improvement and Geosynthetics. American Society of Civil Engineers (ASCE)
33. Shen P, Han J, Zornberg JG, Morsy AM, Leshchinsky D, Tanyu BF, Xu C (2019) Two and three-dimensional numerical analyses of geosynthetic-reinforced soil (GRS) piers. Geotext Geomembr 47(3), 352–368
34. Zheng Y, Fox PJ, McCartney JS (2018) Numerical study on maximum reinforcement tensile forces in geosynthetic reinforced soil bridge abutments. Geotext Geomembr 46(5):634–645
35. Khosrojerdi M, Qiu T, Xiao M, Nicks J (2020) Effects of backfill constitutive behavior and soil–geotextile interface properties on deformations of geosynthetic-reinforced soil piers under static axial loading. J Geotech Geoenviron Eng 146(9):04020072
36. Gebremariam F, Tanyu BF, Christopher B, Leshchinsky D, Zornberg JG, Han J (2020) Evaluation of required connection load in GRS-IBS structures under service loads. Geosynth Int 1–46
37. Mirmoradi SH, Ehrlich M (2015) Modeling of the compaction-induced stress on reinforced soil walls. Geotext Geomembr 43(1):82–88
38. Ambauen S, Leshchinsky B, Xie Y, Rayamajhi D (2015) Service-state behavior of reinforced soil walls supporting spread footings: A parametric study using finite-element analysis. Geosynth Int 23(3):156–170
39. Yu Y, Bathurst RJ, Allen TM (2016) Numerical modeling of the SR-18 geogrid reinforced modular block retaining walls. J Geotech Geoenviron Eng 142(5):04016003
40. Scotland I, Dixon N, Frost M, Fowmes G, Horgan G (2016) Modelling deformation during the construction of wrapped geogrid-reinforced structures. Geosynth Int 23(3):219–232

Negative Skin Friction on Piles: State of the Art

Sujawat Singh Rituraj and B. Giridhar Rajesh

Abstract Pile foundation experiences negative skin friction when the surrounding soil settles more than the pile, which induces an increased compressive force on piles, called drag load, and it also induces settlement in pile called down drag. The drag load acting on the pile reduces shaft capacity to resist the ultimate load. Geotechnical engineers had to deal with the effects of negative skin friction on structural capacity and settlement of piles for the efficient design of pile foundations. The mechanism of negative skin friction on pile nonetheless is still not well understood and needs to be explored. This paper aims to present the state of the art relating to negative skin friction on piles. This review is prepared based on analytical, numerical, experimental and field studies that are of significant contribution in understanding and incorporating the effect negative skin friction in the design. Topics covered include: response of single piles and pile groups to negative skin friction, methods to calculate NSF and incorporation of NSF in the pile design. Finally, the key problems that lie ahead for future research related to negative skin friction on pile and pile groups are addressed briefly.

Keywords Negative skin friction · Down drag · Drag load · Shielding · Interface

1 Introduction

Due to the rapid urbanization in the twenty-first century, construction using pile foundation has become inevitable as it is having more applications in the building of skyscrapers, hydraulic structures, retaining walls, offshore structures, bridge abutments, wharves and various other Civil Engineering problems. So, it is very important to design and analyze the pile foundation efficiently and cost-effectively. Negative skin friction (NSF) due to the settling of surrounding soil more than the pile has been reported by many authors. NSF may occur under the application of a surcharge load,

S. S. Rituraj · B. Giridhar Rajesh (✉)
Department of Civil Engineering, National Institute of Technology Andhra Pradesh, Tadepalligudem, Andhra Pradesh, India
e-mail: rajesh@nitandhra.ac.in

A. K. Choudhary et al. (eds.), *Advances in Geo-Science and Geo-Structures*,
Lecture Notes in Civil Engineering 154,
https://doi.org/10.1007/978-981-16-1993-9_34

withdrawal of the groundwater table, soil consolidation, remolding of clay during pile installation and liquefaction of loose sand. Ignorance of the NSF while accessing the ultimate capacity of deep foundations can lead to excessive settlement of foundation and structure above it. Negative skin friction on piles has been studied for more than 70 years [1]. Various strategic solutions have been investigated to understand their complex behavior [2–9]. With advancements in computer capacity to handle large computations, FEM studies were performed which paved a way for complex situations such as calculations of NSF in a single pile and pile groups in both single-layered and multi-layered soil [8, 10–20]. Numerical models were developed using different FEM platforms for better assessment of NSF behavior in different conditions using various soil models and pile models. This paper reviews the response of down drag and dragload under different pile-soil conditions. The effect of different pile tip locations, axial load, surcharge loads and stiffness of bearing soil on negative skin friction is detailed. Conditions such as single pile under surcharge loading, the effect of different pile tip locations under axial load. Various methods to reduce drag load were also discussed. Finally, a note is presented which highlights the different values of α, β and μ to characterize the soil-pile interface properties for different NSF conditions. The note also includes some provisions which describe the various parameters associated with the interface properties. These provisions are quite useful for practicing engineers and for numerical modeling of pile foundations. Although there are several studies available which are centering on the design and modeling aspects of the pile foundations, but to the best of authors' knowledge, there is no specific state-of-the-art study pertaining to negative skin friction. Hence, in this paper an attempt has been made to summarize the major findings, current research trends and future prospects related to negative skin friction.

2 Response of Single Piles and Pile Groups to Negative Skin Friction

This section discusses the NSF under different pile-soil conditions such as single pile under surcharge loading, the effect of different pile tip locations, the effect of axial load, the influence of stiffness of bearing stratum and soil-pile interface.

2.1 Behavior of Single Pile Under Surcharge Loading to Negative Skin Friction

One of the major causes which is responsible for the drag load is the generation of NSF from the settlement of surrounding soil under the loading of surcharge. The surcharge can be due to the fill from the excavation or some construction on surrounding soil. This settlement increases the relative settlement of pile and surrounding soil and hence induces negative skin friction on the pile. Liu et al. [19] developed a 2-D axisymmetric FEM model using ABAQUS and suggested that the depth of the neutral plane decreases with the increase in surcharge intensity. The author had shown that the depth of neutral plane lowers down from the 0.62 to 0.67 L under the same surcharge increment from 25 to 400 kPa. The unit skin friction at all elevations increases with the increase in the intensity of surcharge loading which results in an appreciable increase in the drag load, and also, the comparison is made for two types of bearing stiffness layers where one layer is 1000 times stiffer than another layer. As expected, the drag load is more in stiffer layer than the softer layer case which is due to the fact that the higher stiffness prevents the settlement of the pile hence inducing larger relative displacement between piles and surrounding soil. Lv et al. [11] conducted a numerical study for the NSF behavior on the pile for different tip locations for $Y = 1.00D, 0.25D, 0.00D$ and $-1.00D$ under the surcharge (where $Y =$ distance between bottom tip of the pile and bearing stiff soil layer; $D =$ diameter of the pile). The soil at the ground surface is directly exposed to surcharge loading, thus maximum settlement of soil can be observed at that point which is also responsible for maximum relative settlement between soil and pile at that point. As expected, the drag load is more for the case where the pile tip moves closer to the bearing soil because it prevents the further settlement of the pile, hence increasing the relative soil-pile displacement. But against the expectation, as the pile tip is started to insert into the bearing soil, the drag load again starts decreasing; this is due to reduction in relative soil-pile displacement. Hence, it was concluded that for reducing the drag load on floating pile, the pile tip should be at a larger distance from bearing soil, but for reducing drag load on end-bearing pile, the pile tip should be embedded as deep as possible into the bearing soil. Yao et al. [21] performed an analytical study for study of negative skin friction on super long piles under surcharge loading by developing a three-dimensional numerical model. The authors concluded that the skin friction above the neutral plane along the pile can be divided into two zones. One zone is where the negative skin friction first increases, and the second zone is where it subsequently decreases with the further increment of depth till neutral point. After neutral point, it is positive skin friction which continuously increases with depth.

2.2 *Effect of Pile Tip Locations on Negative Skin Friction*

The location of the pile tip above the hard stratum greatly influences the behavior of negative skin friction. Several studies were conducted to understand the effect of pile tip location considering various load configurations. Abdrabbo and Ali [12] discussed the 3-D numerical analysis by considering the two cases for the pile tip location considering $Y = 0.25D$ and $Y = -1.00D$, where Y is the distance between pile tip location and the bearing layer and D is the diameter of pile. It was found that for later case, the drag load induced is more compared to the former due to higher soil–pile relative displacement. Lv et al. [11] also observed the behavior of

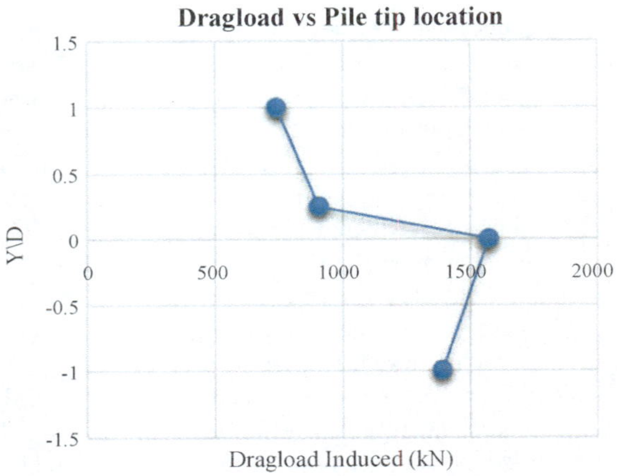

Fig. 2.1 Drag load for different pile tip locations. Adapted from Lv et al. [11]

pile tip locations from 3-D numerical analysis for four different locations of piles and reported the trend for drag load as shown in Fig. 2.1.

The trend discontinued after further embedment of pile tip greater than $Y = -1.00D$ as the relative displacement of pile head and top of soil layer decreases. The pile capacity is not simply the sum of maximum drag load and axial load acting on the pile, and a reduction factor needs to be introduced for drag load at any given axial load which is given by the following equation (Lv et al. [11]).

$$R = \frac{P_{\text{max}} - P_{\text{max,axial}}}{P_{\text{max}}} \times 100\% \quad (1)$$

where R is the rate of maximum drag load reduction, P_{max} is the maximum drag load before the application of an axial load, and $P_{\text{max, axial}}$ is the maximum drag load at a given axial load.

Ng et al. [22] compared the negative skin friction for end-bearing pile and floating pile by keeping the tip locations as $-1.00D$ and $0.25D$, respectively. It is observed that the depth of the neutral plane is much higher for end-bearing pile as compared to the floating pile. As the depth of pile increases, for both the cases, the drag load distribution is common up to 60% length of the pile, but with the further increase, the difference in drag load between the end-bearing pile and floating pile becomes greater. This is because near the neutral plane for floating pile, the mobilization of NSF decreases due to the increment of positive skin friction. It is also observed that the drag load on end-bearing pile is 60% greater than the floating pile. Lam et al. [23] studied the behavior of drag load and negative skin friction for tip locations $1.00D$ and $0.25D$ and come up with the conclusion that the magnitude of axial load required to fully eliminate the maximum drag load is $1.25P_{\text{max}}$ and $1.75P_{\text{max}}$ (P_{max} is the maximum drag load acting on the pile at 90% consolidation of soil before the action of the axial load). The required axial load for the piles which are shielded from sacrificing pile for negative skin friction with spacing 5.00*d* and 6.00*d* are 1.5 P_{max} and 1.75 $P_{max,}$ respectively, where *d* is the diameter of sacrificing pile.

2.3 Effects of Axial Load on Negative Skin Friction

Pile foundations are having a wide application in the construction of skyscrapers, so it is essential to study the behavior of negative skin friction under the load coming from the superstructure in the form of axial loads. When the piles are subjected to negative skin friction due to the settlement of surrounding soil under surcharge load, then by imposing some extra-axial load on the pile will reduce the relative settlement between pile and soil, hence reduces the drag load on the pile.

Lv et al. [11] had calculated the allowable axial load on the pile for a given pile tip location using Davisson's criterion [24] following equation:

$$\Delta = \frac{\text{PL}}{\text{AE}_P} + \frac{D'}{120} + 4 \tag{2}$$

The allowable load is calculated using the $\Delta = 1.24D$ from the performed centrifuge test. Here, Δ is the settlement of pile in mm, L is the length of pile (mm), A denotes the cross-sectional area of the pile (mm^2), E_p denotes the young modulus for the material of the pile (kN/mm^2), and D' denotes the least lateral dimension of the pile (mm). The allowable load for different pile tip locations is presented in Fig. 2.2:

The axial loads as shown in Fig. 2.2 are plotted after considering 63% of the actual allowable load. As expected, the settlement of pile decreases with the decrease tip location (*Y*/*D*). From the results, it was concluded that the tip location controls the pile settlement and influences the pile capacity. The maximum drag load reduction

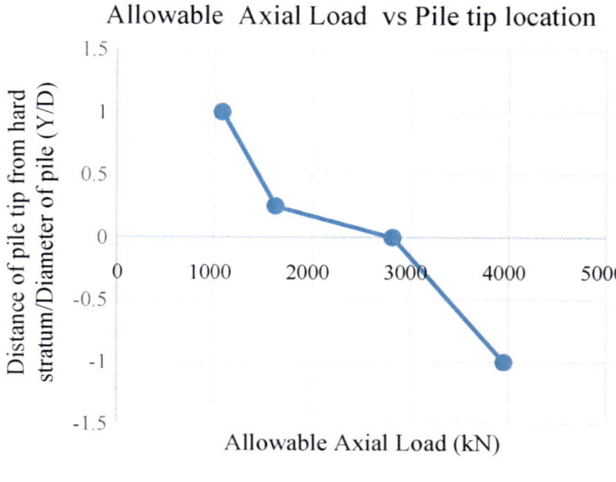

Fig. 2.2 Allowable axial load versus pile tip location. Adapted from Lv et al. [11]

reaches 100% at approximately $P/P_{max} = 2.0, 2.5, 3.5$ and 4.5 for the piles with $Y = 1.00D$, $0.25D$, $0.00D$ and $-1.00D$, respectively. Jeong et al. [25] also reported that an axial load of approximately 1.25 to 3.25 P_{max} is required for full elimination of drag load. Fellenius [2] and Bozozuk [26] observed that the axial load applied after development of negative skin friction would be totally resisted by the reversed locked-in NSF (the amount of negative skin friction developed on the pile before the application of external axial load). This implies that the axial load applied and the locked-in drag load would not act simultaneously on the piles. The author concluded that only the drag load and the load applied prior to development of NSF should be considered in calculating the axial capacity of a pile, as long as the applied loads were smaller than twice the maximum drag load.

Lam et al. [23] had performed the experimental analysis to determine the amount of axial load required for the complete elimination of negative skin friction. Four centrifuge tests were performed to investigate the load transfer mechanism where two cases are of single floating piles with pile tip location $Y = 1.00D$, $0.25D$ piles, and another two cases are of center piles shielded in 3X3 sacrificing pile groups having locked-in NSF with pile spacing of 5.0d and 6.0d (d and D are the diameter of sacrificial piles and center pile, respectively). The pile tip location for both the cases of unshielded piles is 0.25D. The axial load was applied after the maximum drag load had developed. With the further increment of the drag load, the pile will start moving downwards, inducing positive skin friction. It was observed that when the axial load reaches 1.25 P_{max}, the drag load is eliminated. The authors have also studied the hang-up effects for the soil between the piles and sacrificial piles. Hang-up effects are investigated in terms of vertical effective stress for the unshielded piles and pile shielded by sacrificial piles for spacing of 5.0d and 6.0d. The values for the effective stress before and after the action of axial loads are given in Tables 2.1 and 2.2, respectively.

From Tables 2.1 and 2.2, it was observed that hang-up effects on the center piles with the sacrificial piles around and having a spacing of 5.0d are maximum because hang-up effects on the soil between the pile increases as the pile spacing decrease. Also, upon axial load application, it was observed that pile head settlement is maximum for the case of center pile shielded with sacrificial piles which are having a spacing of 5.0d. This is due to the fact that the lower value of effective stress induces a smaller value of positive shaft friction acting on the pile. Thus, it was suggested

Table 2.1 Before the action of axial loads, the average effective vertical stress $(\sigma)'_v$

Type of piles	Average effective vertical stress
Unshielded piles	σ'_v
Shielded pile (with sacrificial piles spacing 5.0d)	74% of σ'_v
Shielded pile (with sacrificial piles spacing 6.0d)	88% of σ'_v

After Lam et al. [23]

Table 2.2 After the action of axial load with magnitude, the average effective vertical stress $(\sigma)'_v$

Type of piles	Average effective vertical stress
Unshielded piles	102.5% of σ'_v
Shielded pile (with sacrificial piles spacing 5.0*d*)	81% of σ'_v
Shielded pile (with sacrificial piles spacing 6.0*d*)	93% of σ'_v

After Lam et al. [23]

that positive skin friction on the pile shielded with sacrificial piles decreases with the increase in hang-up effects which results in a larger-pile head settlement.

2.4 Effect of Stiffness of Bearing Soil on Negative Skin Friction

Liu et al. [19] conducted a detailed parametric study on negative skin friction with the help of numerical simulations by developing a 2-D axisymmetric FEM model using ABAQUS. In end-bearing piles, it is observed that the neutral plane is located near the tip of the pile for most of the cases. The stiffness of the bearing layer is considered in terms of the ratio E_b/E_c (where E_c is the modulus of soil in consolidating layer and E_b is the modulus of soil of bearing layer). The stiffer is the bearing layer, the greater will be the mobilization of negative skin friction. Influence of bearing layer on the drag load is almost negligible for the upper-half part of the pile, but for the lower half, it changes drastically. Abdrabbo et al. [12] reported that the location of the neutral plane is inappreciably affected by the embedment depth of the pile tip within the range of $-2D$ to $-4D$. The author concluded that, irrespective of the magnitude of bearing stiffness or embedded depth of pile, the limiting relative displacement between soil and pile for mobilization of NSF is 5 mm.

2.5 Effect of Soil-Pile Interface on Negative Skin Friction

The interaction between soil and pile plays a very important role in the stress transfer mechanism at the interface of the pile, and it also affects the bearing capacity of the pile. The interfaces are usually considered through the coefficient of friction (μ) which further depends on friction angle (δ) (Liu et al. [19]), i.e.,

$$\mu = \tan \delta \tag{3}$$

Liu et al. [17] studied the impact of soil-pile interface on NSF. From the obtained results, it was observed that by increasing the value of μ from 0.05 to 0.4 (in terms of bitumen coating), the drag load is reduced by about 70%. It is also observed that the neutral plane shifts slightly upwards with an increase in the friction coefficient. Kamal et al. [27] performed a series of direct shear tests on the concrete blocks coated with the bitumen and plastic sleeves to investigate the shielding effect. The author found that the drag load is reduced by about 78–95% through the bitumen and about 80–98% by plastic sleeves.

3 Methods to Calculate NSF and Incorporation of NSF in the Pile Design

In this section, few basic methods for calculating the negative skin friction will be discussed. There are two popular methods to calculate the negative skin friction: One is undrained shear strength method (α-method) or total stress analysis, and another one is effective stress method (β-method) or effective stress analysis. The overview of current codal provisions on considering NSF is also presented in this section.

3.1 *Total Stress Analysis (α-method)*

Terzaghi and Peck [28] proposed the relations for calculation of negative skin friction on pile embedded in undrained clay. The empirical adhesion factor α value depends upon the material of the pile and the undrained shear strength of the clay. It should be noted that the use of high α values for negative skin friction results in a very conservative design [29, 30]. The skin friction acting on the pile can be evaluated as

$$f_s = \alpha S_u \tag{4}$$

where

α = empirical adhesion factor.

$\alpha = 0.5 - 1.0$ (for negative skin friction [31]).

S_u = average undrained shear strength of clay.

3.2 *Effective Stress Analysis (β-method)*

For long-term stress analysis, the negative skin friction is evaluated with the help of β method. Whenever there is a change in excess pore water pressure of soil, the effective stresses also change along the pile length. This change in effective stresses is time-dependent [2, 26, 32–35], and in such cases, the effective stress method is useful in the evaluation of shaft friction [36]. The mobilized skin friction at any depth cannot exceed the limiting skin friction which is given by:

$$f_s = \beta\sigma_v' \tag{5}$$

Here, σ'_v is the vertical effective stress at any depth, and β is the empirical factor that depends on soil type, pile material and method of pile installation [36]. The various values of proposed in the literature are summarized below.

For normally consolidated clay, $\beta = (1 - \sin\varphi')\tan\delta$.

For over-consolidated clay, $\beta = (1 - \sin\varphi')\,\mathrm{OCR}^{0.5}\tan\delta$.

where δ = friction angle of the pile and soil interface.

Values of β as per NAVFAC [37] are:

$$\beta = 0.2 - 0.25(\text{for clay})$$
$$\beta = 0.25 - 0.35(\text{for silt})$$
$$\beta = 0.35 - 0.50\ (\text{for sand})$$

3.3 The Florida Department of Transportation (FDOT) Method

Structure design guidelines (2007) of Florida Department of Transportation determined negative skin friction effect by calculating the Required Driving Resistance (RDR). RDR is calculated using load and resistance design factor methodology.

$$\mathrm{RDR} = \frac{(\mathrm{FPDL} + 2 \times \mathrm{NSF})}{0.65} \tag{6}$$

where FPDL = factored permanent design load (excludes transient loads).

3.4 Briaud and Tucker Method

The bearing capacity approach by the National Cooperative Highway Research Program Report (Briaud and Tucker 1997) incorporates a safety factor of 1.7 for down drag, i.e.,

$$\mathrm{RDR} = \frac{(\mathrm{FPDL} + 2.35 \times \mathrm{NSF})}{0.75} \tag{7}$$

4 Present Research Trends

Due to the complex behavior of negative skin friction, researchers are working on the various other possibilities of soil-pile conditions that may be subjected to negative skin friction. The present researches on negative skin friction are focusing more upon the development of accurate design parameters that can allow the field engineers to determine the drag load on the piles precisely [10, 12, 38, 39]. Recent study of Sun et al. [10] performed a coupled consolidation analysis on piles embedded in consolidating ground and proposed a design chart to estimate the skin friction distribution for long-term cases, where the pile head load is absent. Ha et al. [13] developed a finite element model for calculation of drag load on a batter pile with an inclination angle of 0° and 30°. Lv et al. [8] investigated the drag load and negative skin friction of different geometric shapes of the pile with the help of centrifuge modeling and numerical simulation. It is observed that the neutral planes of rectangular, circular and *H* piles are located at the normalized depths of 0.86*L*, 0.82*L* and 0.75*L* from the upper end of the pile, respectively (where *L* is the length of the pile). The depth of the neutral plane is minimum for the *H* pile as it is subjected to "hang-up" effects because of vertical shearing inside the flanges and Web. Chow et al. [40] presented a study of soil–structure interaction for a rigid inclusion column which was embedded for a monitoring period of 3.5 years.

Another aspect of the recent studies is to develop an efficient shielding method to reduce the drag load on piles. Lam et al. [7] conducted a comparative study between the sheet piles and sacrificial piles around the center pile in consolidating ground for reducing the negative skin friction. The authors have conducted centrifuge and numerical studies and concluded that the sacrificial sheet pile sleeve is an effective method to reduce the drag load than sacrificial pile. The sacrificial sheet pile sleeve method is also more economical, environment-friendly and more convenient as compared to application of bitumen coating on pile. Feng et al. [41] investigated the impact of different coatings on concrete for reducing the negative skin friction. Four types of coating treatments, namely no coating, paraffin oil mixture coating, coating with polymer nanomaterial and coating with paint, are investigated. It was observed that the performance of paraffin oil mixture coating was best among all the considered treatments.

5 Future Prospects

The complex nature of negative skin friction in different soil-pile conditions demands more research in this arena. For different soil conditions, it is required to investigate both the accurate design parameters to evaluate the pile capacity and the different shielding methods to reduce drag load. The following enlists the future scopes involved in the research related to negative skin friction:

- Majority of the studies of negative skin friction are analyzed by considering static loading. Research in the area of time-varying load along with the problem of negative skin friction needs to be investigated thoroughly, and the proper design methods should be developed.
- There are very limited case studies available focusing on the failure of foundations due to negative skin friction. More back investigation studies from the field should be conducted for the complete understanding of negative skin friction.
- Bitumen is the material which is largely used for shielding of piles from negative skin friction. Yeung et al. [42] explained the contamination of groundwater and surrounding soil medium due to direct contact with bitumen coated pile. Bitumen is obtained from volatile constituents having a large concentration of polycyclic aromatic hydrocarbons which is considered as a very toxic substance. Considering the sustainable construction practices, it is a very urgent need for researchers to find the alternatives for bitumen coating.
- A large number of three-dimensional coupled consolidation analyses for negative skin friction on piles for various soil conditions can help to get a better understanding of NSF.

6 Conclusion

Deep foundations are having a vital application in the construction industry. Accurate estimation of pile capacity is required for the efficient design. The negative skin friction effects the deep foundation design considerably. The present study reviewed the literature related to the negative skin friction effect on piles in different soil conditions. The past studies, current state of the art and the future scope related to NSF were summarized. Various design methods were also presented for accessing the negative skin friction at soil-pile surface. From the literature, it is observed that the negative skin friction is having diversified behavior in different conditions of soil and pile. Hence, generalizing its design procedure may not be a correct approach. Currently, the researches are centering upon the establishing accurate design parameters and design charts based on advanced constitutive models for precise drag load estimation. It was observed from the past studies that there exist certain research gaps in the area of negative skin friction as detailed in future prospects. If these gaps are addressed, it can improve the overall understanding of negative skin friction and paves the way for sustainable development of pile foundations.

References

1. Chellis RD (1961) Pile foundations. McGraw-Hill, New York
2. Canadian Geotechnical Society (2006) Canadian foundation engineering design manual, 4th edn, p 273

3. District W, City T, District W (2014) Innovative and sustainable use of geomaterials and geosystems, pp 87–96
4. Koerner RM, Mukhopadhyay C (1972) Behavior of negative skin friction on model piles in medium-plasticity silt. Highway Research Record P No. 34–50
5. Poorooshasb HB, Alamgir M, Miura N (1996) Negative skin friction on rigid and deformable piles. Comput Geotech 18:109–126.https://doi.org/10.1016/0266352X(95)00026-7
6. Leung CF, Liao BK, Chow YK, Shen RF, Kog YC (2004) Behavior of pile subject to negative skin friction and axial load. Soils Found 44(6):17–26
7. Lam SY, Ng CWW, Poulos HG (2013) Shielding piles from downdrag in consolidating ground. J Geotech Geoenviron Eng 139:956–968. https://doi.org/10.1061/(ASCE)GT.1943-5606.0000764
8. Lv YR, Ng CWW, Lam SY (2017) Geometric effects on piles in consolidating ground: centrifuge and numerical modeling. J Geotech Geoenviron Eng 143. https://doi.org/10.1061/(ASCE)GT.1943-5606.0001714
9. Huang T, Zheng J, Gong W (2015) The group effect on negative skin friction on piles. Procedia Eng 116:802–808. https://doi.org/10.1016/j.proeng.2015.08.367
10. Sun TK, Yan WM, Su D (2015) Fully coupled consolidation analysis of shear strength mobilization and dragload of a pile subject to negative skin friction. Int J Geomech 15. https://doi.org/10.1061/(ASCE)GM.1943-5622.0000381
11. Lv Y, Ding X, Wang D (2013) Effects of the tip location on single piles subjected to surcharge and axial loads. Sci World J. https://doi.org/10.1155/2013/149706
12. Abdrabbo FM, Ali NA (2015) Behaviour of single pile in consolidating soil. Alex Eng J 54:481–495. https://doi.org/10.1016/j.aej.2015.05.016
13. Ha MMH, Hassanlourad M (2015) Numerical modeling of the negative skin friction on single vertical and batter pile. Acta Geotechnica Slovenica 12:47–55
14. El-Mossallamy YM, Hefny AM, Demerdash MA, Morsy MS (2013) Numerical analysis of negative skin friction on piles in soft clay. Housing Build Natl Res Center J 9:68–76. https://doi.org/10.1016/j.hbrcj.2013.02.006
15. Comodromos EM, Bareka SV (2005) Evaluation of negative skin friction effects in pile foundations using 3D nonlinear analysis. Comput Geotech 32:210–221. https://doi.org/10.1016/j.compgeo.2005.01.006
16. Lee CJ, Lee JH, Jeong S (2006) The influence of soil slip on negative skin friction in pile groups connected to a cap. Geotechnique 56:53–56. https://doi.org/10.1680/geot.2006.56.1.53
17. Kong GQ, Yang Q, Luan MT (2008) Study of loading rate effects on characteristics of negative skin friction for pile groups. Electr J Geotech Eng 13 L:1–12
18. Dijkstra J, Broere W, Heeres OM (2011) Numerical simulation of pile installation. Comput Geotech 38:612–622. https://doi.org/10.1016/j.compgeo.2011.04.004
19. Liu J, Gao H, Liu H (2012) Finite element analyses of negative skin friction on a single pile. Acta Geotech 7:239–252. https://doi.org/10.1007/s11440-012-0163-x
20. Nazir R, Momeni E, Gofar Nurly N, Maizir H (2013) Numerical modeling of skin resistance distribution with depth in piles. Electron J Geotech Eng 18 L:2477–2488
21. Yao W, Liu Y, Chen J (2012) Characteristics of negative skin friction for superlong piles under surcharge loading. Int J Geomech 12:90–97. https://doi.org/10.1061/(ASCE)GM.1943-5622.0000167
22. Ng CWW, Asce M, Poulos HG (2009) Effects of tip location and shielding on piles in consolidating ground. J Geotech Geoenviron Eng 134:1245–1260
23. Lam SY, Ng CWW, Leung CF, Chan SH (2009) Centrifuge and numerical modeling of axial load effects on piles in consolidating ground. Can Geotech J 46:10–24. https://doi.org/10.1139/T08-095
24. Davisson M (1972) High capacity piles. Innov Found Constr 81–112
25. Jeong S, Lee J, Lee CJ (2004) Slip effect at the pile-soil interface on dragload. Comput Geotech 31:115–126. https://doi.org/10.1016/j.compgeo.2004.01.009
26. Bozozuk M (1981) Bearing capacity of pile preloaded by downdrag. National Research Council Canada, Division of Building Research

27. Tawfiq KS, Caliendo JA (1995) Bitumen coating versus plastic sheeting for reducing negative skin friction. J Mater Civ Eng 7:69–81. https://doi.org/10.1061/(ASCE)0899-1561(1995)7:1(69)
28. Terzaghi K, Peck RB, Mesri G (1996) Soil mechanics in engineering practice. Wiley, 433
29. Neill MWO, Ecturer L, Oston B (2001) The Thirty-Fourth Terzaghi Lecture Presented at the American Society of Civil Engineers, pp 3–16
30. Meyerhof GG (2002) Bearing capacity and settlement of pile foundations. Geotechnical Special Publication, pp 1684–1715. https://doi.org/10.1016/B978-0-444-98929-1.50006-0
31. Canadian Geotechnical Society (2006) Canadian foundation engineering manual, pp 143–149
32. Fellenius H (1999) Bitumen selection for reduction of downdrag on piles. J Geotech Geoenviron Eng 125:341–344. https://doi.org/10.1061/(ASCE)1090-0241(1999)125:4(341)
33. Hm C, Ih S (1970) Bearing capacity of foundation piles state of the art. Highway Res Rec 87–103
34. Garlanger EJ (1974) Measurement of pile downdrag beneath a bridge abutment. In: 53rd Annual Meeting of the Highway Research Board, Transportation Research Board, pp 61–69
35. Burland J (1973) Shaft friction of piles in clay. Ground Eng London 6:30–42. https://doi.org/10.1007/978-94-017-2473-9_30
36. Jeong S, Ko J, Lee C, Kim J (2014) Response of single piles in marine deposits to negative skin friction from long-term field monitoring. Mar Georesour Geotechnol 32:239–263. https://doi.org/10.1080/1064119X.2012.735344
37. Masaki Kubota Y, Ishida KY (2014) 3-dimensional atom probe method and the Auger depth analysis of multilayer thin films by electron spectroscopy (Surface Analysis Society 42nd Society lecture material). J Surf Anal 21(1): A-1
38. Sutman M, Speranza G, Ferrari A et al (2020) Long-term performance and life cycle assessment of energy piles in three different climatic conditions. Renew Energy 146:1177–1191. https://doi.org/10.1016/j.renene.2019.07.035
39. Hanna A, Mashhour I (2018) Drag load on piles in partially saturated collapsible soils. Geotechnical Special Publication, Nov 2017, pp 290–302. https://doi.org/10.1061/9780784481707.029
40. Chow YK, Lim CH, Karunaratne GP (1996) Numerical modelling of negative skin friction on pile groups. Comput Geotech 18:201–224. https://doi.org/10.1016/0266-352X(95)00029-A
41. Feng Z, Hu H, Zhao R et al (2019) Experiments on reducing negative skin friction of piles. Adv Civil Eng. https://doi.org/10.1155/2019/4201842
42. Yeung AT, Viswanathan R, Briaud JL (1996) Field investigation of potential contamination by bitumen-coated piles. J Geotech Eng 122:736–744. https://doi.org/10.1061/(asce)0733-9410(1996)122:9(736)